Lecture Notes
in Physics

Edited by J. Ehlers, München, K. Hepp, Zürich,
H. A. Weidenmüller, Heidelberg, and J. Zittartz, Köln
Managing Editor: W. Beiglböck, Heidelberg

43

Laser Spectroscopy

Proceedings of the Second International Conference,
Megève, June 23–27, 1975

Edited by
S. Haroche, J. C. Pebay-Peyroula, T. W. Hänsch, and
S. E. Harris

Springer-Verlag
Berlin · Heidelberg · New York 1975

Editors

Prof. S. Haroche
Université de Paris VI
Ecole Normale Supérieure
24, rue Lhomond
75231 Paris/France

Prof. J. C. Pebay-Peyroula
Lab. de Spectrométrie Physique
Université de Grenoble
B.P. 53
38041 Grenoble/France

Prof. T. W. Hänsch
Dept. of Physics
Stanford University
Stanford, CA 94305/USA

Prof. S. E. Harris
Microwave Laboratory
Stanford University
Stanford, CA 94305/USA

The Second Laser Spectroscopy Conference has been sponsored by:

- Société Française de Physique,
- Union of Pure and Applied Physics,
- Centre National de la Recherche Scientifique,
- Délégation Générale à la Recherche Scientifique et Technique,
- Direction des Recherches et Moyens d'Essais,
- Commissariat à l'Energie Atomique,
- Chromatix,
- Coherent Radiation,
- Compagnie Générale d'Electricité,
- International Business Machines Corporation,
- Instruments S.A.,
- Molectron,
- Ugine-Kuhlmann,
- Spectra Physics.

The Physics Laboratory of the Ecole Normale Supérieure and the Laboratory of Physical Spectroscopy of the University of Grenoble, have contributed materially to the organization of the conference.

ISBN 3-540-07411-2 Springer-Verlag Berlin · Heidelberg · New York
ISBN 0-387-07411-2 Springer-Verlag New York · Heidelberg · Berlin

This work is subject to copyright. All rights are reserved, whether the whole or part of the material is concerned, specifically those of translation, reprinting, re-use of illustrations, broadcasting, reproduction by photo-copying machine or similar means, and storage in data banks.
Under § 54 of the German Copyright Law where copies are made for other than private use, a fee is payable to the publisher, the amount of the fee to be determined by agreement with the publisher.

© by Springer-Verlag Berlin · Heidelberg 1975
Printed in Germany

Offsetprinting and bookbinding: Julius Beltz, Hemsbach/Bergstr.

PREFACE

This volume contains the invited papers presented at the Second Laser Spectroscopy Conference at Megeve, France, in June 1975. The goal of this Conference, as that of its predecessor at Vail, Colorado, in June 1973, was to bring together physicists, spectroscopists, chemists, and device engineers to review the current status of research and development in the rapidly advancing field of laser spectroscopy.

The organization of the conference and the preparation of the proceedings has been a stimulating task for the editors of this book who, as conference directors and chairmen of the program committee, have cordially and efficiently collaborated from both sides of the Atlantic Ocean. We have received the invaluable help of a Program Committee of nineteen experts from eight different countries. The Conference benifited from the generous support of numerous sponsors, to whom we express our gratitude and appreciation.

The Conference Program reflected the dramatic impact of the development and refinement of tunable laser sources. Powerful new linear and non-linear laser spectroscopic techniques are beginning to complement traditional spectroscopic methods and in some cases to improve their resolution and sensitivity. Applications of laser spectroscopy are now spanning a wide range from fundamental physics to chemical research. An entire session was devoted to the timely subject of laser isotope separation. Unfortunately, the free flow of information was here still severely hampered by secrecy and classification, and the wisdom of including such a "political" topic in the program has since been seriously questioned.

The manuscripts of the invited papers in this book appear in the same sequence as presented during the Conference. The titles and abstracts of 19 contributed post-deadline papers have been included. Thanks to the cooperation of the authors and the energetic efforts of the publisher it has been possible to achieve a short publication time and to make this volume available in less than four months after the Conference.

The Second Laser Spectroscopy Conference brought together almost 270 invited participants from Europe, USA, Canada, USSR, and Japan. It was held in Megeve, France, an alpine resort in the Haute-Savoie, some 30 km from the Mont Blanc. The technical sessions in the morning and evening hours were held in the Palais des Congrès of Megeve. The afternoons remained unscheduled, enabling the participants to seek personal contacts and discussions, or simply to enjoy some of the many possible walks in the mountains, which are particularly beautiful in late June. An excursion to Chamonix and the Aiguille du Midi on Wednesday, June 25, gave an inspiring impression of the majestic mountain world of the French Alps. The Conference banquet on Thursday evening, near the end of the Conference, was enlightened by the humerous and witty remarks of Professor A.L. Schawlow on lasers and technical progress.

The large number of participants and the tight program document the vigorously growing interest in the field of Laser Spectroscopy, and it is hoped that it will be possible, despite this intense pressure, to keep the next Conference of this series at a moderate size, inductive to stimulating discussions and personal contacts.

S. Haroche	*T.W. Hänsch*
J.C. Pebay-Peyroula	*S.E. Harris*
Conference Directors	Chairmen of the Program Committee

TABLE OF CONTENTS

Preface III

SPECTROSCOPY I

La préhistoire de la découverte des lasers. Absorption négative
et dispersion négative
    A. KASTLER 1

High Resolution Studies with Doppler Free Resonances; Recent
Works at MIT
    A. JAVAN 439

A New Measurement of the Relativistic Doppler Shift
    J.J. SNYDER and J.L. HALL 6

Laser - Nuclear Spectroscopy
    V.S. LETOKHOV 18

Nonlinear Spectroscopy
    N. BLOEMBERGEN 31

TUNABLE LASERS I

Recent Developments in Dye Lasers
    F.P. SCHÄFER 39

Generation of Vacuum Ultraviolet Radiation by Nonlinear Mixing
in Atomic and Ionic Vapors
    P.P. SOROKIN, J.A. ARMSTRONG, R.W. DREYFUS, R.T. HODGSON,
    J.R. LANKARD, L.H. MANGANARO and J.J. WYNNE 46

Tunable VUV Lasers and Picosecond Pulses
    D.J. BRADLEY 55

Rotation-Vibration Spectroscopy of Gases by Coherent Anti-Stokes
Raman Scattering: Application to Concentration and Temperature
Measurements
    F. MOYA, S.A.J. DRUET and J.-P.E. TARAN 66

Stratospheric Studies Using Tunable Laser Spectroscopy
　　C.K.N. PATEL 71

SPECTROSCOPY II

Spectroscopy with Spin-Flip Raman Lasers: Mode Properties and
External Cavity Operation
　　S.D. SMITH and R.B. DENNIS 79

New Laser Measurement Techniques for Excited Electronic States
of Diatomic Molecules
　　R.E. DRULLINGER, M.M. HESSEL and E.W. SMITH 91

Excimer and Energy Transfer Lasers
　　D.C. LORENTS and D.L. HUESTIS 100

Laser Fluorimetry
　　R.N. ZARE 112

Selective Photochemistry in an Intense Infrared Field
　　R.V. AMBARTZUMIAN, N.V. CHEKALIN, Yu.A. GOROKHOV,
　　V.S. LETOKHOV, G.N. MAKAROV and E.A. RYABOV 121

Laser Magnetic Resonance (LMR) Spectroscopy of Gaseous
Free Radicals
　　P.B. DAVIES and K.M. EVENSON 132

SPECTROSCOPY III

High Resolution Laser Spectroscopy of the D-Lines of on-Line
Produced Radioactive Sodium Isotopes
　　H.T. DUONG, G. HUBER, P. JACQUINOT, P. JUNCAR, R. KLAPISCH,
　　S. LIBERMAN, J. PINARD, C. THIBAULT and J.L. VIALLE 144

Comparison of Saturation and Two-Photon Resonances
　　V.P. CHEBOTAYEV 150

High Resolution Two-Photon Spectroscopy
　　B. CAGNAC 165

Optically Induced Atomic Energy Level Shifts and Two-Photon
Spectroscopy
　　J.E. BJORKHOLM and P.F. LIAO 176

Infrared Laser Stark Spectroscopy
    Y. UEDA and K. SHIMODA 186

TUNABLE LASERS II

Recent Advances in Tunable Infrared Lasers
    A. MOORADIAN 198

A Broadly Tunable IR Source
    R.L. BYER, R.L. HERBST and R.N. FLEMING 207

Broadly Tunable Lasers Using Color Centers
    L.F. MOLLENAUER 227

The Oxygen Auroral Transition Laser System Excited by
Collisional and Photolytic Energy Transfer
    J.R. MURRAY, H.T. POWELL and C.K. RHODES 239

Synchronous Mode-Locked Dye Lasers for Picosecond Spectroscopy
and Nonlinear Mixing
    L.S. GOLDBERG and C.A. MOORE 248

LASER ISOTOPE SEPARATION

Photochemistry and Isotope Separation in Formaldehyde
    A.P. BARONAVSKI, J.H. CLARK, Y. HAAS, P.L. HOUSTON and
    C.B. MOORE 259

Separation of Uranium Isotopes by Selective Photoionization
    B.B. SNAVELY, R.W. SOLARZ and S.A. TUCCIO 268

Laser Isotope Separation
    C.P. ROBINSON... 275

Isotopic Enrichment in Laser Photochemistry
    R.D. DESLATTES, M. LAMOTTE, H.J. DEWEY, R.A. KELLER,
    S.M. FREUND, J.J. RITTER, W. BRAUN and M.J. KURYLO 296

Laser Chemistry
    A.N. ORAEVSKY and A.V. PANKRATOV 304

SPECTROSCOPY IV

Atoms in Strong Resonant Fields, Spectral Distribution of the
Fluorescent Light
    C. COHEN-TANNOUDJI 324

Perturbed Fluorescence Spectroscopy
    W. HAPPER 340

Laser Spectroscopy of Small Molecules
    J.C. LEHMANN 346

Atomic Fluorescence Induced by Monochromatic Excitation
    H. WALTHER 358

On the $2P_{2/3}-2S_{1/2}$ Energy Difference in Very Light Muonic Systems
    E. ZAVATTINI 370

SPECTROSCOPY V

Ultrafast Vibrational Relaxation and Energy Transfer in Liquids
    W. KAISER and A. LAUBEREAU 380

Studies of Chemical and Physical Processes with Picosecond
Lasers
    K.B. EISENTHAL 390

Time Resolved Spectroscopy with Sub-Picosecond Optical Pulses
    C.V. SHANK and E.P. IPPEN 408

Quantum Electrodynamic Calculation of Quantum Beats in a
Spontaneously Radiating Three Level System
    A. SCHENZLE and R.G. BREWER 420

Collision Induced Optical Double Resonance
    S. STENHOLM 429

TITLES AND ABSTRACTS OF POST-DEADLINE PAPERS

Some Comments on the Dissociation of Polyatomic Molecules
by Intense 10.6μm Radiation
    N. BLOEMBERGEN 450

Excitation of Highly Forbidden Transitions by Tunable Lasers
and Search for Parity Violation Induced by Neutral Currents
    M.A. BOUCHIAT 450

Collisional Angular Momentum Mixing in Rydberg States of Sodium
    T.F. GALLAGHER, S.A. EDELSTEIN and R.M. HILL 451

Spectroscopy of Highly Excited s and d States of Potassium by
Two Photon Absorption
    M.D. LEVENSON, C.D. HARPER and G.L. EESLEY 452

High-Resolution, Two-Photon Absorption Spectroscopy of Highly-
Excited d States of Rb Atoms
    Y. KATO and B.P. STOICHEFF 452

Two-Photon Molecular Electronic Spectroscopy in the Gas Phase
    L. WUNSCH, H.J. NEUSSER and E.W. SCHLAG 453

Two-Photon Laser Isotope Separation of Atomic Uranium -
Spectroscopic Studies, Excited State Lifetimes, and Photoionization Cross Sections
    G.S. JANES, I. ITZKAN, C.T. PIKE, R.H. LEVY and L. LEVIN 454

Isotope Separation in the Solid State
    D.S. KING and R.M. HOCHSTRASSER 456

Saturated Dispersion by Laser Beam Deviation in a Saturated
Medium
    B. COUILLAUD and A. DUCASSE 456

Progress in Saturated Dispersion Spectroscopy of Iodine
    C. BORDÉ, G. CAMY and B. DECOMPS 458

Magnetic Octupole Interaction in I_2
    K.H. CASLETON, L.A. HACKEL and S. EZEKIEL 458

High-Resolution Raman Spectroscopy With a Tunable Laser
    B. BÖLGER 460

Time Dependence of the Third-Harmonic Generation in Rb-Xe
Mixtures
    H. PUELL, C.R. VIDAL 461

Generation of Tunable Coherent Radiation at 1460 Å in
Magnesium
    S.C. WALLACE and G. ZDASIUK 462

Non-Optical Observation of Zero-Field Level Crossing Effects
in a Sodium Beam
    J.-L. PICQUÉ 462

Dressed Atom Picture of High Intensity Gas Laser
    P.R. BERMAN and J. ZIEGLER 464

Nonlinear Resonant Photoionization in Molecular Iodine
    F.W. DALBY, G. PETTY and C. TAI 465

Infrared - X-Ray Double Resonance Study of $2P_{3/2}$-$2S_{1/2}$ Splitting
in Hydrogenic Fluorine
    H.W. KUGEL, M. LEVENTHAL, D.E. MURNICK, C.K.N. PATEL and
    O.R. WOOD 465

Stark Ionization of High Lying Rydberg States of Sodium
    T.W. DUCAS, R.R. FREEMAN, M.G. Littman, M.L. ZIMMERMAN
    and D. KLEPPNER 466

LA PREHISTOIRE DE LA DECOUVERTE DES LASERS.
ABSORPTION NEGATIVE ET DISPERSION NEGATIVE.

Alfred KASTLER
Ecole Normale Supérieure - Paris

I. Le mémoire d'Einstein de 1917

Le mémoire fondamental pour tout ce qui concerne les propriétés de l'émission lumineuse induite est le mémoire d'Einstein de l'année 1917, paru dans "Physikalische Zeitschrift, tome 18, page 121", intitulé : "Zur Quantentheorie der Strahlung" ([1]). C'est dans ce mémoire qu'Einstein introduit, à côté d'une probabilité d'émission spontanée $A_{m \to n}$ d'un état d'énergie supérieur E_m vers un état inférieur E_n d'un atome, une probabilité d'émission induite $B_{m \to n} \cdot \rho_\nu$, proportionnelle à la densité d'énergie ρ_ν sur la fréquence $\nu_0 = \nu_{nm} = \frac{E_m - E_n}{h}$. Il s'agit donc là d'un mécanisme d'émission provoqué par la présence de radiation auprès de l'atome, émission qui n'est pas isotrope mais qui est emportée par le rayonnement inducteur qui se trouve ainsi amplifié. D'autre part, le champ de radiation entourant l'atome produit, sur les atomes se trouvant au niveau inférieur E_n de la transition, le phénomène d'absorption, processus qui est induit également par le rayonnement et qui est caractérisé par une probabilité d'absorption $B_{n \to m} \cdot \rho_\nu$.

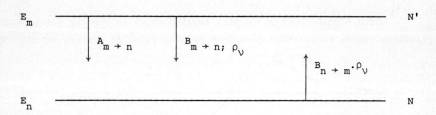

Entre ces coefficients d'Einstein existent les relations suivantes ([2]) :

$$\frac{B_{m \to n}}{B_{n \to m}} = \frac{g_n}{g_m} = \frac{g}{g'}$$

$$\frac{A_{m \to n}}{B_{m \to n}} = \frac{2h\nu_0^3}{c^2}$$

g et g' sont respectivement les poids statistiques du niveau inférieur et du niveau supérieur de la transition.

Enfin, le coefficient d'émission spontané $A_{m \to n}$ est lié à la durée de vie moyenne τ du niveau supérieur de la transition par la relation $\tau = \frac{1}{A}$.

II. Relation entre absorption et dispersion anormale ([2])

Le coefficient d'absorption d'un milieu pour une radiation de fréquence ν, k_ν, est défini par la décroissance du flux lumineux le long du trajet lumineux x dans le milieu absorbant :

$$\phi = \phi_0 \exp(-k_\nu \cdot x)$$

La bande d'absorption est définie par l'intégrale d'absorption totale $\int k_\nu d\nu$ de la bande centrée à la fréquence ν_0 du maximum d'absorption.

A une courbe d'absorption est liée une courbe de variation de l'indice de réfraction du milieu, appelée courbe de dispersion anormale. Si on considère un milieu ayant une seule bande d'absorption, la dispersion au voisinage de ν_0 est caractérisée par la relation :

$$n - 1 = \frac{c}{4\pi\nu_0} \cdot \frac{1}{\nu - \nu_0} \int k_\nu d\nu$$

dans l'échelle de fréquence, c étant la vitesse de la lumière dans le vide.

Transposé dans l'échelle des longueurs d'onde, cette relation s'écrit :

$$n - 1 = \frac{1}{4\pi c} \frac{\lambda_0^3}{\lambda - \lambda_0} \int k_\lambda d\lambda$$

Cette formule peut se déduire aussi bien de la théorie électromagnétique classique que de la relation quantique de dispersion de Kramers-Heisenberg.

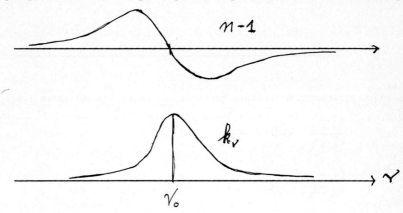

L'intégrale d'absorption $\int k_\nu d\nu$ est liée aux coefficients d'Einstein par la relation :

$$\int k_\nu d\nu = \frac{h\nu_0}{4\pi} \left[B_{n \to m} \cdot N - B_{m \to n} \cdot N' \right]$$

où N et N' sont les nombres d'atomes par unité de volume dans les niveaux inférieur et supérieur. Tenant compte des relations précédentes, cette formule peut se mettre sous la forme :

$$\int k_\nu d\nu = \frac{\lambda_0^2}{8\pi} \frac{g'}{\tau} \left[\frac{N}{g} - \frac{N'}{g'} \right]$$

Ici $\frac{N}{g}$ représente le nombre d'atomes, par unité de volume, par sous-niveau Zeeman de l'état inférieur E_m et $\frac{N'}{g'}$ représente le nombre équivalent par sous-niveau Zeeman de l'état supérieur de la transition spectrale. Cette formule montre que l'absorption de rayonnement à partir du niveau inférieur est partiellement compensée par l'émission induite qui s'exprime par le 2e terme de la parenthèse. Pour cette raison, on appelle ce terme "terme d'absorption négative".

Nous pouvons encore donner à la formule précédente la forme :

$$\int k_\nu d\nu = \frac{\lambda_0^2}{8\pi} \frac{g'}{g} \cdot \frac{N}{\tau} \left[1 - \frac{N'/g'}{N/g} \right]$$

La formule de dispersion, donnant n - 1 en fonction de la fréquence, s'écrit à son tour :

$$n - 1 = \frac{\lambda_0^3}{32\pi^2} \frac{1}{\nu - \nu_0} \frac{g'}{g} \cdot \frac{N}{\tau} \left[1 - \frac{N'/g'}{N/g} \right]$$

On a l'habitude de définir une grandeur f appelée force d'oscillation et liée à une transition spectrale ν_0 par la relation suivante :

$$f = \frac{1}{\tau} \cdot \frac{\lambda_0^2}{8\pi^2} \frac{mc}{e^2} \frac{g'}{g}$$

C'est donc une grandeur inversement proportionnelle à la durée de vie du niveau supérieur.

La formule de dispersion peut alors s'écrire en fonction de f :

$$n - 1 = \frac{e^2}{4\pi mc^2} \frac{\lambda_0^3}{\lambda - \lambda_0} Nf \left[1 - \frac{N'/g'}{N/g} \right]$$

On appelle quelquefois l'expression

$$\mathcal{N} = Nf\left[1 - \frac{N'/g'}{N/g}\right]$$

le "nombre d'électrons de dispersion" associé à la transition de l'atome.

Au terme négatif d'absorption correspond ainsi aussi un "terme négatif de dispersion".

III. Les travaux de l'équipe de recherche de Ladenburg sur la dispersion négative

L'existence de ce terme négatif de dispersion a été mis en évidence par Rudolf Ladenburg et ses élèves dans une série de recherches s'échelonnant entre les années 1926 et 1930 et publiées essentiellement dans les tomes 48 (1928) et 65 (1930) de la Zeitschrift für Physik ([3]). La plus remarquable de ces études fait l'objet de trois mémoires publiés par Kopfermann et Ladenburg sur l'étude de la dispersion du gaz néon au voisinage des raies d'émission rouge dans un tube de néon, siège d'une décharge électrique, en fonction de l'intensité du courant de décharge. Ils ont très nettement observé, en poussant le courant de décharge, la diminution des amplitudes de la courbe de dispersion anormale lorsque l'intensité du courant augmente, mettant ainsi en évidence l'influence grandissante des termes négatifs de dispersion due à l'accroissement de population des niveaux supérieurs en fonction de l'intensité du courant de décharge. S'ils avaient eu l'audace de continuer leurs expériences et d'employer des courants de décharge encore plus intenses, ils auraient sans doute réussi à obtenir des inversions de population, et la découverte des lasers aurait été avancée d'une trentaine d'années.

Il est en tout cas évident que l'inversion de population entre les deux niveaux d'une transition spectrale ne transforme pas seulement l'absorption du rayonnement en amplification, mais inverse également la courbe de dispersion anormale, et ce phénomène mériterait de faire l'objet d'investigations expérimentales systématiques ([4]).

Références

(¹) A. EINSTEIN - Physikal. Zeitschrift 18 (1917), p. 121
(²) Voir par exemple A.C.G. MITCHELL a. M.W. ZEMANSKY - Resonance Radiation and Excited Atoms, chap. III
(³) R. LADENBURG - Zeitschr. f. Physik, 48 (1928), p. 15
   H. KOPFERMANN et R. LADENBURG - ibid. p. 26 et p. 51
   A. CARST et R. LADENBURG - ibid. p. 192
   H. KOPFERMANN et R. LADENBURG - Zeitschr. f. Physik 65 (1930), p. 167
   R. LADENBURG et S. LEVY - ibid. p. 189
(⁴) A. KASTLER - Annales de Physique 7 (1962), p. 57

A NEW MEASUREMENT OF THE RELATIVISTIC DOPPLER SHIFT

J. J. Snyder[*] and J. L. Hall[†]

Joint Institute for Laboratory Astrophysics
National Bureau of Standards and University of Colorado
Boulder, Colorado 80302

It is widely believed that laser techniques will make possible a new generation of interesting tests of the fundamental concepts underlying contemporary physical thought. Given the recent progress in laser frequency stabilization and in the achievement of ever higher spectral resolution, one may imagine laser devices ultimately serving as quantum frequency standards in a number of interesting and fundamental experiments. Several such experiments being actively considered are: more precise measurements of the gravitational redshift, more sensitive tests for spatial anisotropy, and frequency comparison experiments designed to look for a secular drift of the frequency ratio of atomic clocks based on different physical principles. But happily enough, sometimes the available techniques are sufficient to make interesting measurements even before the great Laser Millenium arrives.

In this paper we report on our high precision measurements of the relativistic (or "transverse") Doppler effect using laser saturated absorption techniques on a high speed atomic beam. Previous optical measurements of the effect[1] have used comparable beam speeds, but have been limited by normal Doppler broadening to a few percent accuracy. The Mössbauer experiments[2] obtained similar accuracy by using very high spectral resolution, but were limited by the relatively low speed attainable with a mechanical rotor. Meson experiments[3] have wonderful v/c values, but extreme precision is hard to achieve in measuring the time of flight and the decay length.

Our experiment is based on the observation that the particles observed in saturation spectroscopy are free of first-order Doppler shifts and broadening. The transverse effect, however, is more persistent. It

---

[*]NRC-NBS Postdoctoral Fellow.
[†]Staff Member, Laboratory Astrophysics Division, National Bureau of Standards.

arises physically from the relativistic dilation of the apparent time interval between events occurring in a moving frame, as measured by an observer in the laboratory frame. A suitable clock for these studies is provided by a quantum transition between states of a moving atom. Because of the relativistic time dilation effect, we will find that resonance excitation of the moving atoms occurs for an applied frequency which -- in the laboratory frame -- is red-shifted from the natural atomic frequency. We note that Ives and Stilwell[1] measured the effect in emission rather than absorption as in the present case, and were therefore severely limited by collection-angle broadening.

To summarize our experiment, a beam of metastable $1s_5$ neon atoms with velocity up to $\simeq 10^{-3}$ c is produced by charge transfer of a ≤50 keV Ne^+ beam in an oven containing Na vapor. The upward 588 nm neon $1s_5 \rightarrow 2p_2$ transition is excited by a 2 cm diameter diffraction-limited optical standing wave, produced by our 100 mW frequency-stabilized dye laser.[4] The nonlinear absorption resonance, observed in the $2p_2 \rightarrow 1s_2$ (660 nm) fluorescence channel, is used to identify atomic beam particles which have zero first-order Doppler shift. Thus we can in principle have a resonance linewidth approaching the natural linewidth of ~10 MHz. Since the total relativistic time dilation for the neon atomic clocks moving with 50 keV energy corresponds to about 1368 MHz frequency shift, we are optimistic that an important accuracy improvement can be achieved -- perhaps approaching 10^{-4} of the effect itself.

To obtain the available high accuracy and to identify potential systematic errors, the experiment has been cast into the form of a differential measurement: We establish successively in time an evenly-spaced comb of stable optical dye laser frequencies. We then make high accuracy measurements of the set of atomic beam acceleration voltages which shift the atomic transition into resonance with that set of laser frequencies. Essentially we use the relativistic time dilation -- through the relation $\nu_\perp = \nu_0 \sqrt{1-(v/c)^2}$ -- to tune the atoms into resonance. A 100 MHz frequency increment to the red corresponds to a neon beam kinetic energy increase of approximately 3.7 keV. The comb of precise optical frequencies is established by sequentially locking the dye laser to consecutive orders of a stable high-finesse optical reference cavity. The necessary long-term frequency stability of the reference cavity is obtained by servo-controlling its length to the radiation of a CH_4-stabilized 3.39 μm laser. The entire system concept is illustrated schematically in Fig. 1. Before proceeding to discuss these sub-

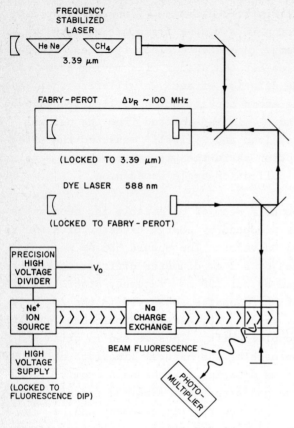

Fig. 1. Schematic diagram of the experimental apparatus. See text for description.

systems in detail, it is useful to turn our attention to the question of the basic measurement equation, the required optical alignment tolerances, and, of course, to possible residual systematic errors.

A general expression for the laboratory frequency of a planewave which resonantly excites a beam of moving absorbers of velocity $\bar{v}$ and rest frequency ω_o is given by

$$\omega_L = \omega_o \sqrt{1-\beta^2} \Big/ (1 - \beta \cos\theta) \tag{1}$$

Here $\beta \equiv v/c$ and θ is the angle between $\hat{v}$ and the optical wavevector $\hat{k}$. We are interested in the saturated absorption geometry, where the light beam is reflected back on itself to produce the standing wave and non-linear resonance dip. The angle between the two k-vectors is taken to be $(180° + 2\alpha)$, where 2α measures the deviation from perfect retroreflection. It is easy to show that there is no first-order frequency shift of the resonance due to small deviations of θ away from 90°. However in our case of a unidirectional atom beam, the finite optical angular error $\alpha \simeq 0$ does introduce a residual first-order frequency shift of the nonlinear resonance peak. One finds that the peak occurs at the frequency:

$$\omega_L = \omega_o \sqrt{1-\beta} / (1 + \alpha\beta) \qquad (2)$$

Thus the burden of precision is shifted from the atom beam to the optical beam(s). They, however, can be very sensitively studied and tested.

To minimize the linear term introduced by $\alpha \neq 0$, we use a "cat's-eye" retroreflector formed by a very high quality lens of 20 cm focal length with a highly reflecting mirror at its focus. As shown in a study[5] of this type of retroreflector, the angular field is maximized when the mirror radius is equal to the lens focal length. Techniques have been developed to set the focus (and so the average retro-angle α) to about 1/5 of the diffraction limit. The laser beam-diameter expansion factor-- and so the scale of the diffraction effect -- is still to be chosen. As a compromise between matching the time-of-flight and natural decay broadenings (at $w_o = 1.4$ cm for 50 keV) and the high saturation parameter, high signal/noise regime of smaller beam diameters, we have chosen $w_o = 1$ cm. This gives a diffraction angle $\alpha_d = (\lambda/\pi w_o) = 19$ μrad. At 50 keV the neon beam has $\beta = v/c = 2.3 \times 10^{-3}$. Thus the scale of the first-order term in Eq. (2) is about 2% of the interesting velocity-squared term, but its net influence is strongly reduced by careful centering of the laser beam on the cat's-eye symmetry axis. Finally, in the data analysis we will have to carefully account for a possible non-zero value for α. We turn now to the choice and properties of the atomic beam.

The desire for a high velocity beam of accurately known kinetic energy leads one first to consider electrostatic acceleration of light ions such as $(Li^+)^*$. However, analysis revealed unacceptable transverse acceleration due to mutual electrostatic repulsion and so led to consideration of fast ion beams which could be resonantly charge-transferred back to the neutral state. For a variety of reasons, we have chosen to work with a fast metastable neon beam produced by charge transfer of Ne^+ in a cell containing Na vapor. The spectroscopy of the relevant neon levels is indicated in Fig. 2. The figure also shows that the Na ionization potential is nearly identical to the binding energy of $Ne(1s_5)$, leading us to expect a large charge-transfer efficiency. The dye laser, tuned to 588 nm, pumps atoms upward from the metastable level, $1s_5(2p^53s[3/2])$, to the $2p_2(2p^53p^1[1/2])$ level. The transition rate $A_{ij} = 0.82 \times 10^7$ s^{-1} leads to an intensity requirement of 6 w/cm^2 for

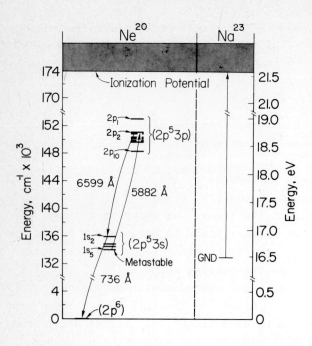

Fig. 2. Partial term diagram for Ne, showing the 5882 Å laser-pumped transition $1s_5 \rightarrow 2p_2$ (Paschen notation) and the 6599 Å fluorescence channel $2p_2 \rightarrow 1s_2$.

saturation by the dye laser at 588 nm. We monitor the strong decay fluorescence at 660 nm, $2p_2 \rightarrow 1s_2 (2p^5 3p^1 [1/2] \rightarrow 2p^5 3s [1/2])$ which has an Einstein $A_{ij} = 2.49 \times 10^7$ s^{-1}.[6] The $2p_2$ level lifetime is 18.8 ns which leads to a natural linewidth of 8.5 MHz FWHM for the 588 nm transition.

In the experiment, a few μA of neon positive ions are extracted from a low power discharge, and then focused with an Einzel lens into the sodium vapor oven about 130 cm distant. A one parameter beam energy control is approximated by relating the focus and steering voltages to the main acceleration voltage. The 3 μA typical primary beam is essentially quenched by charge transfer within the 10 cm oven. Strong fluorescence at 660 nm from the $2p_2$ level was observed with laser pumping at both 588 nm ($1s_5$-$2p_2$) and at 616 nm ($1s_3$-$2p_2$). State-resolved charge-transfer information is in principle available from these experiments, but has not yet been investigated in detail. However, it is clear from consideration of the observed fluorescent intensity that a large fraction -- probably most -- of the incident neon ions are converted to metastable neon atoms in the charge-transfer region.

We note that for the present experiment the existence of a large charge-transfer cross section is rather more than an experimental convenience: large impact parameter collisions transfer only a small amount of momentum to the target and thus preserve the velocity definition of the primary ion beam. To study this crucial factor further we measured the actual size of the neutral beam by monitoring the 660 nm fluorescence as the softly-focused 588 nm laser beam was scanned vertically. The neutral beam size did not exceed 4 mm for acceleration voltages above 5 kV, and reached $\simeq 1$ mm at 40 kV. The ion focal spot sizes could be viewed on a downstream fluorescent screen when the Na oven was not heated and were essentially the same as quoted above for the neutral beam. Geometrically, the divergence implied by a 4 mm spot size is 2.3 mrad. The angular divergence of the neutral beam was further studied by sweeping the direction of a collimated 588 nm laser beam in the horizontal plane. The first-order Doppler shift served to selectively excite only those neon beam atoms within the <<1 mrad angular cone of the laser beam. The measured angular width of the atom beam varied from 2.4 to 3 mrad as a function of the acceleration voltage. Considerations of the elementary momentum-conservation triangle of the charge-transfer collision with such well-collimated and coaxial input and output beams leads to the conclusion that the transverse impulse given to the Na target atom is appreciably less than 1% of the longitudinal momentum of the incident ion. Joint application of momentum and energy conservation then shows that to within 1×10^{-4} the neutral atom kinetic energy is equal to the kinetic energy of the accelerated positive ions. Having now considered the bottom line of the experiment's schematic plan, given on Fig. 1, we turn now to the dye laser and then to the frequency measurement problems.

The dye laser optical design is derived from the usual three mirror astigmatically-compensated design[7] and is indicated in Fig. 3. Some of the laser's salient output characteristics are also indicated. While these rather good results can be reproduced essentially every day, a certain amount of careful "twiddling" does seem to be required. Especially sensitive is the single-mode power conversion efficiency: the present usual value is about 12% although 15% was readily obtained when the mirrors were fresh. The output amplitude noise of a few percent is reduced below 10^{-3} by the amplitude servo, limited mainly by geometrical effects.

Unfortunately, the free-running stability of our present flowing-jet dye laser, about 20 MHz, is by itself not really adequate for the experiment, and it is therefore necessary to use a two-stage frequency control system. The fast loop serves to narrow the laser's spectral width to 50 kHz while preserving its tuneability; the CH_4-referenced slow loop provides the long term stability. Figure 3 shows how this

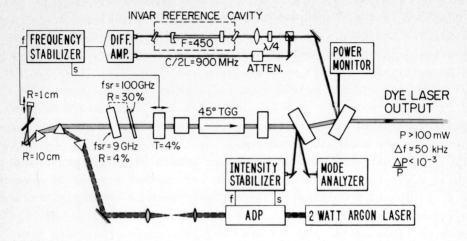

Fig. 3. Dye laser schematic. See text for details.

fast loop works: An attenuated portion of the dye-laser beam is mode-matched into a high finesse (~450) mechanically-stable optical cavity. A pair of photodiodes compare the reflected and transmitted light, allowing us to lock to the side of the transmission fringes as described previously.[4] Conversion of amplitude noise into frequency modulation is minimized by the high gain and high attack speed (>80 kc) of this frequency servo: The control system is always operating very near the balance point so that laser amplitude variations affect only the loop <u>gain</u> and are not converted to frequency noise. This system reduces the laser linewidth to ~50 kHz rms. The use of an adaptive acquisition and relocking algorithm has greatly reduced the need for operator intervention.

The high finesse of the test cavity means the dye laser can be rapidly and tightly controlled to it. Its invar construction and pressure isolation give very high intrinsic stability, thus allowing very slow updating rates (~3 s^{-1}), mainly for drift compensation. Of course the dye laser frequency precisely tracks this cavity so we must supply the desired laser frequency modulation to it, as well as the long-term stabilization information. Separate piezo tranducers are used to introduce the dc control and the ac modulation onto the invar cavity and hence onto the dye laser. This separation eliminates a dc-voltage-induced change in the small nonlinearity of the modulator piezo displacement coefficient (which would otherwise tend to appear as a coherent but spurious shift in the position of the neon atomic beam resonances).

Now we turn to the problem of long-term stabilization and frequency calibration of the dye laser. As indicated in Fig. 1 the frequency shift scale for our relativistic Doppler effect measurement is fixed by the sharp, successive axial-order transmission fringes of the reference cavity. To be free of fringe shape distortion due to admixed higher spatial modes, one should choose a cavity configuration for which the first nearly-degenerate off-axis mode is many finesse-narrowed linewidths away from degeneracy. Excitation into this mode is minimized when its azimuthal quantum number is large and odd. At present we are using a 215 cm radius mirror against a flat spaced 154 cm away, (g = 0.716). The first near degeneracy is for $\ell + m = 3$ and falls near 0.04 (c/2L), about five linewidths away from the axial order $n + 3$. Since the mode admixture is only ~10% and since only changes of the admixture affect the experiment, this mirror configuration was judged to be (just) acceptable.

To long-term stabilize this reference cavity we have chosen for simplicity to illuminate it directly with a part of the CH_4-stabilized laser's output. Without the isolation implicit in our usual two laser, frequency-offset-locked system, it is necessary to consider the question of possible contamination of the CH_4 peak signal by a weak return reflection from the transfer cavity. Thus in addition to the $\lambda/4$ polarization isolation used jointly with the visible laser beam, we use a YIG Faraday isolator to give an extra >26 dB reverse attenuation. As a result of the dual wavelength requirement for these cavity mirrors, a finesse of only 30 was obtained at 3.39 μm. Thus the infrared fringes of the transfer cavity are about four times broader than the 3/4 MHz FWHM width of the CH_4 peak, making it attractive to servo to the CH_4 peak using a third-derivative algorithm. The transfer cavity operates well with a first derivative servo since its baseline signal is zero between the transmission fringes. The laser frequency modulation at 10 kc is about 1.5 MHz pp, not too far from the optimum 1.64 and 0.7 linewidths for the third and first derivative systems respectively. Considering also that the cavity reflection reaches a minimum when it is locked on the transmission fringes, we expect no pulling above 2×10^{-11} and none is observed.

The frequency scale for our differential frequency shift measurements is given by the axial order spacing of the optical reference cavity and was accurately determined by strongly phase modulating the dye laser's output using a $LiNbO_3$ crystal driven by a ≈100 MHz rf power source (~3 W). A "vernier" action between the high order FM sidebands and the comb of cavity frequencies, together with high cavity finesse (~150), led to a very secure knowledge of cavity axial-order frequency spacing. We note that this measurement method intrinsically gives the correct optical frequency measurement even in the presence of a small wavelength variation of the mirrors' phase shift upon reflection. For the present mirror configuration we find the value (97.35 ± 0.01) MHz. This value sets the frequency scale for the experimental determination of the relativistic frequency shift and could probably be refined at least another decade if it were useful.

The frequency modulation amplitude needed to optimally recover the neon beam saturation resonances tends to be ~10-20 MHz -- much wider than the sharp 0.6 MHz transmission peaks of the CH_4-stabilized reference cavity. Thus it is attractive to use a gated, sampling-type servo to slowly control the modulated invar cavity (and thus the average dye laser frequency) so that the transmission peak of the CH_4-controlled reference cavity is centered in the 1 kHz modulation waveform. A centering precision and stability well below 10^{-3} modulation widths were readily obtained.

Thus we are finally equipped to stabilize the modulated dye laser to a "picket fence" of optical frequencies with a precision and reproducibility better than 50 kc (10^{-10}). The axial frequency mesh interval of ~100 MHz is known to 10 kc and may be divided into four-fold more known intervals by shifting the _infrared_ axial quantum number by 1, 2, or 3, using the known wavelength ratio, 5.765644. We now turn to a presentation of the data now available.

In the first experiment the reality of the shift was dramatically demonstrated by observing non-linear resonance peaks simultaneously from the fast neon beam and from a weak low pressure discharge in pure neon. Figure 4a shows the signal achieved with about one minute of integration. The reader may judge for himself how lasers will be affecting spectroscopic investigations generally by comparison of Figs. 4a and 4b, where 4b has been reproduced from the original paper of Ives and Stilwell. In their classic experiment, atomic beam fluorescence emitted in both the forward and backward directions was spectroscopically analyzed for the relativistically-predicted asymmetry between the parallel and anti-parallel Doppler shifts. The asymmetry of about 0.03 Å in a splitting of 38 Å was only marginally observable, as can be seen from Fig. 4b.

After the entire laser stabilization system was functional, a set of experiments was made in which the neon ion beam voltage was scanned

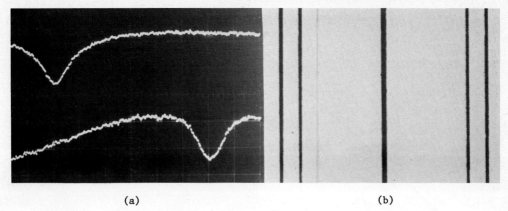

Fig. 4. Experimental evidence for the transverse Doppler shift. a) Fluorescence curves from a Ne discharge cell (upper trace) and from a 5 keV Ne atomic beam (lower trace). The 137 MHz red-shift of the atomic beam transition frequency displaces the line-center fluorescence dip of the atomic beam several linewidths from the Lamb dip in the discharge cell. b) Fluorescence lines emitted in both the forward and backward direction by a fast beam of H_2^+ and H_3^+. The center line is due to background gas fluorescence. The Doppler splitting of the beam fluorescence is slightly asymmetric due to relativistic effects. Reproduced from H.E. Ives and G.R. Stilwell, "An Experimental Study of the Rate of a Moving Atomic Clock," J.O.S.A. 28, 215 (1938).

slowly over the resonance region for the particular dye laser stabilization frequency chosen. The saturation resonances in derivative form were stored on an analog storage oscilloscope. The first data were taken with a sophisticated, adaptive servo link (JJS) between the lockin output and the high voltage power supply control. We made ten or so independent settings to the estimated center of the stored derivative trace, recording the corresponding values of the beam acceleration voltage. This method allowed an effective though subjective correction to be made for baseline tilt (voltage-dependence of the ion beam current or steering). The various lock frequency/beam voltage pairs were interleaved and rechecked occasionally to check for unexpected drifts: none were found.

These 80 points were subsequently analyzed by blocks, yielding 8 discrete acceleration voltages for 8 dye laser frequencies locked to 8 adjacent axial orders of the CH_4-stabilized reference cavity. The standard deviation of these voltage averages was typically 30 V whereas the interval between pairs was 3554 V. These data are plotted in Fig. 5.

A large number of analysis techniques have been investigated. The presently-preferred method chooses one point as the reference, and fits

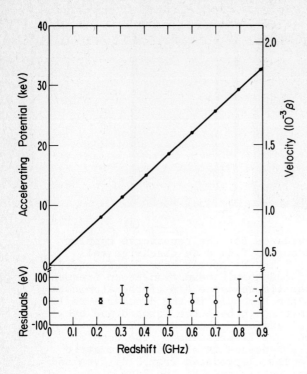

Fig. 5. Experimental results. The solid line is a least-squares fit to the eight data points. The residuals, and one-sigma error bars are shown on an expanded scale in the lower part of the figure.

the resulting set of voltage-differences to a finite difference form of Eq. (2). We regard the frequency as precisely known, with the noise appearing in the dependent variable, the beam voltage. Unfortunately the algebraic simplicity of Eq. (2) is not preserved when inverted, and it becomes convenient to use an iterative form of weighted least-squares analysis. For the present purposes it is convenient to report the frequency shift with velocity by expansion of Eq. (2)

$$\omega = \omega_0 \left(1 - \frac{2eV}{Mc^2}\right)^\gamma + \omega_0 \sqrt{\frac{2eV}{Mc^2}} \cdot \alpha \tag{3}$$

where the speed parameter β has been written in terms of the acceleration voltage V using $\beta^2 = 2eV/Mc^2$. For ^{20}Ne we have the value $2e/Mc^2 = 1.073946 \times 10^{-10}$ V^{-1}. We find with this first data set that $\gamma = .502 \pm .003$ and $\alpha = (2.5 \pm 10)$ μrad. This precision for γ, ~1/2%, is already competitive with the best previous experiment.[3] A factor ~30 improvement in the precision of the voltage points should result from closing the loop so that the atomic beam resonance derivative signal controls the beam voltage to the line center condition.

The discovery of the universal 3° blackbody radiation has made it again interesting to consider new tests[8] for the possible existence of a "preferred frame." Following Robertson,[9] we can show how certain types of preferred-frame effects would be manifest in our experimental results. Also with certain obvious modifications of our experimental techniques, it should be possible to make a very sensitive test for a potential anisotropy in the speed of light.

This work was supported in part by the National Science Foundation through Grant 39308X to the University of Colorado.

References

1. H. E. Ives and G. R. Stilwell, J. Opt. Soc. Am. **28**, 215 (1938); and H. I. Mandelberg and L. Witten, J. Opt. Soc. Am. **52**, 529 (1962).
2. D. C Champeney, G. R. Isaak, and A. M. Khan, Proc. Phys. Soc. **85**, 583 (1965).
3. A. J. Greenberg, D. S. Ayres, A. M. Cormack, R. W. Kenney, D. O. Caldwell, V. B. Elings, W. P. Hesse, and R. J. Morrison, Phys. Rev. Lett. **23**, 1267 (1969).
4. R. L. Barger, M. S. Sorem, and J. L. Hall, Appl. Phys. Lett. **22**, 573 (1973).
5. J. J. Snyder, "Paraxial Ray Analysis of a Cat's-eye Retroreflector," Appl. Opt. to appear August, 1975.
6. All lifetimes and A-values quoted here are from W. R. Bennett, Jr. and P. J. Kindelman, Phys. Rev. **149**, 38 (1966).
7. H. W. Kogelnik, E. P. Ippen, A. Dienes, and C. V. Shank, IEEE J. Quant. Electron. **QE-8**, 373 (1972).
8. J. L. Hall, to be published; and L. W. Alvarez, "New Ether Drift Experiments. Motion through the Cosmic Black-Body Radiation Field," Seminar presented Nov. 6, 1974 at Colorado State University, Fort Collins, Colorado.
9. H. P. Robertson, Rev. Mod. Phys. **21**, 378 (1949); and H. P. Robertson and T. W. Noonan in _Relativity and Cosmology_ (W. B. Saunders Co., Philadelphia, 1968), Ch. 3.

Laser - Nuclear Spectroscopy
V.S. Letokhov
Institute of Spectroscopy, Academy of Sciences USSR
Akademgorodok, Podolskii rayon, 142092, USSR

Introduction. One of the fixed ideas, which have been haunting me for a few last years, is a connection between atomic and molecular quantum transitions in the optical region of spectrum and quantum transitions in the γ-region of nuclei present in these atoms and molecules. Instead of bare nuclei we always have to deal with compound quantum systems:

$\ll$ atom $\gg$ = $\ll$ nucleus + electron shell $\gg$

$\ll$ molecule $\gg$ = $\ll$ a group of atoms $\gg$ =

$\ll$ a group of nuclei + electron shell $\gg$

Nevertheless, we are used to treat atomic or molecular spectra neglecting nuclear transition spectra, and vice versa. This is a very good approximation, but since it is an approximation it is achieved at the expense of some effects lost. I have had some papers published in which I try to reveal these lost effects. In the present report I want to summarize the results of these papers. It should be noted, that despite such an abstract statement, this problem is directly concerned with the subject of our conference, because the most effects are based on using atomic and molecular quantum transitions under laser radiation.

Basic ideas. There are at least two effects which enable us in a way to combine atomic and molecular transitions to nuclear ones. They are the recoil effect and the Doppler effect.

For a bare nuclei the lines of nuclear emission and absorption transition are intershifted by the value of recoil energy

$$R = E_o^2 / 2Mc^2, \tag{1}$$

where M is the nuclear mass and $E_o \ll Mc^2$. The emission and absorption frequency shift is caused by changes in nuclear translational state when a γ-quantum is emitted or absorbed due to the recoil effect. If a nucleus is located in an atom or in a molecule, the law of conservation of momentun and angular momentum orders not only change in translational state, for instance, of the molecule but also in its internal (electronic, vibrational and rotational) state.

Laws or conservation of momentum and energy for the system "nucleus in atom or molecule + γ-quantum" have the following form in a nonrelativistic case:

$$M\vec{v_o} \pm \hbar\vec{\kappa_\gamma} = M\vec{v},$$
$$\pm \hbar\omega_\gamma + \mathcal{E}_i + \tfrac{1}{2}Mv_o^2 = \mathcal{E}_f \pm E_o + \tfrac{1}{2}Mv^2, \qquad (2)$$

where $\vec{v_o}$, $\vec{v}$ denote the initial and final velocities of translational motion of a particle (an atom or a molecule), E_o is the energy of the nuclear transition under consideration, E_i and E_f are the initial and final internal energies of the particle; the signs «+» and «-» correspond to absorption and emission of γ-quantum. It follows from (2) that the energy of absorbed or emitted γ-radiation is determined by the expression:

$$\hbar\omega_\gamma^\pm = E_o \pm R + \hbar\vec{\kappa_\gamma}\vec{v_o} \pm (\mathcal{E}_f - \mathcal{E}_i), \qquad (3)$$

where the first term corresponds to the nonshifted transition, the second one gives recoil shifts due to change of particle translational state, the third one gives frequency shift for emission and absorption lines due to the Doppler effect, and the last term gives line shifts caused by changes in atomic or molecular internal states.

Fig.1 ahows the spectrum of γ-transitions in absorption or emission for the nucleus in an initially excited atom or molecule ($E_i > 0$). During γ-quantum emission a part of nuclear excitation energy may be transferred to the internal state of particle ($E_f > E_i$), and then a γ-satellite appears which is "red" - shifted about the emission energy $E_o - R$ for which internal state of particle remaining the same. In a like manner, the particle excitation energy together with the nuclear excitation energy may be transferred to the γ-quantum, and then a satellite appears "blue" - shifted about the line $E_o - R$. An analogous situation takes place during γ-quantum absorption as well.

For the nucleus in an atom additional satellites of the γ-line are conditioned by electron - nuclear transitions. In the case of nucleus in a molecule changes may be in the electronic, vibrational and rotational energies of the molecule, and because of this electron - vibrational - rotational - nuclear transitions occur. Naturally, the intensity of additional satellites depends on the probability of such composite transitions for the system "nucleus in an atom" or "nucleus in a molecule".

Changing the population of atomic or molecular excited states by laser radiation we can, firstly, control the intensity of composite γ-transitions and, secondly, set up new γ-transitions shifted to the long wavelengths about the γ - absorption line $E_o + R$ and shifted to the short wavelength side about the γ - emission line $E_o - R$ (Fig.1).

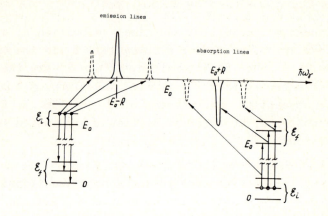

Fig.1 Spectrum of nuclear γ-transitions in an excited atom or a molecule (on the left - γ - emission lines, on the right - γ - absorption lines).

The frequency of nuclear γ-transition is shifted by the value $\vec{K}_\gamma \vec{V}_o$ due to the Doppler effect. If the distribution of nuclear velocities, that is of atomic and molecular velocities, is thermal (equilibrium), the term $\vec{K}_\gamma \vec{V}_o$ in expression (3) gives the Doppler broadening of γ-lines. By laser ratiation we can excite atoms or molecules with a certain projection of the velocity on the chosen direction (the laser wave direction), that is we can change the velocity distribution of particles at the levels connected by the laser field (Fig.2 a). For example, it is possible to have excited atoms (molecules) with the velocity $\vec{V}_{res}$ determined by the optical resonance condition:

$$\vec{K}_o \vec{V}_{res} = \omega - \omega_o , \qquad (4)$$

where $\vec{K}_o$ is the laser wave vector, ω is the laser field frequency, $E_i = \hbar \omega_o$ is the atomic (molecular) transition energy. It is clear

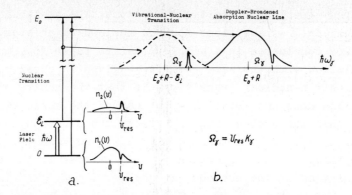

Fig.2 Formation of narrow resonances of γ-absorption when an atom or a molecule is excited by a coherent light wave in low--pressure gas.

that the absorption spectral line of the composite γ-transition, in which excited atoms (molecules) with a nonequilibrium velocity distribution participate, will have a narrow resonance peak (Fig.2 b) rather than an ordinary Doppler profile. The frequency of this peak is shifted about the centre of the line ($E_o + R - E_i$) by the value

$$\Omega_\gamma = \vec{K}_\gamma \vec{v}_{res} = \frac{\omega - \omega_o}{\omega_o} \omega_\gamma . \qquad (5)$$

It can be tuned within the whole of the Doppler contour of γ-line when tuning the laser field frequency along the Doppler line if optical transition absorption occurs (!).

The idea to obtain narrow tunable γ-resonances of absorption and emission was proposed in papers /1,2/ in 1972. The occurence of vibrational sattelites of nuclear γ- transitions in a molecule was considered in the simplest classical model in these papers as well. It is evident, that γ-lines free of Doppler broadening can be obtained not only at the frequencies of composite γ-transitions but also in any case, when the nuclear velocity distribution is changed in some way by laser radiation /3,4/. Therefore, the both approaches to the change of the γ-transition spectrum under laser radiation (additional satellites and narrow resonances arising in the Doppler profile of γ-lines) may be used both together and separately.

Let's consider now the specific quantum systems (an atom, a diatomic molecule, a polyatomic molecule, a positronium) where these ideas can be realized, and calculate for them the probabilities of such composite quantum transitions.

Electron - nuclear transitions in atoms. The possibility of electron-nuclear γ-transitions for the nucleus in an atom and their intensity are considered in a simple model in paper /6/, and a rigorous calculation with the same results is given in paper /10/. The cause of electron-nuclear transitions is that the centre of inertia of the nucleus does not coincide with that of the whole atom, and because of this the nuclear recoil affects the electron motion, and vice versa.

The optical electron coordinate $\vec{r}$ is related to the coordinate of centre of mass of the nucleus $\vec{R}$ as:
$$\vec{R} + \frac{m}{M}\vec{r} = 0, \qquad (6)$$
where the coordinates begin at the centre of mass of the atom, m and M are electron and atomic masses respectively. The probability of the γ-transition "a" $\to$ "b" with the change in optical electron quantum state "i" $\to$ "f" is given by the expression
$$W_{fi}^{ba} = A_{ba}|\langle \Psi_f^*(\vec{r})|e^{-i\vec{K}_\gamma \vec{R}}|\Psi_i(\vec{r})\rangle|^2 = A_{ba} P_{fi}, \qquad (7)$$
where A_{ba} is the probability of the γ-transtition "a" $\to$ "b" between two levels of the bare nucleus, $\vec{K}_\gamma$ is the wave vector of γ-quantum, $\Psi_{i,f}(\vec{r})$ denotes wave functions for the electron state, and the coordinates $\vec{r}$ and $\vec{R}$ are connected by expression (6). The vibration amplitude for the centre of mass of the nucleus in an atom is much smaller than $\lambdabar_\gamma$, that is $\vec{K}_\gamma \vec{R} \ll 1$, and the expression for the electron transition probability P_{fi} reduces to /6/:
$$P_{fi} = (k_\gamma \frac{m}{M})^2 |\langle \Psi_f^*(\vec{r})|\vec{n}_\gamma \vec{r}|\Psi_i(\vec{r})\rangle|^2 = (k_\gamma \frac{m}{M})^2 (\vec{r}_{if} \vec{n}_\gamma)^2, \qquad (8)$$
where $i \neq f$, $\vec{n}_\gamma$ is the unit vector in the $\vec{K}_\gamma$ direction, $e\vec{r}_{if}$ is the matrix element of dipole momentum of the transition "i" $\to$ "f". The probability of initial atomic state conservation $P_{ii} \simeq 1$. Evidently owing to the condition $\vec{K}_\gamma \vec{R} \ll 1$ probabilities for $i \neq f$ will be $P_{fi} \ll 1$.

The γ-transition frequency with a change in electron state is determined by expression (3). If before emission a γ-quantum the atom is in the ground state ($E_i = 0$), all electron satellites are distributed on the long-wave side of the basic emission line $E_o - R$ and spaced at intervals equal to the excitation energy of the corresponding states of electron shell. The intensity of different satellites, according to (8), is proportional to the square of electron transition dipole momentum and, hence, decreases in proportion to n_f, where n_f denoted the principal quantum number of final state.

The relation between the intensities of the satellite and the basic line can be estimased by the formula

$$K_{on} = \mathcal{E}_{on} [eV] (f_{on} E_\gamma [MeV]/A)^2, \qquad (9)$$

where E_{on} is the energy (eV) of the electron transition "o" → "n", f_{on} is the oscillator strength for the transition "o" → "n", A is the atomic mass (in atomic units). For instance, for a nuclear transition in the Ne^{21} isotope with the energy $E_o = 6$ MeV the intensity of the satellite, which corresponds to the excitation of the resonance level Ne I with $E_{o1} = 16.7$ eV, will be $K_{o1} \simeq 5 \cdot 10^{-3}$. Atomic excitation in γ-emission can be detected with subsequent fluorescence.

The relative intensity of electronic satellites is small in view of a weak connection between electron motion and nuclear motion during resoil.

Vibrational - nuclear transitions in a polyatomic molecule.
It is clear from simplest considerations that, when the nucleus in a molecule absorbs or emitts a γ-quantum, it feels a instantaneous shock which excites the molecular vibrations where this nucleus takes part in. If the emitting or absorbing nucleus «n» is off the centre of mass of the molecule, such a shock gives rise to molecular rotation inevitably. The probability of changes in the vibrational or rotational molecular states, unlike changes in the atomic electron state, is not small and should be accounted even in a zero -order approximation. Vibrational - nuclear transitions in a polyatomic molecule with the nuclear «n» at the centre of mass of the molecule (the nuclear coordinate $\vec{r} = o$ about the centre of mass of the molecule), when the molecular rotational state remains constant, are studied in detail in papers /5,7/. Here only some results are given.

The probability of vibrational - nuclear transition is determined by the expression like (7), where $\Psi_{i,f}$ means wave functions of molecular vibrational states depending on the coordinate of nuclear inertia centre $\vec{R}$. The nuclear inertia centre $\vec{R}$ in a molecule is positioned so:

$$\vec{R} = \vec{R}_o + \vec{r} + \vec{u}, \qquad (10)$$

where $\vec{R}_o$ is the position of the centre of mass of the molecule, $\vec{r}$ is the vector connecting the centre of mass of the molecule with the equilibrium position of the nucleus, $\vec{u}$ is the vibrationsl displacement of the centre of mass of the nucleus about equilibrium position. With $\vec{r}=o$, the matrix element giving the change in molecular state reduces to:

$$\langle \Psi_f^*(\vec{R})|e^{-i\vec{k}_\gamma \vec{R}}|\Psi_i(\vec{R})\rangle = \langle \Psi_f^*(\vec{u})|e^{-i\vec{k}_\gamma \vec{u}}|\Psi_i(\vec{u})\rangle \quad (11)$$

where the wave function $\Psi(\vec{u})$ gives the vibrational displacement of the nucleus «n». The nuclear vibrational displacement can be given as a shift superposition owing to different normal molecular vibration:

$$\vec{u}_n = \sum_e Q_e \vec{P}_{en}, \quad (12)$$

where Q_e is the normal l-th molecular coordinate, $\vec{P}_{en}$ is the shift of the n-th nucleus due to the l-th normal vibration.

In a harmonic approximation it is easy to deduce an expression for probability of the γ-transition which is folowed by the transition $0 \to \nu_e$ of an initially-unexcited molecule /5,7/:

$$P_{\nu_e,0} = \frac{1}{\nu_e!} Z_e^{\nu_e} e^{-Z_e}, \quad (13)$$

as well as for the vibrational transition $1 \to \nu_e$ of the molecule initially excited to the first vibrational level:

$$P_{\nu_e,1} = \frac{1}{\nu_e!} Z_e^{\nu_e-1}(Z_e - \nu_e)^2 e^{-Z_e} \quad (14)$$

Parameter Z_e is the average vibrational energy (in terms of $\hbar\omega_e$) assumed by an unexcited molecule due to emission or absorption a γ-quantrum. This parameter depends on the relations of the recoil energy R to the energy of vibrational quantum $\hbar\omega_e$, and the mass of the nucleus «n» to that of the whole molecule. The position of vibrational - nuclear transitions is determined by expression (3), where E_i and E_f denote the initial and final vibrational energies of the molecule.

And now, without deducing other formulas (averaging over molecular orientation, inclusion of vibration anharmonizm, etc.), we shall list the numerical data for vibrational - nuclear transitions of the nucleus of ^{188}Os in the OsO_4 molecule obtained in paper /8/. The ^{188}Os nucleus can be excited to a state with E_0 = 155 KeV, and the vibration ν_3 of the OsO_4 molecule with $\hbar\omega_3$ =0.119 eV can be excited by CO_2 - laser radiation (experimental studies on exciting the vibrations of the OsO_4 molecule by CO_2 - laser radiation are performed in paper /11/). At emission or absorption a γ-quantum the shock of the Os nucleus may change the state of the triple - degenerated vibration ν_3. The parameter Z_e for this vibration will be /8/

$$Z_3 = \frac{R}{\hbar\omega_3} \frac{4m_0}{m_{Os}} = 0.146 \quad (15)$$

The recoil energy R = 0.0512 eV and, hence, the molecule assumes a comparatively small vibrational energy due to recoil. Table 1 gives values of probabilities averaged over molecular orientation for the most intensive vibrational - nuclear transitions $\bar{P}_{v_f v_i}$ of the $^{188}Os^{16}O_4$ molecule.

Table 1.

Probabilities $\bar{P}_{v_f v_i}$ for $^{188}OsO_4$

$v_f \backslash v_i$	0	1	2	3	4	5
0	0,865	0,0956	0,0057	0,00023		
1	0,0956	0,684	0,160	0,050		
2	0,0057	0,160	0,532	0,273	0,031	
3	0,00023	0,050	0,273	0,405	0,360	0,043
4			0,031	0,360	0,297	0,444
5				0,043	0,444	0,250

Fig.3 shows Doppler - broadened lines of γ-absorption of vibrational - nuclear transitions of ^{188}Os in the OsO_4 molecule, as well as the position of γ-radiation line of Os (the dashed line) in a solid state source of γ-radiation.

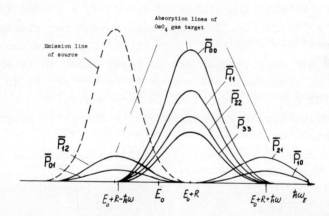

Fig.3 Doppler-broadened lines of γ-absorption of vibrational-nuclear transitions for ^{188}Os in OsO_4 molecule. The dashed line shows the emission of solid state Os-source of γ-radiation.

Exciting the vibration ν_3 of the OsO_4 molecule by CO_2-laser radiation we may control the intensity of long-wave vibrational satellites $\bar{P}_{01}$ and $\bar{P}_{12}$ of γ-absorption line and, hence, change the absorption coefficient of γ-radiation by gaseous $^{188}OsO_4$ target. The ^{188}Os nucleus, after having absorbed a γ-quantum, transits to the ground state by emitting a quantum and owing to internal conversion, that is the energy of nuclear excitation is transferred to the electron shell. There are two ways to observe the effect of γ-radiation absorption modulation by: detecting of scattered γ-quanta and detecting of conversion electrons. The main problem of such an experiment is that, for a good signal-noise ratio to be schieved with an intensity of γ-radiation source of 1 Curie, a gaseous OsO_4 target at a pressure of some torr at least should be used. At such a pressure the CO_2- laser radiation induces thermal heating of the gas /11/. This leads to thermal population of vibrational levels and to thermal broadening of the Doppler profile of γ-absorption line that mask the effect of γ-absorption modulation by vibrational satellites. The necessity of high intensity of γ-radiation sources and small γ- radiation absorption coefficient makes it difficult to conduct experiments at a low pressure of OsO_4 when without Doppler-broadening narrow resonances of γ-emission and absorption can be obtained.

<u>Diatomic molecule.</u> When a nucleus in a diatomic molecule emitts or absorbs a γ-quantum there may occur a simultaneous change in the electronic, vibrational and rotational states. As opposed to polyatomic molecules with the centre of symmetry, a change in rotational state here is inevitable. The full calculation for electron - vibrational - rotational - nuclear transitions of the nucleus in a diatomic molecule has been carried out in paper /12/. As it shows, the probability of electron state change is determined by an expression similar to (8) for the atom, that is very small.

The probability of vibrational state change is given by the same expressions as for one normal vibration of a polyatomic molecule. The expression for probability of the vibrational transition $v_i \rightarrow v_f$ takes the form /12/:

$$P_{v_f v_i} = \frac{v_i!}{v_f!} e^{-z} z^{v_f - v_i} \left[L_{v_i}^{v_f - v_i}(z) \right]^2, \qquad (16)$$

where $z = z_o \cos^2 \nu$, $z_o = \frac{R}{\hbar \omega_o} \frac{m_2}{m_1}$, ν is the angle between $\vec{K}_\gamma$ and the molecular axis, $\hbar \omega_o$ is the vibration frequency, m_1 and m_2 denote the atomic masses, L is the Laguerre polynomial. With $v_i = 0,1$ expression (16) reduces to (13) and (14), respectively.

The probability of rotational state change in a rigid rotator approximation can be also calculated exactly. In an case of the transition $J_i = 0 \rightarrow J_f = J$ the expression for transition probability takes the form:

$$\bar{P}_{J,0} = \frac{\pi}{2a}(2J+1)\mathcal{J}_{J+\frac{1}{2}}(a),\qquad(17)$$

where $a = K_{\gamma}r_0 m_2/(m_1+m_2)$, r_0 is the internuclear spacing, m_1 is the mass of the nucleus interacting with γ-radiation, $\mathcal{J}$ is the Bessel's function, the line above means orientation averaging. The parameter a is equal to the average value of the angular momentum (in units of $\hbar$) transferred to the molecule.

It is worthy of notice that the vibrational and rotational degrees of freedom of a diatomic molecule derive of give up the same energy which may be both more (with $m_1 < m_2$) and less (with $m_1 > m_2$) than the recoil energy R. It is apparent that, when the nuclear masses differ greatly and γ-radiation interacts with the lighter nucleus, vibrational - rotational satellites will absorption (emission) lines by distances larger than the value of the R energy of recoil to molecular translational motion.

<u>Atom of positronium</u>. Let's consider now the method of producing narrow and tunable lines of positronium annihilation radiation at $\hbar\omega_{\gamma}$ = 0.511 MeV. This method is based on selective conversion of ortho-positronium atoms to parapositronium under laser radiation and magnetic field /9/.

As known /13/, a considerable part of slow positrons annihilate through the formation of positronium atoms, 25% through the formation of para-positronium atoms (p-Ps) and 75% through that of ortho--positronium atoms (o-Ps). The p-Ps atoms reach quickly the ground singlet state $^1S_0$, where they annihilate in a time $\tau_s^o = 1.25 \cdot 10^{-10}$ sec emitting two 0,511 MeV γ-quanta (Fig.4). The o-Ps atoms reach the ground triplet state $^3S_1$ also quickly, where they line 10^3 times longer than the p-Ps atoms and annihilate in a time $\tau_T^o = 1,4 \cdot 10^{-7}$ sec emitting three γ-quanta with total energy of 1,022 MeV. Thus, the positronium atoms emit a line of 2γ-annihilation in a short time τ_s^o and continuous spectrum of 3γ- annihilation in a longer time τ_T^o. The line of 2γ- annithlation is inhomogeneously broadened due to the Doppler effect. The relative value of Doppler width will be $\Delta\omega_D^{\gamma}/\omega_{\gamma} = (W_0/m_0c^2)^{1/2} = 10^{-3} - 10^{-4}$, where W_0 is the positronium kinetic energy. Natural width annihilation line is only $\Gamma = 1/\tau_s^o \simeq 10^{10} \text{sec}^{-1}$.

Narrowing and tuning of the positronium 2γ- annihilation line by laser radiation consists in converting the o-Ps atoms, for which the 2γ - annihilation is forbidded, to p-Ps atoms, and the conversion mechanism should make possible a selective conversion of the o-Ps atoms with a definite projection of their velocity to the chosen direction. This may be accomplished by the following method.

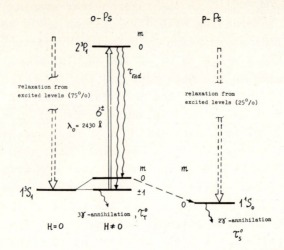

Fig. 4 The scheme of velocity-selective ortho-positronium conversion from the state 1^3S_1 to the state 1^1S_0 under the action of magnetic field and laser radiation.

In the ground triplet state the o-Ps atoms are distributed over three magnetic sublevels (m = 0, $\pm$ 1). On switching on a stationary magnetic field of several KG the sublevels m=o of the states $^1S_0$ p-Ps and $^3S_1$ o-Ps get mixed. Due to this mixing the o-Ps atoms on the sublevel m=o undergo 2γ-annihilation /13/. As a result, only o-Ps atoms on states m= $\pm$ 1 stay in a long-lived triplet state, which must be velocity-selectively converted to the sublevel m=0.

Assume that the Zeeman δ - components of the line L_α o-Ps at λ_0=2430 Å are interacted with laser wave that performs stimulated transitions from the states $1\ ^3S_1$, m=$\pm$1 to the state 2^3P_1, m=o. Only those atoms, the Doppler-shifted transition frequency of which coincides with the laser wave frequency ω, go to an excited state, i.e.

$$\omega = \omega_0 \frac{(1-v^2/c^2)^{1/2}}{1-(\vec{v}/c)\vec{n}} \quad \text{or} \quad (\omega - \vec{K}\vec{v}) = \omega_0 \left(1 - \frac{v^2}{c^2}\right)^{1/2}, \qquad (18)$$

where ω_0 is the L_α line frequency for fixed o-Ps atoms, $\vec{K} = \vec{n}\frac{\omega}{c}$ is the wave vector, $\vec{v}$ is the velocity of positronium.

The excited atoms o-Ps during the time τ_{rad} = 3.2·10^{-9} sec spontaneously return to the sublevels of the 1^3S_1 ground state, one third of the atoms falling on the sublevel m=o, where 2γ-annihilation occurs. Two thirds of the atoms return to the m=$\pm$ 1 initial sublevels from which they may again be excited by the laser field to the state 2^3P_1, m=o. For conversion of all atoms o-Ps laser intensity should be about $0,1 W/cm^2$.

The radiation of 2γ-annihilation of the fields converted o-Ps atoms with the resonant velocity $\vec{v}$ determined by condition (18), that is observed in the laser wave direction, has the frequency:

$$\omega_\gamma = \omega_{\gamma o}\left(1-\frac{v^2}{c^2}\right)^{1/2} \Big/ \left(1-\frac{\vec{v}}{c}\vec{n}\right) = \omega_{\gamma o}\frac{\omega}{\omega_o}, \qquad (19)$$

where $\hbar\omega_{\gamma o} = m_o c^2$. Thus, tuning the laser frequency along the Doppler profile of the ortho-positronium line L_α we can tune the annihilation radiation frequency along the Doppler profile of the 2γ-annihilation line.

The width of annihilation radiation narrow line is determined by the width of resonance region of condition (18), that is by the natural width of the L_α line:

$$\Delta\omega_\gamma/\omega_\gamma = (\omega_o \tau_{rad})^{-1} = 4 \cdot 10^{-8} \qquad (20)$$

on condition that second order Doppler broadening ($W_o/m_o c^2$) = (v_o^2/c^2) is less, v_o is average positronium velocity. The interval of tuning is determined by the relative width of Doppler profile (the same for L_α - line and 2γ- annihilation line), i.e.

$$\Delta\omega_{tun}^\gamma/\omega_\gamma = \Delta\omega_D^{opt}/\omega_o = (W_o/m_o c^2)^{1/2} = 10^{-3} - 10^{-4} \qquad (21)$$

To conduct of this experiment there should be gas of positronium with thermal velocities in the vacuum. This can be accomplished by injecting a positron beam to the volume in a porous shell (a cell with its walls of high-dispersion medium). In the walls the positrons slow down, positronium atoms appear, and diffusion to vacuum and to the region of laser and magnetic fields occurs.

- - -

So, in this report I took an attempt to discuss some effects in the region of γ-radiation caused by nuclear transitions and annihilation and which are connected with optical atomic and molecular transitions; hence, its can be controlled by laser radiation. To this problem we have a reverse approach as well, that is occurence of nuclear interactions in optical molecular and atomic spectra. As an example, we can refer to recently suggested experiments, which violence of parity during weak interactions in atomic /14/ and molecular spectra /15/. We hope that such ingenious effects will be revealed with progress in laser spectroscopy methods.

References

1. V.S.Letokhov. Pis'ma Sh.Eksper. I.Teor. Fiz. <u>16</u>,428 (1972)
2. V.S.Letoknov. Phys.Rev. Letters. <u>30</u>, 729 (1973)
3. V.S.Letokhov. Phys.Letters. <u>43A</u>, 179 (1973)
4. V.S.Letokhov. Proceedings of Intern.Colloquium "Methodes de Spectroscopie sans Largeur Doppler de Niveaux Excites de Systemes Moleculaires Simples", Aussois, France, 23-26 May 1973 (Publ. N217, CNRS, Paris, 1974), p.127
5. V.S.Letokhov. Phys.Letters. <u>46A</u>, 257 (1974)
6. V.S.Letokhov. Phys.Letters. <u>46A</u>, 481 (1974)
7. V.S.Letokhov. Phys.Review (in press); Preprint of Spectroscopy Institute, USSR, N167 (1973)
8. O.N.Kompanetz, V.S.Letokhov, V.G.Minogin. Preprint of Spectroscopy Institute, USSR, N6/138 (1974)
9. V.S.Letokhov. Phys.Letters. <u>49A</u>, 275 (1974)
10. L.N.Ivanov, V.S.Letokhov. Zh.Eksper. I.Teor. Fiz. <u>68</u>, 1748 (1975)
11. O.N.Kompanetz, V.S.Letokhov, V.G.Minogin. Kvantovai Electronika. <u>2</u>, 370 (1975)
12. V.S.Letokhov, V.G.Minogin. Zh.Eksper. I.Teor.Fiz. (in press)
13. V.I.Goldanskii. Physical Chemistry of Positron and Positronium. (Publ. "Nauka", Moscow, 1968)
14. M.A.Bouchiat, C.C.Bouchiat. Phys.Letters. <u>48B</u>, 111 (1974)
15. V.S.Letokhov. Phys.Letters (in press).

NONLINEAR SPECTROSCOPY

N. Bloembergen
Division of Engineering and Applied Physics
Harvard University
Cambridge, Massachusetts 02138, U.S.A.

I - HISTORICAL INTRODUCTION

With the formulation of quantum mechanics by Dirac[1] and others in the late twenties, it became clear that the interactions of electromagnetic radiation and matter also involved processes in which several photons are involved. An early example is provided by Raman scattering. In this process one photon is absorbed, another is emitted and the difference in photon energy is taken up by a transition in the material system from an initial state $|i\rangle$ to a final state $|f\rangle$. The symbolic Hamiltonian describing this event is proportional to $a_1^+ a_2 c_i c_f^+$, where a_1^+ represents a creation operator for photons in mode 1, a_2 represents an annilation operator for photons in mode 2, while c_i takes an electron out of state $|i\rangle$, c_f^+ puts one electron in state $|f\rangle$. The closely related two-photon absorption process is described by a term in the Hamiltonian proportional to $a_1 a_2 c_i c_f^+$. The theory for this letter process was developed by Maria Goeppert-Mayer in her Ph.D. thesis at the University of Gottingen.[2] While the Raman effect was discovered in 1927, the two-photon absorption process was first demonstrated[3] in 1961, about thirty years after the theoretical paper of Mayer. This difference in the experimental development can be explained by the fact that the Raman scattering involves the spontaneous emission of a Stokes-shifted photon. The spontaneously scattered light intensity at ω_2 is proportional to the intensity of the incident field at ω_1. The power absorbed in the two-photon absorption process is, however, proportional to the square of the incident intensity. Its demonstration consequently requires a high light intensity, and had to await the advent of lasers. Furthermore, the frequency of the laser must be adjustable if the two-photon transition occurs between sharp energy levels and no accidental coincidences between laser frequency and energy separations exist. It is therefore understandable that two photon spectroscopy could really develop fully only after high power tunable dye lasers became available.[4]

The current interest in two-photon absorption spectroscopy is based on the following characteristics:

1. The initial and final states have the same parity.

2. The final state may have an excitation energy in the far UV, while the incident light beam has a frequency in the near UV or blue part of spectrum.

3. It is possible to eliminate momentum transfer between the electromagnetic field and the atom or molecule, and consequently to eliminate Doppler broadening.

The absence of Doppler broadening, if two photons of equal energy and opposite momentum are absorbed, was first analyzed by Chebatoyev and coworkers[5] in 1970, and experimentally demonstrated by Cagnac et al.[6] and by Levenson[7] in 1974. Numerous results have subsequently been published and Doppler-free two photon spectroscopy promises to be an important new spectroscopic tool. In saturation "Lamb-dip" spectroscopy, a small segment of the inhomogeneous Doppler distribution of resonant frequencies is selected to contribute to the signal. In two photon absorption without momentum transfer all atoms or molecules contribute to the signal. A simple way to see this is to consider the case of an atom at rest with the frequency ν of the light wave adjusted so that $2h\nu$ corresponds exactly to the energy difference between two sharp energy levels with the same parity. Now consider an atom moving with a velocity component v parallel to one of the light beams. The apparent frequency of this beam is down-shifted by an amount $-(v/c)\nu$. The apparent frequency of the light beam, propagating in the opposite direction, is up-shifted by the same amount, $+(v/c)\nu$. Thus the two linear Doppler shifts cancel each other exactly, and for any atom the resonant condition is fulfilled, if quadratic Doppler shifts are ignored.

II - DOPPLER-FREE TWO-PHOTON SPECTROSCOPY

The first transition to be investigated was the 3s-5s two-photon line of the Na^{23} atom. The frequency falls in the range of the Rhodamine-6G dye laser, which has a very high efficiency, and the strength of the transition is enhanced by the presence of the 3p level almost half-way between the 3s and 5s level, as shown in Figure 1a.

The experimental resolution is determined by the spectral width and stability of the dye laser. For transitions with strong second order matrix elements, the power of cw dye lasers is adequate for detection. An instrumental width of a few MHz is not uncommon in this case. The 1s → 2s transition in hydrogen gas, where the life time for spontaneous two-photon decay of the metastable 2s-level is 0.14 sec., could be determined with unprecedented spectroscopic precision if dye

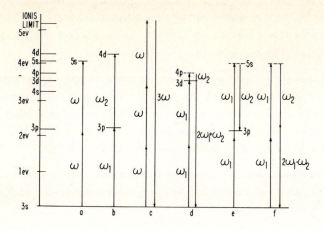

Figure 1: Nonlinear Processes in the Na atom
  a. Two-photon absorption
  b. Two-photon absorption with resonance of intermediate level
  c. Third harmonic generation
  d. Sum-frequency three wave mixing
  e. Hyper-raman three-photon process
  f. Parametric mixing of light waves at ω_1 and ω_2 with generation of the combination frequency $2\omega_1 - \omega_2$

lasers of sufficient power and frequency stability could be developed. Hansch[8] has already determined the Lamb shift of the 1s state of hydrogen and deuterium from the observation of this two-photon transition. Second order Doppler broadening and interaction time of atoms with the beam would, of course, set certain limits.

Doppler-free two-photon spectroscopy has permitted the following observations on high-lying s- and d-states of alkali atoms, such as Na, K and Rb.

   a. hyperfine structure of s-states[9]
   b. fine structure of d-states[10,11]
   c. Zeeman splittings[12,13]
   d. Stark splittings and shifts[14]
   e. Isotope shifts[15]
   f. Light-induced power-dependent shifts and broadening[16]
   g. Collision induced shifts and broadening.[17]

The selection rules for electric-dipole two-photon transitions may be derived in a straightforward manner. If the intermediate levels are sufficiently far from resonance, $\hbar\lambda \ll \omega_{gi} - \omega$, so that the summation over intermediate levels gives equal weight to all sublevels in a multiplet, the selection rules are determined purely by the symmetry of the initial and final states. The selection rules and the Zeeman effect for $S \to nS$ transitions have some unique features,[12] because of the spatial isotropy of initial and final states.

Two photon transitions can be enhanced to the strength characteristic of one-photon allowed transitions, by tuning the frequencies of two dye laser beams so that they correspond to two one-photon transitions in cascade. The virtual intermediate level tends to become a real one. Such experiments have been performed by Bjorkholm and Liao[18] in the Na-atom as sketched in Figure 1b.

The theoretical situation is analogous to resonant Raman scattering and the cascade process of resonant absorption with subsequent emission of fluorescent light. The distinction between a two-photon process and a cascade of two one-photon processes is based on the role played by the intermediate state $|n\rangle$, represented by the diagonal element of the density matrix ρ_{nn}, in the calculation of the expectation value of polarization. A general calculational framework exists, which encompasses a wide variety of nonlinear phenomena. Besides two photon absorption and raman processes it is capable of describing saturation spectroscopy, coherent quantum beats, parametric frequency up and down conversion, and the interrelationships between these various processes.

III - NONLINEAR SUSCEPTIBILITIES

The general evolution of a quantum mechanical system is described by the equation of motion for the density matrix

$$\dot\rho = -i\hbar^{-1} [\mathcal{H}_o + \mathcal{H}_1, \rho] \tag{1}$$

where $\mathcal{H}_o$ is the Hamiltonian for the unperturbed atom or material system. The interaction with the electromagnetic fields which for the present purpose will be considered as classical quantities, is in the electric dipole approximation described by,

$$\mathcal{H}_1 = -\Sigma \, e \, \underset{\sim}{r} \cdot \underset{\sim}{E} \tag{2}$$

where the summation is over the particles in the system. For one-electron systems the expectation value for the electrical polarization is given by

$$\underset{\sim}{P} = \langle \Sigma e \underset{\sim}{r} \rangle = N \, \mathrm{Tr}\,(e \underset{\sim}{r} \rho) \tag{3}$$

where N is the number of systems per unit volume. The set of equations must be made self-consistent by having $\underset{\sim}{P}$ and $\underset{\sim}{E}$ satisfy Maxwell's wave equation

$$\nabla \times \nabla \times \underset{\sim}{E} + \frac{1}{c^2} \frac{\partial^2 \underset{\sim}{E}}{\partial t^2} = -\frac{1}{c^2} \frac{\partial^2 \underset{\sim}{P}}{\partial t^2} \tag{4}$$

In order to proceed further with solutions often only a very small number of modes of the electromagnetic field and a very small number of energy levels and eigenfunctions of the Hamiltonian $\mathcal{H}_o$ are considered. The case of one traveling electromagnetic mode and a two-level material system leads to a description of laser action and self-induced transparency. It is always possible to find solutions which are correct for arbitrarily large amplitudes in one of the electric fields by transforming to a rotating coordinate system. In saturation spectroscopy, two photon and raman spectroscopy, involving only one pair of energy levels or three energy levels, numerous solutions have been discussed in the literature.

If one wishes to focus attention on the transition from non-resonant to resonant processes and give a comparative survey of a large variety of nonlinear processes, a perturbation expansion in terms of powers of the amplitudes of the externally applied electromagnetic fields is indicated.[19] This procedure permits to retain interference effects between non-resonant parametric processes, and dissipative processes with power absorption or emission at one or more frequencies. The steady state nonlinear response of media with inversion symmetry (atomic vapors, molecular gases and fluids, crystals with a center of symmetry) can often be adequately described by the lowest nonvanishing term. This polarization cubic in the applied field amplitudes may be written phenomenologically as

$$P_i^{NL}(\omega_4) = \chi_{ijk\ell}^{(3)}(-\omega_4, \omega_1, \omega_2, \omega_3) E_j(\omega_1) E_k(\omega_2) E_\ell(\omega_3) \tag{5}$$

A perturbation expansion of Eqs. (1-3) leads to the following expression of the xxxx-component of the nonlinear susceptibility tensor,

$$\chi_{xxxx}^{(3)}(-\omega_4, \omega_1, \omega_2, \omega_3) = N e^4 \hbar^{-3} \sum_{n, n', n''} x_{in} x_{nn'} x_{n'n''} x_{n''i}$$

$$\left[(\omega_{ni} - \omega_1 + i\Gamma_{ni})(\omega_{n'i} - \omega_1 - \omega_2 + i\Gamma_{n'i})(\omega_{n''i} - \omega_4 + i\Gamma_{n''i}) \right]^{-1}$$

$$+ \text{Perm}(1, 2, 3, 4). \tag{6}$$

The other 23 terms correspond to permutations of the order in which the photons at ω_1, ω_2 and ω_3 are absorbed and the photon at the sum frequency ω_4 is emitted. Similar expressions can be written down for the other components of the third-rank susceptibility tensor. Resonances occur when one of the frequencies ω_1, ω_2, ω_3 and ω_4 or linear combinations, such as $\omega_1 + \omega_2$, corresponds to the energy difference between a pair of states of the system. It is clearly possible to get multiple resonant enhancement.

Different nonlinear processes are distinguishable by the choice of frequencies ω_1, ω_2 and ω_3, which may be taken equal to each other, or may be negative. Near resonance the nonlinear susceptibility obviously becomes complex. The perturbation expression retains its validity provided

$$\hbar \left| \omega_{ni} - \omega + i\Gamma_{ni} \right| \gg \left| e x_{ni} E \right| \tag{7}$$

This is always satisfied off resonance. At resonance, the damping parameter measured in energy, $\hbar\Gamma$, must be larger than the product of the electric dipole moment and the applied field amplitude.

Two photon absorption is described by taking $\omega_1 = -\omega_3$, $\omega_4 = \omega_2$ and $\omega_1 + \omega_2 = \omega_{fi}$ in Eqs. (5) and (6). One retains only the resonant term with $\omega_{n'i} = \omega_{fi}$ in the summation over n'. The nonlinear polarization at ω_2 is 90° out of phase with E_2 and proportional to intensity $|E(\omega_1)|^2$. The imaginary part of $\chi^{(3)}$ is thus proportional to the two-photon absorption. The frequencies ω_1 and ω_2 may, or may not, be equal.

Third harmonic generation is obtained by taking $\omega_1 = \omega_2 = \omega_3$ and $\omega_4 = 3\omega_1$. The intensity at $3\omega_1$ is proportional to the absolute square $|\chi^{(3)}(\omega_4)|^2$. Momentum matching conditions are important. For $\omega_1 = \omega_3 \neq \omega_2$ one obtains the generation of the sum frequency $\omega_4 = 2\omega_1 + \omega_2$. These parametric processes have been observed in Na vapor by Harris and coworkers,[20] and are depicted in diagrams c and d of Figure 1. Resonant enhancement from intermediate levels is again possible.

The frequency ω_2 may also be replaced by its negative, as the real light beam at ω_2 contains both positive and negative frequencies. In this case the generation of the combination frequency $\omega_4 = 2\omega_1 - \omega_2$ is described. The process sketched in diagram f has been observed by Matsuoka[21] in Rb vapor. When $2\omega_1$ is close to a two-photon absorption resonance, this parametric generation is enhanced.

The description of all these processes is quite general and applies equally to atoms, molecules and to condensed matter.

IV - LIGHT MIXING SPECTROSCOPY IN CONDENSED MATTER

Excitations in condensed matter can also exhibit sharp resonances and they may be studied with advantage by nonlinear techniques. Consider, for example, the light generation at the frequency $2\omega_1 - \omega_2$, when two light beams at ω_1 and ω_2 respectively, with $|\omega_1 - \omega_2| \ll \omega_1$, are incident on a crystal. The specimen may be transparent, with very small dispersion, in the entire visible range containing the frequencies ω_1, ω_2 and $\omega_4 = 2\omega_1 - \omega_2$. The intensity at this combination frequency will, however, reveal dispersive characteristics from resonances at $2\omega_1$, which may lie in the ultraviolet, or from resonances at $\omega_1 - \omega_2$ in the infrared. The resonant terms interfere in a characteristic fashion with the non-resonant contribution, which arises from the excited electronic levels in the far ultraviolet.

It has been possible to study the properties of the sharp longitudinal exciton resonance in a CuCℓ crystal,[22] by observing the intensity in the infrared at $2\omega_1 - \omega_2$, as $2\omega_1$ is tuned through the exciton resonance. The red and infrared light beams penetrate through a crystal several millimeters thick, while one photon absorption processes at the exciton frequency would limit the absorption depth to a few hundred angstroms. The nonlinear technique makes it possible to study the excitons far from crystal boundaries and surfaces. The damping of the excitons could be measured as a function of temperature without surface contribution from physical and chemical impurities. The infrared resonances at $\omega_1 - \omega_2$ often arise from raman-active optical phonons. In this case $2\omega - \omega_2$ may be considered as the antistokes frequency. Because of the interference with the non-resonant term, this electronic nonlinear contribution to $\chi^{(3)}$ may be calibrated in terms of the known raman cross section. The optical phonon line shape in diamond has been observed out to more than one hundred Lorentzian half-widths from the center of resonance.[23] The antistokes generation also provides a good analytical tool for the identification of characteristic vibrations. Unlike the Stokes component it cannot be masked by impurity fluorescence. Rather detailed reports on this type of nonlinear spectroscopy have appeared in the literature.

In CuCℓ one could adjust the dye laser frequencies in such a manner that $2\omega_1$ corresponds to the exciton resonance while at the same time $\omega_1 - \omega_2$ corresponds to the polariton resonance. The two resonant terms interfere with each other and with the nonresonant terms in a manner described by Eq. 6. In condensed matter the resonances are often more damped and the relative importance of the non-resonant terms is increased, compared to the situation in atomic vapor. There is, however, no difference in principle.

The nonlinear spectroscopic techniques described here supplement the linear techniques and saturation spectroscopy techniques in a number of significant ways. The dispersion around states with very high (in the ultraviolet) or very low (in the infrared) energy may be probed while restricting the observations to tunable dye lasers in the visible region of the spectrum. States of either parity can be investigated with high resolution. Doppler and other inhomogeneous broadening mechanisms can be eliminated, without selecting a small subgroup out of the inhomogeneous distribution.

REFERENCES

1. P. A. M. Dirac, The Principles of Quantum Mechanics, (Oxford, Clarendon Press). First edition (1930), fourth edition (1958).

2. M. Goeppert-Mayer, Ann. d. Physik 9, 273 (1931).

3. W. Kaiser and C. G. B. Garrett, Phys. Rev. Lett. 7, 229 (1961).

4. See, for example, Dye lasers, Ed. F. P. Schafer. Topics in Applied Physics, vol. 1, Springer, Berlin, 1973.

5. L. S. Vasilenko, V. P. Chebotaev and A. V. Shishaev, JETP Letters 12, 113 (1970).

6. B. Cagnac, G. Grynberg and F. Biraben, Phys. Rev. Lett. 32, 643 (1974).

7. M. D. Levenson and N. Bloembergen, Phys. Rev. Lett. 32, 645 (1974).

8. T. W. Hansch, S. A. Lee, R. Wallenstein and C. Wieman, Phys. Rev. Lett. 34, 307 (1975).

9. M. D. Levenson and M. M. Salour, Phys. Lett. 48A, 331 (1974).

10. T. W. Hansch, K. Harvey, G. Meisel and A. L. Schawlow, Opt. Comm. 11, 50 (1974).

11. F. Biraben, B. Cagnac and G. Grynberg, C. R. Acad. Sc. Paris 279, B51 (1974).

12. N. Bloembergen, M. D. Levenson and M. M. Salour, Phys. Rev. Lett. 32, 867 (1974).

13. F. Biraben, B. Cagnac and B. Grynberg, Phys. Lett. 48A, 469 (1974).

14. K. C. Harvey, R. T Hawkins, G. Meisel and A. L. Schawlow, Phys. Rev. Lett. 34, 1073 (1975).

15. D. E. Roberts and E. N. Fortson, Opt. Comm. (1975).

16. P. F. Liao and J. E. Bjorkholm, Phys. Rev. Lett. 34, 1 (1975).

17. F. Biraben, B. Cagnac and G. Grynberg, J. de Phys. Lett. 36, L 41 (1975), and C. R. Acad. Sc. Paris, B280, 235 (1975).

18. J. E. Bjorkholm and P. F. Liao, Phys. Rev. Lett. 33, 128 (1974).

19. See, for example, N. Bloembergen, Nonlinear Optics, W. A. Benjamin, Inc., New York, 1965.

20. R. B Miles and S. E. Harris, IEEE J. QE-9, 470 (1973); D. M. Bloom, J. T. Yardley, J. F. Young and S. E. Harris, App. Phys. Lett. 24, 427 (1974).

21. H. Nakatsuka, J. Okada and M. Matsuoka, J. Phys. Soc. Japan, (1974).

22. S. D. Kramer, F. G. Parsons and N. Bloembergen, Phys. Rev. B9, 1858 (1974); S. D. Kramer and N. Bloembergen, Proceedings of the Conference on Optical Properties of Highly Transparent Solids, ed. S. S. Mitra, Waterville (N.H.), 1975.

23. M. D. Levenson and N. Bloembergen, Phys. Rev. B10, 4447 (1974).

Recent developments in dye lasers

F. P. Schäfer
Max-Planck-Institut für Biophysikalische Chemie, Göttingen

Progress in the dye laser field has been manifold recently:

1. New dyes have been found and new experimental findings allow a better understanding of the photodegradation process in dye lasers.

2. Vapor phase dye lasers have been operated.

3. New resonator configurations and tuning methods have been employed.

4. Higher average power and higher reliability have been achieved.

5. UV- and IR-wavelength regions that can be generated with dye lasers by nonlinear methods have been extended and the efficiencies of the conversion processes have increased.

These points will be reviewed in the following.

It is relatively easy to find new dyes that are useful with nitrogen laser pumping and we have published recently a list of more than 70 dyes that lase well with nitrogen laser pumping [1]. It is somewhat more difficult to find good dyes for flashlamp pumping since in this case one has to find dyes with little triplet-triplet absorption in the spectral region of the fluorescence band. This restriction is even more severe for cw dye lasers. Recently 16 new dyes of the cyanine class of dyes have been found by the Kodak group to operate satisfactorily in flashlamp-pumped dye lasers in the near infrared [2]. We have found a new class of dyes, namely the phenoxazones, which exhibit large solvatochromic shifts [3]. They are closely related to the well-known phenoxazines as shown in the Fig. 1. One observes that in contrast to the phenoxazines, like nileblue A, shown here as an example, the phenoxazones have an unsymmetrical structure which creates a dipolemoment already in the ground state, which changes by 5 Debye with excitation and is responsible for the large solvatochromic shifts seen with this class of dyes. We have not yet found a convenient way of synthesizing these dyes and thus have only studied the analog of nileblue A shown here. But we have made molecular orbital calculations of new dyes of this class and have found that some of these dyes should be capable of dye laser action to beyond

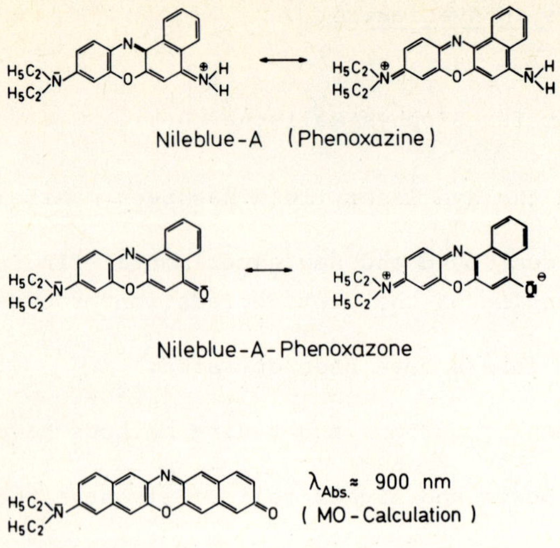

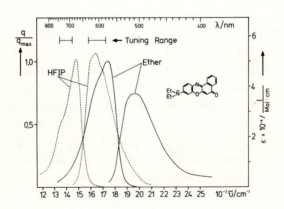

Nileblue-A (Phenoxazine)

Nileblue-A-Phenoxazone

$\lambda_{Abs.} \approx 900$ nm
(MO-Calculation)

Fig. 1

one micron. For example the dye shown in the third line should have its peak absorption around 900 nm and correspondingly show fluorescence at about 1000 nm. Absorption- and fluorescence spectra and the tuning range for the nileblue-A-phenoxazone is shown in Fig. 2 for the two

Fig. 2

solvents ether and hexafluoroisopropanole. As one can see the tuning range in ether is from 580 to 631 nm and in HFIP from 690 to 743 nm. Many other solvents with intermediate polarity can be used and we have seen that any intermediate tuning range can also be obtained with this dye.

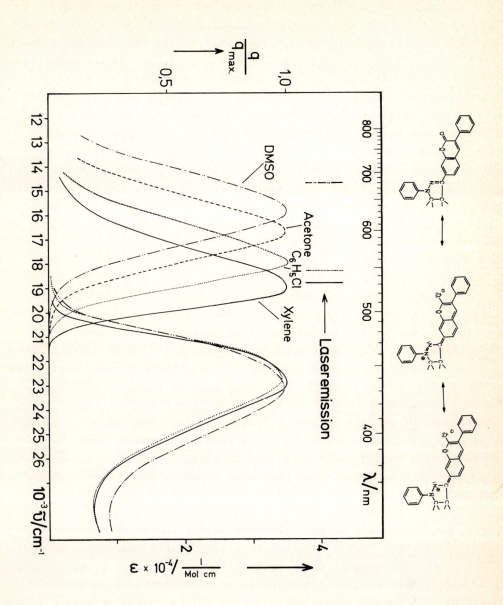

Such an extended tuning range without a gap is not at all trivial as can be seen in another example which we have investigated of a dye that shows a large solvatochromic shift. The chemical formula and the spectra of this dye are shown in Fig. 3. Two of the mesomeric formula are highly polar so that the molecule possesses a certain dipolmoment already in the ground state. The large change in dipolmoment with excitation (for which we measured the very high value of 23 Debye) causes the large solvatochromic shifts of the fluorescence spectra with change of the solvent which is shwon here for a few solvents only. Pumping with a nitrogen laser to avoid triplet-triplet absorption, we could not find laser emission between 550 and 680 nm in any of the many solvents we tried. This is indicative of a strong excited singlet state absorption in this region, which was further corroborated by the spectral shifts observed with mixtures of benzene and DMSO as solvents.

Another important recent development is the vapor phase dye laser which holds promise for the future of direct electrical excitation. We first obtained dye laser emission in the vapor of the scintillator dye POPOP with nitrogen laser pumping, a result that was independently obtained by a group of the Bell Laboratories. Similarly, a Russian group reported dye laser operation of POPOP at 250 ^{o}C with pentane "under high presse" being present. Since the critical temperature of pentane is 190 ^{o}C this means that here we have the dye in a supercritical solution and not a true vapor phase laser. Meanwhile we have used 3 additional dyes in a vapor phase laser and found a considerable number of dyes that will probably work in the near future [4].

As a next step we plan two experiments using the experimental arrangement shown in Fig. 4 which is still under construction but almost completed now. In both experiments we will use a molecular beam 3 mm wide and 80 mm long generated in an oven with a glass channel array of 10 μ bore diameter which should give a molecular beam with a collimation of better than 1 : 100. The beam is condensed at a cold trap. In the first experiment pumping will be done with two flashlamps

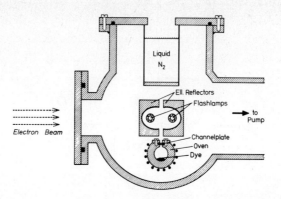

Fig. 4

in a double elliptical cavity. In this configuration we hope to be able to use also dyes that have a lower vapor pressure and insufficient absorption at the nitrogen laser wavelength. In the second experiment the flashlamps and reflectors will be removed as well as the blind flange at the side and the apparatus will be flanged to an electron beam machine so that we can try electron beam pumping at various electron energies. It could well happen, that too many molecules are cracked by the electron collisions so that no laser action is possible, but since no one to my knowledge has done such an experiment with large molecules, it is worth trying. Another possibility would be to deposit the electron energy in a suitable buffer gas and then use energy transfer to the dye molecules. In this respect it is of interest that we have found that 1 atmosphere of helium or argon as a buffer gas does not reduce the laser output noticeably and the Russian group has found similar results for nitrogen as a buffer gas. If these electron beam pumping experiments are successful we are probably one step nearer to a more efficient dye laser.

Two new prism dye ring laser arrangements have recently been developed in our lab and are promising because of easy alignment and low threshold [5].

Our efforts to increase the average power and the reliability of pulsed dye lasers for photochemical purposes started with work described in [6]. We have recently completed a novel design shown in Fig. 5 using four flashlamps in an aplanatic lens-mirror system which collects the light very efficiently into an $80°$ sector and images the flashlamp into the cuvette with only 1.5 mganification and very little aberrations. The fourfold symmetry and focussing gives a very uniform

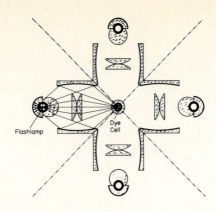

Fig. 5

pumping over the whole dye volume which is most important in dye lasers and the coupling efficiency is much higher than in a fourfold elliptical or close-coupling cavity.

We will now use four large lamps and thus hope to increase the output to at least 50 Watts within the next few weeks. When one remembers that 50 Watts in the visible is roughly one mole light quanta in two hours one can imagine that this type of dye laser can be very useful in photochemistry.

Of the various possibilities to generate ultraviolet and infrared wavelengths with dye lasers by nonlinear optical methods I would only like to point out to you the very fine work done by Werner Schmidt and his group at Zeiss. In this approach a powerful flashlamp-pumped dye laser is focussed into a cell with pressurized hydrogen. As the threshold of stimulated Raman emission is passed, the first Stokes line is generated, which simultaneously serves as pump light for the second Stokes line, which appears at somewhat higher input powers and so on in a cascade process. Since the vibrational frequency of hydrogen is about 4000 wavenumbers, a normal tuning range of the dye laser gives larger wavelength tuning ranges for the Stokes lines, the third Stokes line being already tunable from 2 to 12 μ. The main difficulty with this approach is the necessity to use a confocal resonator with relatively high reflectivity broadband mirrors in order to get a low threshold of Raman laser action and it is very difficult to make good broadband dielectric mirrors for the infrared. An efficiency of over 1 % for the 3rd Stokes line has already been achieved [7]. Further development of this dye laser pumped Raman laser will certainly result in a very handy tunable coherent infrared light source.

References

[1] D. Basting, F. P. Schäfer, B. Steyer: Appl. Phys. <u>3</u> (1974) 81 - 88.
[2] J. P. Webb, F. G. Webster, and B. E. Plourde: IEEE J. Quant. Electron. <u>QE-11</u> (1975) 114.
[3] D. Basting, Dissertation, Marburg, 1974.
[4] B. Steyer, F. P. Schäfer: Appl. Phys. <u>7</u> (1975) 113 - 122.
[5] G. Marowsky: Appl. Phys. Lett. <u>26</u> (1975) 647.
[6] J. Jethwa, F. P. Schäfer: Appl. Phys. <u>4</u> (1974) 299.
[7] W. Schmidt: private communication.

GENERATION OF VACUUM ULTRAVIOLET RADIATION BY NONLINEAR MIXING IN ATOMIC AND IONIC VAPORS

P. P. Sorokin, J. A. Armstrong
R. W. Dreyfus, R. T. Hodgson
J. R. Lankard, L. H. Manganaro
and
J. J. Wynne

Introduction

The method introduced by HODGSON et al[1] for vacuum ultraviolet (VUV) generation differs from the earlier work of HARRIS' group[2] in that two input dye laser beams are employed. One laser (ν_1) is always tuned to two-photon resonance with an excited state of the atom in order to achieve resonance enhancement of $\chi^{(3)}$, the nonlinear susceptibility governing the 4-wave mixing process. The other laser (ν_2) is then free to be tuned, resulting in tunable sum frequency generation of light at $2\nu_1 + \nu_2$. In order to prevent the generation of a resonantly enhanced signal at $3\nu_1$ the optical configuration shown in Fig. 1 is used. During the past year and a half further studies on VUV generation have been pursued by our group. Most of our recent work involves Sr, the atomic vapor originally studied in Ref. 1, although a few other systems have been successfully tried, as we shall relate below. These notes summarize the main results achieved by our group since the publication of Ref. 1.

Resonance Enhancement from Autoionizing States

Speculation was presented in Ref. 1 that autoionizing states are involved in the resonances seen in sum frequency generation as ν_2 is tuned. This has now been confirmed. In Fig. 2 a photometric scan of the VUV absorption of Sr, measured a number of years ago by GARTON and co-workers[5] with a one meter normal incidence vacuum spectrometer is plotted. By comparison of Fig. 2 with the traces obtained in sum frequency generation in Sr, it becomes apparent that there is a one-to-one correspondence between the resonances seen by the two different methods. This is evident, for example, in Fig. 3, which shows the variation in the VUV sum mixing signal generated in a spectral region that spans the resonances marked "X" and #5 in Fig. 2. The importance of resonance enhancement due to autoionizing states is dramatically shown in Fig. 3, since the ratio of the heights of the two peaks is ~10:1, and the smaller peak itself is roughly ten times greater than the VUV signal in the range 1896-1917A. For Fig. 3 the 5s5d $^1D_2$ state was employed as the two-photon resonant state, and the beam at

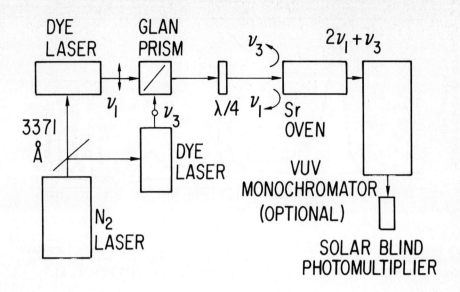

Figure 1. Method of combining beams from two dye lasers to produce a single frequency, tunable, coherent VUV source.

ν_2 was supplied by a Na-fluorescein laser. The distortion which appears on the low frequency side of the smaller peak in Fig. 3 is attributable to the presence of a small amount of Ba impurity in the Sr vapor cell (both being alkaline earths).

Although the above-mentioned one-to-one correspondence of resonances is apparent in Figs. 3 and 3, it is also clear that the <u>shapes</u> of corresponding absorption and generation resonances are entirely different. This again can be seen by comparing Fig. 2 with Fig. 4. The latter trace covers a range that both overlaps and extends beyond the right-hand portion of Fig. 2. For Fig. 4 the $5p^2\ ^1D_2$ state was chosen as the two-photon resonant state (λ_1 = 5409A), and the scanning laser was a coumarin laser. Strong, well-defined resonances in the VUV signal are observed. In particular, the two peaks at the shortest wavelengths (1712.3A and 1715.5A) are seen to be well resolved. The authors of the work from which Fig. 2 has been extracted remark that below 1722A the spectrum was too crowded for reliable data to be taken with the dispersion available. Thus, at short wavelengths the technique of sum frequency mixing can evidently be used to study autoionizing states with higher accuracy than with state-of-the-art spectrometers. In the original Sr high resolution photographic work of GARTON and CODLING[6], a three meter normal incidence grating was used, and the plates presented in that work do show fully resolved autoionizing lines out to ∼1650A.

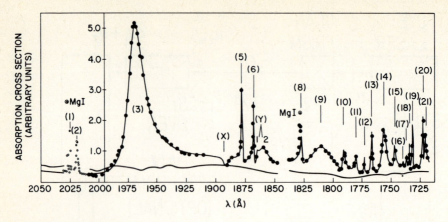

Figure 2. Autoionization resonances seen in absorption in Sr. Plotted points are measured values of relative absorption cross-section; curve is summation of profiles determined by q_g parameters deduced in Ref. 5. The 1st. ionization limit of Sr is off to the left, at 2177 Å. (After GARTON et al.[5])

If one compares Fig. 4 with Fig. 2 and with various measurements listed in Refs. 5 and 6, one arrives at the following conclusions: The largest peak at the longest wavelength shown in Fig. 4 corresponds to peak #17 in Fig. 2. The much smaller peak on its high frequency wing corresponds to peak #18 in Fig. 2. Note that in Fig. 2 peak #18 is larger than peak #17. Even more remarkable is the fact that there is absolutely no sign of a peak in the VUV generation spectrum that corresponds to peak #19 of Fig. 2, which in that figure is seen to be larger than #17 or #18! In Fig. 4 the off-scale central peak is actually a double peak (measured wavelengths 1722.9Å, 1722.3Å) which would appear to correspond to the pair of peaks #20, #21 in Fig. 2, although there is some discrepancy in wavelengths. (The separation between the peaks is greater in the case of Fig. 2 than in Fig. 4.) Finally, in Ref. 6 the next two shorter wavelength autoionizing lines listed are the $5s^2\ {}^1S_0 - 4d\ (^2D_{5/2})7f[1/2]_1^\circ$ transition at 1715.65Å, and the $5s^2\ {}^1S_0 - 4d(^2D_{3/2})\ 10p\ [1/2]_1^\circ$ transition at 1712.44Å. These are seen to correlate well with the measured values of 1715.5Å and 1712.3Å for the two shorter wavelength lines in Fig. 4. The relative sharpness of these two lines as seen in VUV generation, particularly the one at 1712.3Å, is also apparent in absorption in the plates of GARTON and CODLING[6]. As a general rule, both in absorption and in VUV generation, the resonances get sharper for the higher members of a sequence approaching a series limit. The two transitions described above fall in this category. Further examples of observed VUV generation spectra in Sr showing autoionization resonances are given in Ref. 7.

There has been a surprisingly swift development of a theory[8,9] which appears to be very adequate in explaining the observed shapes of the VUV generation resonances.

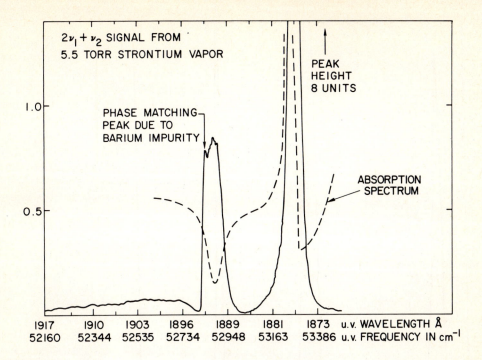

Figure 3. Intensity variation of VUV generated in Sr in a range that spans the Fano resonances "X" and #5 of Fig. 2 (solid curve). Part of the absorption curve of Fig. 2 has been superimposed (dashed curve) in order to show the correspondence between VUV resonances observed in absorption and in sum frequency generation.

FANO[10] originally developed a theory to explain the shapes of autoionizing resonances seen in absorption. ARMSTRONG and WYNNE[8] have recently obtained a simple expression for the shape of the VUV generation peaks by substituting into $\chi^{(3)}$ matrix elements from FANO's theory. The theory essentially depends upon only two parameters: (a) q_g, the same parameter Fano introduced (see below) to characterize the absorption from the ground state to an autoionizing state and (b) $q_{j'}$, an analogously defined quantity characterizing the transition from the two-photon resonant state to the autoionizing state. FANO[10] showed that the ratio of the actual absorption cross-section in the vicinity of an autoionizing line to that of the unperturbed photoionization continuum can be written $|\mu_{g\Psi_\nu}|^2/|\mu_{g\psi_\nu}|^2 = (q_g + \varepsilon)^2/(1+\varepsilon^2)$, i.e., as one of a single family of curves specified by only one parameter q_g. Here ε is the frequency offset of the probing radiation from the center of the resonance, normalized to the halfwidth of the line. As amended by ARMSTRONG and BEERS[9], the expression for the square magnitude of $\chi^{(3)}$, to which the VUV generation signal is proportional, now becomes

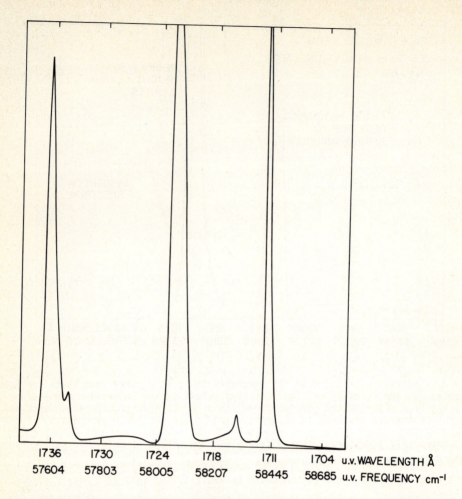

| 1736 | 1730 | 1724 | 1718 | 1711 | 1704 u.v. WAVELENGTH Å |
| 57604 | 57803 | 58005 | 58207 | 58445 | 58685 u.v. FREQUENCY cm⁻¹ |

Figure 4. Intensity variation of VUV generated in Sr in a range that includes the extreme right-hand portion of Fig. 2.

$$|\chi^{(3)}|^2 \sim |\mu_{g\psi_\nu}|^2 |\mu_{\psi_\nu j'}|^2 \left\{ \frac{q_g^2 \, q_{j'}^2 + [x + (q_g + q_{j'})]^2}{1 + x^2} \right\}$$

Note that the expression for $|\chi^{(3)}|^2$ given above reduces to $|\mu_{g\psi_\nu}|^2 \cdot |\mu_{\psi_\nu j'}|^2$ as x, the normalized frequency offset of the VUV, becomes large. This product is a measure of the VUV signal strength away from the autoionizing resonance. For large values of q_g, q_j, the VUV signal becomes enhanced at resonance ($x = 0$) by a factor $\sim q_g^2 \, q_{j'}^2$, which may easily reach a value $\sim 10^4$.

Excellent results have been achieved[9] fitting the lineshapes observed in VUV sum mixing[8] with the above expression for $|\chi^{(3)}|^2$. The q_g are available from fits to VUV absorption data, such as given in Ref. 5. From an adequate computer fit to the

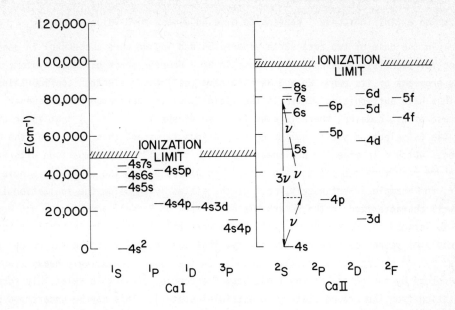

Figure 5. Energy level diagram of Calcium I and II, showing observed resonantly enhanced THG process.

VUV generation spectra the q_j, can thus be derived. Recently we have succeeded in obtaining resonantly enhanced photoionization spectra from which the q_j, can also be directly measured. In these experiments the solar blind photomultiplier is replaced by an ionization probe. As in the case of VUV generation, one laser is tuned to a double-quantum resonance, and the other laser ν_2 is scanned over the autoionizing states. It can be shown that the rate of three-photon ionization is proportional to $|\mu_{\psi_\nu j'}|^2 (q_j, + \varepsilon)^2/(1 + \varepsilon^2)$. Thus, there is provided an independent measure of q_j, from the variations in ionization observed as ν_2 is scanned.

VUV Generation in Other Systems

With the same techniques used in Sr we were unable to generate <u>any</u> detectable VUV signal in Na vapor. Two particular schemes were tried. In the first scheme a single linearly polarized Rhodamine 6G laser was tuned to two-photon resonance with the 5d states. In the second, two dye lasers were employed. One laser (ν_1) was tuned between the $p_{1/2}$, $p_{3/2}$ components of the 3p state, roughly twice as close to the $p_{1/2}$ component as to the $p_{3/2}$ component in order to achieve phase-matching. The other laser (ν_2) was then adjusted so that the sum frequency $\nu_1 + \nu_2$ again was equal to the frequency of the 5d states. We attribute the lack of signal in both cases to the fact that at the wavelengths of the expected VUV generation (~1900A) the smooth Na photoionization extinction curve[11] happens to have an almost zero value, i.e., $|\mu_{g\psi_\nu}| = 0$.

In the case of two rare earth vapors, Eu and Yb, we were successful in generating strong VUV signals, comparable with those in Sr. However, each of these systems introduces problems of its own. Eu and Yb both have ($6s^2$) outer electron configurations and thus have autoionizing states in the vicinity of the generated VUV frequency. In the case of Eu, however, there is a half-filled 4f shell ($4f^7\ {}^8S_{7/2}$) that interacts with the $6s^2$ electron shell, splitting its various terms and thus "diluting" the matrix elements which determine the nonlinear susceptibility $\chi^{(3)}$. By comparison with Sr, there are many more two-photon resonances encountered during the scan of a single dye laser, for example. Fortunately, most of the states lying below the ionization limit are well characterized in the remarkable early work of Russell and King[12], who were the ones to "crack" the Eu spectrum. With Eu we were able to demonstrate still another principle of resonance enhancement, namely, that the intensity of the VUV output generated by sum mixing can be further enhanced in the case of relatively heavy (two-electron) atoms by tuning one of the input beam frequencies close to a relatively forbidden transition from the ground state g to a triplet state j. This may be understood by considering the general expression for $\chi^{(3)}$. For example, if it is assumed that the oscillator strength f_{jg} characterizing the transition from state g to state j is $\sim 1/100$, instead of ~ 1, then $\nu_{jg} - \nu_1$ can be reduced by ~ 100 without increasing the difficulty if phase-matching. Since $\mu_{jg} \sim f_{jg}^{1/2}$, $\chi^{(3)}$ will be increased by ~ 10 and the VUV power output, which varies as $|\chi^{(3)}|^2$, will be increased by ~ 100. In this argument it is assumed that the other matrix elements retain their same order of magnitude when the relevant states lie in the triplet manifold. This assumption can be justified from the known data. Using the above principle, we observed in Eu VUV power increases of roughly 2 orders of magnitude by tuning one laser (ν_1) very close to a triplet state (e.g., $6s6p\ {}^6P_{7/2}$ at 17,340.65 cm^{-1}) while the other laser (ν_2) was tuned so that $\nu_1 + \nu_2$ equalled the frequency of a two-photon resonant state (e.g. $6s6d\ {}^6D°_{9/2}$ at 36,566.6 cm^{-1}). Of course, to take advantage of this technique and still have a tunable output would require the introduction of a third input beam. Finally, this technique should work even better in Sr, whose levels are not diluted by the interaction present in Eu. In Sr one would tune the first laser close to the $5s5p\ {}^3P°_1$ state. The required wavelength (~ 6892.6Å) can be provided by a nitrogen-laser-pumped dye laser using a mixture of Rhodamine B and Nile Blue Perchlorate A.

Yb has a completely filled 4f shell and so does not have the complex spectrum that Eu has. The difficulty in its use is one of a technical nature. Its melting temperature is sufficiently close to the temperature at which suitable vapor pressure for VUV generation is developed that operation of the vapor cell for any but the shortest period of time is impossible without cooling and recharging it. In contrast to the case of most atoms including Sr, very few of the states lying <u>below</u> the first ionization limit of Yb are known. Illustrative of this point is the fact that the authors believe they have located the first excited s state of Yb. We observe a strong two-photon resonantly enhanced THG signal at $2\nu_1 = 34{,}360$ cm^{-1}. By comparing its position with the few other states known for Yb, and deducing that it is a J = 0 state, we arrive at the conclusion that the two-photon resonant state is the 6s7s $^1S_0$ state.

Third Harmonic Generation from Ions

An analysis presented in Ref. 7 indicated that an ion density as high as 10^{16}/cm^3 could be produced by multiphoton ionization in Sr at the focus of one of the dye lasers used for generation of VUV by sum mixing. In calcium vapor we have recently observed relatively strong third harmonic generation (THG) signals at 1278A when the nitrogen-laser-pumped dye laser was tuned very close to 3832.5A. This wavelength setting exactly corresponds to two-photon resonance with the 5s ^{2}S state of Ca II located at 52167 cm^{-1}. Evidently, since only two input violet laser photons are needed to ionize Ca atoms (Fig. 5), enough ions are created by two-photon ionization to provide sizable THG signals when the light wave is also tuned to be in two-photon resonance with the ions. The fact that there exist known, sharp levels of Ca II considerably higher in energy than 52167 cm^{-1} makes it appear feasible to detect <u>6-wave parametric generation</u> (i.e., five waves up, one down) in this system. For such an experiment one would keep $2\nu_1$ at 52167 cm^{-1} and tune ν_2 to 13541.5 cm^{-1} (7330.6A), for instance. Then $2\nu_1 + 2\nu_2$ would be resonant with the 7s ^{2}S state and one would expect to be able to detect radiation at either $2\nu_1 + 3\nu_2$ (93091.5 cm^{-1} or 1074A) or $3\nu_1 + 2\nu_2$ (105,533.5 cm^{-1} or 947.6A). This experiment is currently being tried.

In Sr, also, we have observed VUV generation connected with the presence of ions, but it appears to involve an effect other than THG. Specifically, with a dye laser tuned within a 100 cm^{-1} wide band centered at 23715.2 cm^{-1} (4215.5A), there is produced VUV light in a series of bands (1778.4, 1769.6, 1620.3, 1613, 1537.9 and 1531.3A) that appear to be transitions nf-4d, (with n ranging from 5-7) of the ion The wavelength 4215.5A corresponds exactly to <u>single photon</u> resonance with 5p $^2P°_{1/2}$ state of SrII, but the width of the observed resonance is at least two orders of magnitude greater than the width of the two-photon resonance enhancements observed in the case of THG. Again only two photons are needed initially to ionize the Sr atoms, but the subsequent mechanisms remain to be clarified. The most likely possibility is that the ion is excited to the 7f state by a resonantly enhanced three-photon absorption process. Superfluorescence from this state and from states populated by subsequent cascading processes then occurs.

References

1. R.T. Hodgson, P.P. Sorokin and J.J. Wynne, Phys. Rev. Lett. __32__, 343 (1974).

2. S.E. Harris and R.B. Miles, Appl. Phys. Lett. __19__, 385 (1971).

3. J.F. Young, G.C. Bjorklund, A.H. Kung, R.B. Miles and S.E. Harris, Phys. Rev. Lett. __27__, 1551 (1971).

4. A.H. Kung, J.F. Young, G.C. Bjorklund and S.E. Harris, Phys. Rev. Lett. __29__, 985 (1972).

5. W.R.S. Garton, G.L. Grasdalen, W.H. Parkinson, and E.M. Reeves, J. Phys. B(Proc. Phys. Soc.) __1__ (Ser. 2), 114 (1968).

6. W.R.S. Garton and K. Codling, J. Phys. B (Proc. Phys. Soc.) __1__ (Ser. 2), 106 (1968).

7. P.P. Sorokin, J.J. Wynne, J.A. Armstrong, and R.T. Hodgson, Proc. 3rd. Conf. Laser N.Y. (1975) to be published in Annals N.Y. Acad. of Sciences.

8. J.A. Armstrong and J.J. Wynne, Phys. Rev. Lett. __33__, 1183 (1974).

9. L. Armstrong and B. Beers, Phys. Rev. Lett. __34__, 1290 (1975)

10. U. Fano, Phys. Rev. __124__, 1866 (1961).

11. R.D. Hudson, Phys. Rev. __135__, A1212 (1964).

12. H.H. Russell and A.S. King, Ap. J __90__, 155 (1939)

TUNABLE VUV LASERS AND PICOSECOND PULSES

D. J. Bradley
Optics Section
Physics Department
Imperial College
London SW7 2BZ

Summary

Recent developments in coaxial electron-beam pumped VUV lasers are reported. The production of tunable-frequency sub-picosecond pulses by transient stimulated Raman scattering and the extension of electron-optical chronoscopy to subpicosecond time-resolution and to XUV and X-ray wavelengths are described.

Introduction

The last year has seen sigificant advances in the development of electron-beam pumped, high-pressure VUV lasers, including frequency narrowing and tuning, and in the expansion of the spectral range of picosecond pulses. Simultaneously electron-optical streak-cameras have been further improved in performance to extend time-resolution into the sub-picosecond domain and to extend the range of spectral sensitivity to cover photon energies from 1 keV to 1 eV. With coaxial diode pumping systems efficient, narrow-band, tunable-frequency, low-divergence, megawatt power at $\sim$170nm is obtained from table-top Xe_2 lasers. This performance is already comparable to that of dye lasers at longer wavelengths, and by nonlinear mixing tunable XUV coherent sources ($\sim$50nm) should be available. The broad bandwidth of Xe_2 and other noble gas lasers should also permit the amplification of picosecond pulses, shifted in frequency from the visible. The spectral range of tunable picosecond pulses can also be extended to longer wavelengths by stimulated Raman scattering in organic liquids, with considerable pulse shortening to produce sub-picosecond pulse durations. The purpose of this paper is to briefly review these latest developments in VUV lasers and ultra-short pulse generation and measurement.

The Coaxial Diode Xe_2 Laser

The first quasi-molecular ultra-violet lasers were produced by pumping with relativistic electron-beams typically delivering hundreds of joules, (1-4). Peak powers of 500MW (10J, 20ns) have been obtained (5). Considerable improvements in efficiency and convenience were obtained with the coaxial diode, electron-beam arrangement (6,7) shown in Figure 1. The field-emission diode consists of a thin-

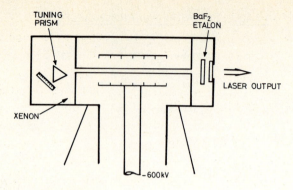

Fig. 1. Coaxial-diode VUV Xenon laser system.

walled ($\sim$ 70μm) stainless steel, tubular anode of $\sim$ 4mm internal diameter, maintained at earth potential. Both ends of the anode, which is also the container of the laser gas, open into high-pressure chambers which contain the optical elements. The concentric cathode, which has an internal diameter of 3.5cm and a length of 10cm, is constructed from titanium sheet, perforated to produce an array of spikes. The first diode was designed for use with a Febetron 706 pulse generator (Field Emission Corporation) which has a load impedance of 60Ω (10kA, 600kV). The anode tube diameter was matched to the range of the electrons, at a pressure of 10kTorr, to obtain uniform pumping. To obtain good frequency narrowing and tuning it was necessary to drastically modify the Febetron 706 generator to produce a 20J, 5ns pulse of 500 keV electrons. Pumping is obtained over a 14cm anode length and a total energy of 10J is transmitted into the gas, corresponding to a deposition of 5.5J cm^{-3}.

The anode tube is normally evacuated to 10^{-4} Torr. The xenon gas is purified by freezing in a high-pressure bomb cooled by liquid nitrogen and pumped to $<10^{-4}$ Torr. The laser resonator reflectors consist of an Al:MgF$_2$ coated plane mirror of $\sim$ 85% reflectivity, and a BaF$_2$ single-plate resonant-reflector, with an effective reflectivity of $\sim$ 20%. When these two reflectors only are employed in an optical cavity of length 25cm, an output laser peak power of $>$ 3MW (10mJ in a 3ns pulse) is obtained. The laser intensity is reproducible to 10% and the beam divergence half-angle is $<$ 1 mrad. A nett gain of 0.25 cm^{-1} is achieved (6,7). It was orignally necessary to allow a minimum internal of 15 minutes between shots to allow the gas to cool to room temperature. The instantaneous rise in temperature of the xenon is $\sim$ 700°C and the anode tube temperature increases by $\sim$ 30°C, after a few seconds. Absorption by ground-state xenon molecules increases rapidly with increasing temperature (8) and gas heating strongly affects both fluorescence and laser intensities.

This difficulty has now been overcome by circulating the xenon gas through a water-cooled heat-exchanger, which can be seen in the photograph of Figure 2. The repetition rate is now limited by the power supply loading only.

A fused quartz prism is employed as the intra-cavity tuning element. At the operating wavelength of $\sim$172nm the angular dispersion of a prism is comparable to that of a grating of the same area. With the prism edge adjusted parallel to the mirror surface, rotation of the prism produces tuning of the narrowed bandwidth. Figure 3 shows three typical spectra recorded on SC7 film in a 1 metre, normal incidence, vacuum spectrograph. From the microdensitometer traces of Figure 4, a laser bandwidth of 0.13nm was determined, giving a spectral narrowing by a factor of x 100 from the $\sim$15nm fluorescence (9), and by a factor of x 10 from the untuned laser bandwidth of 1.3nm (6).

Continuous tuning is obtained over the spectral range shown in Figure 5. The peak power of 0.7MW is comparable to that obtained from high-power flashlamp pumped dye-lasers (10) while the tuning range of $2500 cm^{-1}$ is considerably greater. Further bandwidth narrowing should be achievable with longer pumping pulses giving a greater number of resonator transits. Multiple-prism arrangements could also be employed. A 50cm long coaxial diode has been constructed in our laboratory to operate with 600 keV electrons and delivering up to 100J in 50 nsec. This diode will be employed to produce narrower bandwidths and higher-powers for third harmonic generation

Fig. 2. Photograph of tunable Xe_2 laser showing gas heat-exchanger and rebuilt commercial 500 kV power supply.

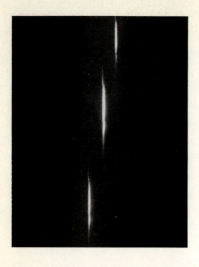
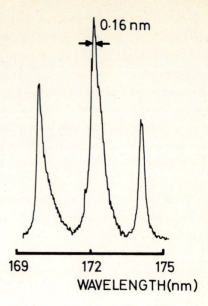

Fig. 3. Frequency narrowing and tuning of Xe_2 laser. (The spectrograph plate was moved vertically between recordings)

Fig. 4. Microdensitometer trace of spectra of Figure 3. (Ordinate is arbitrary linear density scale).

experiments in argon and other gases to produce coherent tunable sources in the XUV spectral region.

It is clear that the Xe_2 laser is likely to play as important a role in VUV spectroscopy as dye lasers are currently playing at longer wavelengths, and that with further development of Ar_2 and Kr_2 lasers it will be possible to employ selective excitation (11) over a wider range of atomic and molecular transitions and to achieve the breaking of bonds by optical means. For all of these applications a repetition rate laser is very desirable, and in collaboration with AWRE Aldermaston Laboratory a 10pps, 600keV, 40ns, 40J power supply has been developed and is undergoing tests (12). The coaxial diode design of high-pressure gas lasers has obvious applications also for longer wavelength lasers employing mixtures of gases (13).

VUV Picosecond Pulse Generation and Amplification

As with dye lasers, the broad bandwidth of the Xe_2, and other noble gas lasers should allow the amplification of picosecond pulses. While the Xe_2 laser has yet to be mode-locked, picosecond pulses at 173.6nm can be produced by four-wave nonlinear

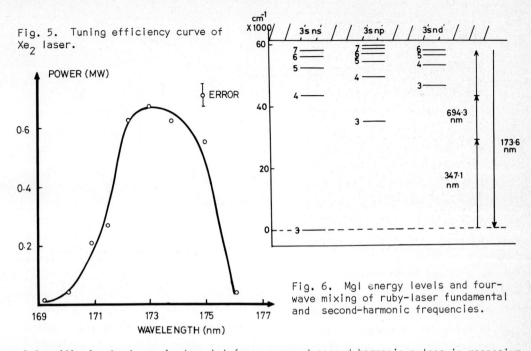

Fig. 5. Tuning efficiency curve of Xe$_2$ laser.

Fig. 6. MgI energy levels and four-wave mixing of ruby-laser fundamental and second-harmonic frequencies.

mixing (4) of ruby-laser fundamental frequency and second-harmonic pulses in magnesium vapour, phase-matched with xenon buffer gas. As can be seen from the magnesium energy level diagram of Figure 6, there is a near two-photon resonance for the $3s^2\ ^1S_0 - 4s^1S_0$, $3s3d\ ^1D_2$ transitions for one ruby second-harmonic photon and one fundamental frequency photon. The train of pulses from a mode-locked ruby laser operating in a low-order single-transverse mode (15), with peak pulse energy of 1mJ (∼50MW), were frequency doubled in ADP, with a conversion efficiency of ∼10%. Both fundamental and second harmonic frequencies were focussed into a magnesium vapour cell, isothermally heated with a sodium/argon heat pipe (16). The output beam was focussed with a B$_a$F$_2$ lens into a vacuum monochromator, with either a CsI photo-multiplier or a CsTe photodiode as detector. Maximum efficiency was obtained at a xenon buffer gas pressure of 8 Torr for a magnesium vapour pressure of 1.6Torr (cell temperature of 615°C) when a peak power of ∼ 200 Watt at 173.6nm was obtained. This corresponds to a power conversion efficiency of 4×10^{-6}. The oscillograms of Figure 7 show the effect of this strongly nonlinear process upon the pulse train profile. Amplification of the oscillator pulses by x 20 should produce VUV megawatt picosecond pulses, provided saturation or other loss mechanisms do not operate. Further amplification could then be achieved in electron-beam pumped Xe$_2$ amplifiers. Provided that breakdown in the laser gas, or two-photon absorption in windows, can be avoided, high-power picosecond pulses could thus be produced for laser plasma generation. The short wavelength will permit penetration of denser plasma for diagnostic studies of high density, high temperature matter produced by laser compression. The ruby fourth-harmonic picosecond pulses can also be employed for studying the detailed time-evolution of the Xe_2 laser pumping mechanisms, and to determine if the laser bandwidth is homogeneously broadened on a picosecond time-scale.

Sub-picosecond Pulse Generation and Measurement

The invention of the extraction-mesh electrode (17,18) for electron-optical streak-tubes permitted the development of camera systems for the direct linear measurement of ultra-short pulse durations, with a time resolution as short as 2 psecs (19,20). For several laser picosecond interaction studies including molecular excited state relaxation rates (21) self-phase modulation (22,23) and transient Raman scattering (24,25) higher time-resolution still is needed. Again for some applications of picosecond light pulses, particularly in photochemistry and laser plasma diagnostics, time-resolution has to be maintained throughout the spectrum of sensitivity of the photo-cathode. To permit a substantial increase in the photo-cathode electric-field strength to obtain sub-picosecond time-resolution (20) we redesigned the electron-optics of the original Photochron tube, at the same time improving the spatial resolution, to double the information content (26). The resultant reduction in magnification in the Photochron II camera also increases the recording speed. To produce a direct demonstration of sub-picosecond resolution it has been necessary to further shorten the pulses from a mode-locked dye laser by transient Raman scattering in ethanol (25,26).

The shortest duration pulses are obtained from mode-locked dye-lasers (27) which have the added advantages of frequency tunability and of being very reproducible in operation, compared with neodymium:glass and other solid state laser systems. Pulses

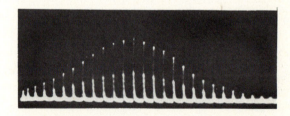

Fig. 7. Oscillogram of ruby laser pulse-train.

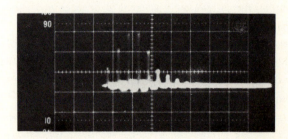

VUV pulses at 173.6nm.

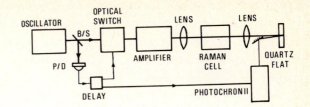

Fig. 8. Arrangement for generation and measurement of stimulated Raman scattered ultra-short pulses.

from a Rhodamine 6G dye laser mode-locked using an ethanolic solution of 1,3'diethyl 4,2'-quinolyoxacarbocyanine iodide (DQOCI) (28) and tuned to operate at 605nm, produced recorded pulse-widths (including the camera instrumental width) as short as 1.5 psec with a S20 streak-tube (26). To directly demonstrate sub-picosecond time-resolution the experimental arrangement shown in Figure 8 was employed. Six pulses from the centre of the mode-locked Rhodamine 6G laser pulse train were selected out by a Pockels cell optical-switch and amplified (28) to peak powers of $\sim$ 300MW. After amplification the pulses were focussed into a cell containing ethanol. Transient stimulated Raman scattering from the C-H stretching vibration generates a Stokes frequency at λ 733.7nm. The transverse relaxation time, T_2, of this vibration is $\sim$ 0.3 psec (spontaneous Raman linewidth of $\sim$ 17.4cm^{-1} (29).) For a laser pulse of duration, tp, $\sim$ 1.5 psec the interaction is then essentially transient in nature. With tp/T_2= 5 the Stokes pulse is delayed and is shorter than the laser pulse (24,30,31).

Figure 10 shows a typical streak record with subpicosecond time-resolution. The Raman Stokes pulse was transmitted through glass filters which removed the dye-laser pumping pulses. Two pulses separated by 60 psec were generated from each Raman pulse by reflection from a quartz flat. From the microdensitometer trace a total recorded duration of 900 fsec (900 x 10^{-15} sec) was measured. It is not sufficient to make the approximation normally used (18,20) in deriving the time-resolution limit of this new camera. For photons of wavelength 733.7nm, the time-dispersion spread between the photocathode and the mesh is 380 fsec. Time-dispersion in the mesh to anode region of the tube adds a further 120 fsec to this to give a total time-dispersion resolution limit of 500 fsec, making a total camera instrumental resolution of 700 fsec (26). Deconvolving this value from the recorded width of 900 fsec gives a Raman pulse duration of 570 fsec. This Photochron II camera system thus permits the study of luminous phenomena throughout the spectrum from the vacuum ultra-violet to the near infra-red with a time-resolution of $\leq$ 2 psec and with sub-picosecond resolution at spectral regions close to the long wavelength response cut-off of the particular type of photcathode employed. With its greater light gain the range of usefulness is extended to weaker light sources, while the improved spatial resolution increases the information capacity. Stimulated Raman scattering also increases the frequency range of tunable picosecond pulses since high conversion efficiencies are obtainable (25).

X-ray and XUV Picosecond Chronoscopy

The study of laser produced plasmas in compression experiments, and the development of short-wavelength sources required the extension of electron-optical chronoscopy to the XUV and X-ray spectral regions, with time-resolution in the picosecond range. Employing a demountable, modified X-ray version (32), of the Photochron streak-tube pulses of $\sim$ 1keV photon energy, with durations of $\sim$ 20 picoseconds have been recorded (Figure 11). A series of 10 psec pulses, separated by 66 psec, generated from a passively mode-locked Nd:glass oscillator, amplifier system, were amplified up to energies of $\sim$ 100mJ and focussed on to a plane copper target to generate a plasma of $\sim$ 100μm diameter, of temperature about 200eV. X-rays in a broadband of energy around 1 keV were selected by an aluminium foil filter. With a 100nm thick, vacuum-evaporated gold photocathode, streaks at a writing speed of 2×10^9 cm sec^{-1} were easily recorded by projecting a shadowgraph of the slit on to the photocathode at a glancing angle of 5°. The shortest recorded pulse-widths

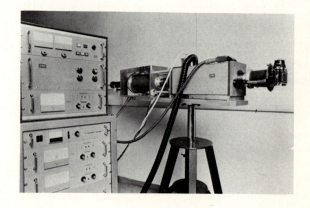

Fig. 9. Photochron II streak-camera.

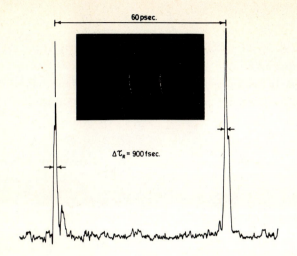

Fig. 10. Streak-record, and corresponding microdensitometer trace, of 500 femtosecond Raman Stokes pulse at ~ 733.7nm.

were 22 psec corresponding to a camera time-resolution limit of ~ 20 psec for 1 keV photons. With slight modifications to the image-tube it should be possible to improve the time-resolution to $\leq$ 10 psec. Picosecond chronoscopy can thus now be carried out with photons covering the energy range 1eV to 10keV.

Fig. 11. Streak photograph of X-ray pulses.

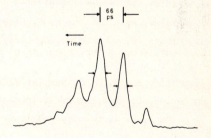

Acknowledgement

The author wishes to thank Dr. M. H. R. Hutchinson, Dr. E. G. Arthurs, Dr. A. G. Roddie, Dr. W. Sibbett and other members of the Imperial College Optics Section Laser Group whose work is described in this paper. Financial support from the Science Research Council, the Paul Instrument Fund and the UKAEA Culham Laboratory is gratefully acknowledged.

References

(1) H. A. Koehler, L. J. Ferderber, R. L. Redhead and P. J. Ebert, Appl. Phys. Letts. Vol. 21, 198 (1972)

(2) W. M. Hughes, J. Shannon, A. Kolb, E. Ault and M. Bhaumik, Appl. Phys. Letts. Vol. 23, 385 (1973)

(3) J. B. Gerardo and A. Wayne Johnson, IEEE J. Quantum Electronics, QE-9, 748 (1973)

(4) P. W. Hoff, J. C. Swingle and C. K. Rhodes, Opt. Commun. Vol. 8, 128, (1973)

(5) R. O. Hunter, J. Shannon and W. Hughes, Maxwell Laboratories Internal Report - MLR - 378 (1974)

(6) D. J. Bradley, D. R. Hull, M. H. R. Hutchinson and M. W. McGeoch. Opt. Commun. Vol. 11, 335 (1974) (UK Patent Application No. 14102/74)

(7) D. J. Bradley, D. R. Hull, M. H. R. Hutchinson and M. W. McGeoch. Opt. Commun. Vol. 14, 1 (1975)

(8) D. A. Emmons, Opt. Commun. Vol. 11, 257 (1974)

(9) D. J. Bradley, M. H. R. Hutchinson and H. Koetser, Opt. Commun. Vol. 7, 187 (1973)

(10) D. J. Bradley, W. G. I. Caughey and J. I. Vukusic, Opt. Commun. Vol. 4, 150 (1971)

(11) D. J. Bradley, P. Ewart, J. V. Nicholas and J. R. D. Shaw, J. Phys. B. Atom. Molec. Phys. Vol. 6, 1594 (1973); Phys. Rev. Lett., Vol. 31, 263 (1973)

(12) C. Edwards, M. D. Hutchinson, J. C. Martin, T. H. Storr. AWRE Report SSWA/JCM/755/99 "Lark - a modest repetive pulse generator".

(13) E. R. Ault, App. Phys. Letts. Vol. 26, 619 (1975)

(14) R. B. Miles and S. E. Harris, IEEE J. Quant. Elect. QE-9, 470 (1973); P. P. Sorokin, J. J. Wynne and R. T. Hodgson, Phys. Rev. Lett., Vol. 32, 343, (1974)

(15) D. J. Bradley, M. H. R. Hutchinson, H. Koetser, T. Morrow, G. H. C. New and M. S. Petty, Proc. Roy. Soc. A, Vol. 328, 97 (1972)

(16) E. G. Arthurs and M. H. R. Hutchinson. Unpublished.

(17) D. J. Bradley, UK Patent 1329977 (1973) US Patent 3761614 (1973)

(18) D. J. Bradley, B. Liddy and W. E. Sleat, Opt. Commun. Vol.2, 39 (1971)

(19) E. G. Arthurs, D. J. Bradley, B. Liddy, F. O'Neill, A. G. Roddie, W. Sibbett and W. E. Sleat. Proc. X Int. Congress on High Speed Photography, (Nice, France 1972), 117.

(20) D. J. Bradley and G. H. C. New, Proc. IEEE, Vol. 62, 313 (1974)

(21) G. Porter, E. S. Reid and C. J. Tredwell. Chem. Phys. Letts. Vol. 29, 469 (1974)

(22) E. G. Arthurs, D. J. Bradley and A. G. Roddie, Appl. Phys. Letts., Vol. 19, 480, (1971)

(23) D. J. Bradley and W. Sibbett, Opt. Commun. Vol. 9, 17 (1973)

(24) D. von der Linde, A. Laubereau and W. Kaiser, Phys. Rev. Lett. Vol. 26, 954 (1971)

(25) R. S. Adrain, E. G. Arthurs and W. Sibbett. Unpublished.

(26) P. R. Bird, D. J. Bradley and W. Sibbett. Proc. XI Int. Congress on High Speed Photography (Chapman and Hall, London) 112, (1975); D. J. Bradley and W. Sibbett. Unpublished.

(27) D. J. Bradley, Opto-Electronics, Vol.6, 25 (1974)

(28) R. S. Adrain, E. G. Arthurs, D. J. Bradley, A. G. Roddie and J. R. Taylor Opt. Commun., Vol. 12, 136 (1974)

(29) D. von der Linde, A. Laubereau and W. Kaiser, Phys. Rev. Lett. Vol. 26, 954, (1971)

(30) R. L. Carman, F. Shimizu, C. S. Wang and N. Bloembergen. Phys.Rev. $\underline{A2}$, 60 (1970)

(31) G. I. Kachen. PhD Thesis, Lawrence Livermore Laboratory, UCRL-53 (1975)

(32) P. R. Bird, D. J. Bradley, A. G. Roddie, W. Sibbett, M. H. Key, M. Lamb abd C. L. S. Lewis. Proc. XI Int. Congress on High Speed Photography (Chapman and Hall, London) 118, (1975) and unpublished work.

ROTATION-VIBRATION SPECTROSCOPY OF GASES BY COHERENT ANTI-STOKES RAMAN SCATTERING : APPLICATION TO CONCENTRATION AND TEMPERATURE MEASUREMENTS

F. Moya, S.A.J. Druet and J-P E. Taran
Office National d'Etudes et de Recherches Aérospatiales (ONERA)
92320 Châtillon (France)

The Raman spectroscopy of gases is greatly facilitated by the use of Coherent anti-Stokes Raman Scattering (CARS). The improvements stem from the parametric nature of this process, as opposed to the incoherent nature of spontaneous Raman scattering :
- the scattered light is well collimated (10^{-3} cone angles are typical)
- its intensity is 5 to 10 orders of magnitude larger in practice ;
- modest powers in the pump pulses are sufficient (1 kW to 1 MW).

These properties have been recognized for some time [1-4]. We have investigated some characteristic features of the effect, in view of its utilization for gas concentration and temperature measurements in aerodynamic flows and flames [5-8].

CARS is a four-wave mixing process. It can be observed in a gas with two intense, collinear optical beams of frequencies ω_1 and ω_2 such that $\omega_1 - \omega_2 \simeq \omega_v$, where ω_v is the frequency of a Raman active vibrational transition ; satisfactory phase matching is obtained for collinear beams since dispersion is negligible in gases : sidebands at the combination frequencies $2\omega_1 - \omega_2$ and $2\omega_2 - \omega_1$ are then generated in the same direction as the incoming beams. The Stokes sideband at frequency $\omega_s = 2\omega_2 - \omega_1$ ($\omega_1 > \omega_2$) is usually less convenient to use, especially if it lies in the red or IR portion of the spectrum because detectors are less efficient, and also when an unwanted fluorescence is likely to be excited either in the gas or in filters. The anti-Stokes sideband at frequency $\omega_a = 2\omega_1 - \omega_2$ is easier to detect and is less prone to fluorescence interference ; all experimental efforts have dealt with that particular sideband so far [1-9].

In many cases, spatial resolution is needed. Instead of parallel beams, focused beams can be used. With focused beams the anti-Stokes power at ω_a is independent of f-number ; it is generated from a narrow region about the focus and in the same cone angle as the pump beams. It is given by [7].

$$P_a \simeq \left(\frac{4\pi\omega_1^2}{c^3}\right)^2 |\chi|^2 P_1^2 P_2 \qquad (1)$$

where we assume $\omega_1 \simeq \omega_2 \simeq \omega_a$, P_1 and P_2 are the powers at ω_1 and ω_2 respectively, and χ is the susceptibility of the gas. One has : $\chi = \chi^{res} + \chi^{nr}$, where χ^{res} is a resonant contribution from the nearby Raman active vibration-rotation resonances and χ^{nr} a nonresonant term independent of $\omega_1 - \omega_2$, contributed by the electrons and the remote resonances.

In a pure gas and on resonance, χ^{res} is 3 to 5 orders of magnitude larger than χ^{nr}. A specific, homogeneously broadened Raman transition j gives a contribution :

$$\chi_j^{res} \simeq \frac{2c^4}{\hbar\omega_1^4} N \Delta_j g_j \left(\frac{d\sigma}{d\Omega}\right)_j \frac{\omega_j}{\omega_j^2 - (\omega_1 - \omega_2)^2 + i\gamma_j(\omega_1 - \omega_2)} \qquad (2)$$

Here, Δ_j is the average population difference per molecule between the lower vibration rotation level

($\vec{v}_j$, J_j) and the upper one, $\hbar\omega_j$ is the energy jump between these levels, γ_j the transition linewidth, g_j the weighting factor (e.g. $g_j = v_j + 1$ for a Q-line in a non degenerate mode), and $(\frac{d\sigma}{d\Omega})_j$ the spontaneous Raman scattering cross section of the mode ; N is the molecular number density. The actual resonant susceptibility of the gas is the algebraic sum $\Sigma_j \chi_j^{res}$ of all the terms such that $\omega_j \approx \omega_1 - \omega_2$. The other resonances in the gas, which are too far to produce an appreciable variation of χ over the spectral domain of interest, are small and real ; therefore they can be lumped together in the constant χ^{nr}.

The experimental conditions under which equations (1) and (2) can be used for point concentration measurements and spectroscopy in gases are listed below :

<u>Spatial resolution</u> : the focal volume from which the signal radiation is generated is approximately a cylinder [7] of diameter $\phi = 4 \lambda f/\pi d$ and length $\ell = 10 \phi^2/\lambda$, where f is the focal length, d the beam diameter, and $\lambda = 2\pi c/\omega$; ℓ is usually on the order of 1 mm.

<u>Phase matching</u> : good coherence is usually maintained across the focal region, unless gas pressures above 100 atm are used, or strong dispersion is produced by an absorbing species.

<u>Pump saturation</u> : the interaction should not induce significant variations of P_1 and P_2, leading to a practical limit $P_1 < 5$ kW for the most stringent case, e.g. H_2 at 1 atm. excited on resonance, due to stimulated Raman scattering.

<u>Population perturbation</u> : a similar requirement on Δ_j [7] also implies $P_1, P_2 < 5$ kW, with f = 10 cm ;

<u>Spectral considerations</u> : χ is roughly independant of ω, except near an absorption ; unless a resonant electronic enhancement of this kind is specifically sought for, the choice of ω_1 is not crucial, and should be made according to PM efficiency and availability of good lasers ;

<u>Detection of trace constituants in a mixture</u> : this is the problem of how small a concentration one can measure ; the essential factor here is the uncertainty on the measurements, which in practice is on the order of 30 % in terms of $\Delta N/N$, and improves as $1/\sqrt{n}$ when n shots are fired ; a realistic value for the limiting concentration is thus one for which the χ^{res} of the gas to detect (which is pure imaginary at resonance) is equal to the χ^{nr} of the buffer gas (which is real). This single shot limit lies in the range 10 ppm to 1000 ppm for many gases excited on resonance, and assuming $\Delta_j = 1$; it is independent of pressure if γ_j does not vary with pressure, provided P_1 and P_2 are large enough : for instance one could detect 100 ppm of H_2 in air at a total pressure of 10^{-5} atm. with $P_1 = P_2 = 1$ MW in 20 ns.

<u>Temperature measurements</u> : rotational and vibrational temperatures can be retrieved [8] from the population difference Δ_j ; to this end, a spectral investigation on the Q branch lines similar to that of Lapp et al [10] for spontaneous Raman scattering must be conducted ; for instance, probing the $0 \to 1$ and $1 \to 2$ vibrational transitions gives the number densities $N\Delta_0$ and ΔN_1 between the ground at 1st levels and the 1st and 2nd levels respectively (anharmonicity and rotation vibration coupling must be sufficient to separate the lines unambiguously) ; the vibrational temperature T_v results from $N\Delta_1/N\Delta_0 = \exp(-\hbar\omega_v/kT_v)$.

An extensive experimental program was conducted on H_2 gas to verify these points. Experimental arrangements with [1] and without [8] a reference leg can be used ; the one with a reference leg was preferred because it permits exact normalization of the signal versus laser powers. Concerning the light sources, many types are available. On early designs, Q-switched ruby lasers pumping stimulated Raman oscillators were used : ω_1 was produced by the laser and ω_2 by the oscillator. The big advantages of this scheme are self-alignment, exact tuning for one molecular species, and reliability (however, spectral investigations are inherently limited [3-5] ; we used it to demonstrate the feasibility of concentration measurements [5]. Extensive spectral investigations now can be done with tunable dye lasers : one can either use a fixed frequency laser and a dye laser [8, 9], or two dye lasers [9, 11]. All these lasers must emit near diffraction-limited radiation, with spectral purity and stability better than 0.1 cm^{-1}, in order to match the Raman linewidths in gases.

At the present time, we use a set-up with a single mode ruby laser pumping a tunable dye laser in a near longitudinal configuration (fig. 1). The mode selection in the ruby laser is done by a set of two high index glass (E8840) flats (one as output reflector) and a 2 mm diameter pinhole. The ruby is temperature stabilized and the cavity is mounted on a small Zerodur bench ; the linewidth is 0.01 cm^{-1}, the stability better than 0.1 cm^{-1}, the power 800 kW. The beam is amplified to the 5-10 MW level through 4 passes in an amplifier. The dye laser is classical,

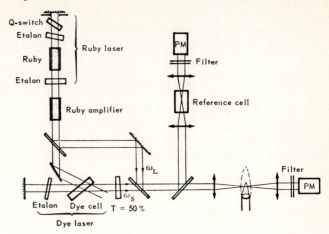

Fig. 1 - Experimental set-up.

with two etalons of 12 and 100 μm thickness, and a 1 mm pinhole ; its spectral width can reach 0.1 cm^{-1}. The advantages of the laser assembly are peak power and frequency stability, the drawback is low repetition rate. The lenses used are achromatic lenses. The reference cell contains 10 atm of N_2 (non resonant at the H_2 vibration frequency) ; we tried to replace this cell by a glass cover slide, but this resulted in poor reproducibility of the signals, certainly because the coherence length is short compared with the dimensions of the focus. Color and interference filters transmit the anti-Stokes pulses to the photomultipliers while blocking the laser and room light. The signal read in the sample leg P_a is divided by that in the reference leg (P_a^{ref}). The ratio P_a/P_a^{ref} is proportional to N^2 for a given gas under the same temperature and pressure conditions ; therefore, if the system is calibrated on a sample of known composition, N can be obtained directly. If the temperature varies, a correction must be made to account for the Boltzmann population changes.

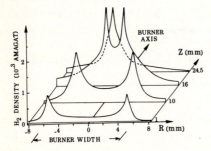

Fig. 2 - H_2 distribution in a horizontal gas flame ; R is the distance from the burner axis, Z the distance along the axis ; the results are deduced from the Q(1) line intensity.

The experiments were done on a Bunsen flame. The next figures present typical data on the H_2 formed by the pyrolysis of hydrocarbon molecules. Figure 2 shows the distribution in the flame (which then was mounted horizontal, the R axis pointing downward). These results were obtained with the first set-up built [5], and were not compensated for temperature ; however, a large maximum in the H_2 concentration is observed in the blue reaction zone ; H_2 vanishes in the diffusion zone.

Figure 3 is the spectral profile of pure H_2 at room temperature, showing a characteristic interference between the real parts of the susceptibilities of the weaker lines and that of the stronger Q(1) line. The solid curve is theoretical ; it was calculated assuming a dye laser linewidth of 0.3 cm^{-1} (the results were obtained at an early stage of the set-up, with mediocre resolution). Figure 4 presents the line maxima obtained at one location in the flame, near the tip of the cone. The lines Q(1) through Q(5) are distributed according to the Boltzmann factors for a temperature of 1350 ± 30 K, in good agreement with a crude thermocouple measurement (1340 ± 20 K). The Q(0) line is stronger than expected ; this result may be an artefact, but it can also be explained by the endothermal nature of the production of H_2 from CH_4, continuously supplying colder molecules into the hot bath. From the absolute Q(1) line intensity, we can also deduce a number density of (5.4 ± 0.5) 10^{16} cm^{-3} for H_2 at that point.

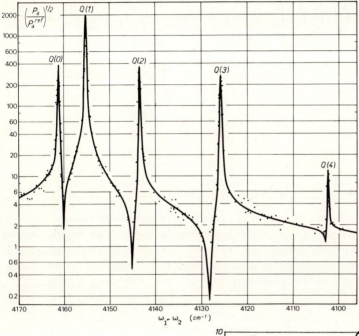

Fig. 3 - Spectrum of the susceptibility of pure H_2 at STP ; the quantity $(P_a/P_a^{ref})^{1/2}$ is plotted in arbitrary units.

Fig. 4 - Lines Q(0) through Q(5) near their maxima, in the flame. The points at line centers represent averages over 10 laser shots ; this and the flame fluctuations explain the irregular shapes and widths. The lines are normalized with respect to Q(3).

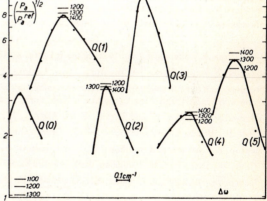

Other gases have also been studied. Figure 5 is the spectrum of the Q branch of O_2 from Q(1) to Q(23), taken with a slightly different dye laser (shortened cavity, with narrowband interference filter plus 75 μm and 500 μm etalons for fine tuning). Simple monochromators made of three highly dispersive Brewster angle prisms were also added in front of the photomultipliers ; this was done in order to prevent the ruby light from entering the colored filters and causing fluorescence at the anti-Stokes frequency. A good fit is seen with the theoretical curve, calculated for a Boltzmann distribution among rotational levels, with a monochromatic ruby laser and a Gaussian 0.14 cm^{-1} dye laser line.

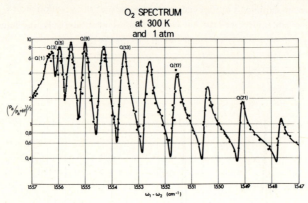

Fig. 5 - CARS spectrum of O_2 at room temperature.

In conclusion, the feasibility of carrying out Raman spectroscopy of gases by CARS is established.

The obvious application of this method at the present time is the measurement of concentrations, rotational and vibrational temperatures in flows, flames and electric discharges ; such measurements are underway in various laboratories. Very promising applications are also opened, such as turbulence measurements in aerodynamic flows with high repetition rate lasers. Finally, one could use the resonant enhancement of $d\sigma/d\Omega$ near electronic absorptions for greatly enhanced sensitivities, or fluorescence free resonance Raman spectroscopy.

REFERENCES
[1] Rado, W.G., Appl. Phys. Letters 11, 123 (1967).
[2] Hauchecorne, G., Kervervé, F. and Mayer, G., J. de Physique, 32, 47 (1971).
[3] De Martini, F., Giuliana, G.P. and Santamato, E., Optics Com. 5, 126 (1972).
[4] De Martini, F., Simoni, F. and Santamato, E., Ibid. 9, 176 (1973).
[5] Régnier, P.R. and Taran J-P E., Appl. Phys. Letters, 23, 240 (1973).
[6] Régnier, P.R. and Taran J-P E., in Laser Raman Gas Diagnostics, edited by M. Lapp and C.M. Penney, Plenum Publishing Corporation, New York, London (1974), p. 87.
[7] Régnier, P.R., Moya, F. and Taran J-P E., AIAA J. 12, 826 (1974).
[8] Moya, F., Druet, S.A.J. and Taran J-P E., Optics Com. 13, 169 (1975).
[9] Begley, R.F., Harvey, A.B., Byer, R.L. and Hudson, B.S., J. Chem. Phys. 61, 2466 (1974).
[10] Lapp, M., Penney, C.M., Goldman, L.M., Optics Comm. 9, 195 (1973).
[11] Levenson, M.D., IEEE J. Quant. Elect. QE 10, 110 (1974).

STRATOSPHERIC STUDIES USING TUNABLE LASER SPECTROSCOPY

C. K. N. Patel
Bell Telephone Laboratories, Incorporated
Holmdel, New Jersey 07733

    This paper reviews recent experiments of spectroscopic determination of stratospheric NO and H_2O using opto-acoustic techniques with a spin-flip Raman (SFR) laser as the source of tunable infrared (IR) radiation. The interest in the measurement of stratospheric NO arises from the important catalytic destruction role it is presumed to play in the stratospheric ozone balance. Recent model calculations[1] have indicated the detrimental effects of additionally introduced NO into the stratosphere in reducing the O_3 concentration and the subsequent increase in the short wavelength ultraviolet (UV) radiation reaching the earth[2]. The increased UV radiance has undesirable biological implications which have received considerable attention recently. In addition, a reduction in O_3 concentration is predicted according to a model which considers the diffusion of chlorofluoromethanes into the stratosphere, their subsequent breakdown in the stratosphere by the short wavelength UV radiation which does not reach ground level, and finally the release of chlorine which, through recombination with ozone and atomic oxygen to form ClO enters into the catalytic destruction of ozone[3] in a manner that is similar to that of NO. While this paper focusses its attention primarily on the NO measurements (with a minor importance placed on H_2O measurements), it will be seen that the spectroscopic nature of the technique described here is general enough to allow quantitative determination of a considerably larger number of stratospheric constituents of interest.

    In contrast to the large number of theoretical models and calculations which have been carried out, experimental determinations of minor constituents of the stratosphere are somewhat scarce. Indeed, O_3 which exists in a concentration of $\sim 10^{12}$ molecules cm^{-3} at an altitude (h) ~ 28 km i.e., a volumetric mixing ratio (VMR) ~ 1 ppM at h ~ 28 km is the only constituent for which detailed measurements are available both for its temporal diurnal as well as long term variations. Constituents such as NO, NO_2, H_2O etc. are only recently beginning to be studied by a variety of techniques. If we look at the ozone creation cycle described by the Chapman reactions[4] and the NO/NO_2 catalytic destruction of O_3 given below we can recognize the importance of measurements that yield not only the absolute concentration of NO (say) but also its temporal variation in determining the role of NO and NO_2 in the proposed catalytic destruction of O_3. The Chapman reactions are given by

$$O_2 + h\nu \; (\lambda < 2400 \, A) \rightarrow 2O \qquad (1)$$

$$O + O_2 + M \rightarrow O_3 + M \qquad (2)$$

$$O_3 + h\nu \; (\lambda < 3300 \, A) \rightarrow O_2 + O \qquad (3)$$

$$O + O_3 \rightarrow 2O_2 \qquad (4)$$

and the NO/NO$_2$ catalytic destruction reactions are described by

$$NO_2 + h\nu \;(\lambda < 4000\,A) \rightarrow NO + O \qquad (5)$$

$$NO_2 + O \rightarrow NO + O_2 \qquad (6)$$

$$NO + O_3 \rightarrow NO_2 + O_2 \qquad (7)$$

where M is any nonreactive molecule. The Eqs. (5) - (7) constitute the catalytic destruction cycle of O_3 because the total NO + NO$_2$ concentration remains unchanged, however, for each cycling of NO and NO$_2$ back and forth, one O_3 molecule and one O atom is destroyed. Without going through details, it is seen that to ascertain the importance of this catalytic cycle the crucial quantities are not only the concentration of NO and/or NO$_2$ but also the rate constants for the various reactions, i.e., the crucial quantity is the total rate at which various reactions proceed. This is best determined by measuring the absolute concentration of NO together with its temporal diurnal variation which is affected by the presence of solar radiation. To expand on the last statement it can be seen that while NO + NO$_2$ concentration is expected to be in the range of $1\text{-}10 \times 10^9$ molecules cm^{-3} at h $\sim$ 28 km and remains unchanged during the catalytic destruction of O_3, the absence of solar UV radiation puts a stop to the reactions described in Eqs. (5) and (6) while the reaction in Eq. (7) converts all (or nearly all) of NO into NO$_2$ i.e., at night time the equilibrium of NO + NO$_2$ is towards almost 100% NO$_2$. With the solar radiation present, generation of NO proceeds through reactions described in Figs. (5) and (6) and the NO + NO$_2$ balance shifts towards somewhere in between 0 and 100% NO. Since the solar radiation makes the NO/NO$_2$ catalytic cycle proceed, the diurnal variation in NO (or NO$_2$) would provide a very crucial check on the importance of the catalytic destruction of O_3 by NO and NO$_2$.

This paper reports measurements of stratospheric NO and H$_2$O obtained by spectroscopic techniques. NO measurements include presunrise, daytime, and sunset data which constitute the first complete information on the diurnal variation in the NO concentration. For the spectroscopic determination of NO and H$_2$O, absorption of tunable IR radiation from a SFR laser[5] by the stratospheric gases is utilized. Since the expected concentration of NO at an altitude, of 28 km is $\sim 10^9$ cm^{-3}, the peak absorption coefficient for the

Table I. Frequency and identification of NO and H$_2$O absorption lines used in the present stratospheric investigations. SFR laser magnetic field shown is that required for the observation of the respective lines using the $P_{10\text{-}9}(17)$ transition of the CO laser at 1893.52 cm^{-1} as the pump source.

SPECIE	LINE IDENTIFICATION	POSITION	SFR LASER* MAGNETIC FIELD
NO	$v=0 \rightarrow 1$, $m=3.5$ $\Omega=3/2$	1887.63 cm^{-1}	2545 G
NO	$v=0 \rightarrow 1$, $m=3.5$ $\Omega=1/2$	1887.55 cm^{-1}	2605 G
H$_2$O	ν_2 fundamental $5_{3,2} \rightarrow 6_{4,3}$	1889.58 cm^{-1}	1742 G

Doppler broadened NO lines (see Table I) is expected to be only $\sim 10^{-8}$ cm^{-1}. For a reasonable path length, say 10 cm, for *in situ* measurements of the NO absorption, it can be seen that a conventional double beam spectroscopy would require measurement capability of detection of changes in the input radiation intensity of 10^{-6} for a S/N ratio of 1 assuming a NO concencentration of 10^9 cm^{-3}. To be able to determine this value with reasonable accuracy, perhaps we will be required to measure an absorption coefficient as small as 10^{-9} cm^{-1}, and a corresponding change of 10^{-7} in the transmitted beam intensity will have to be measured. These measurements are extremely difficult. Thus we use a calorimetric technique which, instead of measuring the small expected change in the IR transmitted intensity, measures the energy left behind in the absorption cell by having excited the NO molecules to the upper state. We have shown earlier that such a calorimetric technique is best exploited by using an opto-acoustic absorption cell and that a NO concentration as small as 1×10^8 cm^{-3} can be measured by using a SFR laser as the source of tunable radiation[5]. An experimental set up consisting of a SFR laser together with a CO pump laser, an opto-acoustic absorption cell, a minicomputer (a Data General Nova) for controlling the experiment and for data handling and the associated electronics was enclosed in a pressurized, temperature controlled vessel for balloon flight to a nominal altitude of 28 km in the stratosphere. To date, two balloon flights of the experiment have taken place. Figure 1 shows a schematic diagram of the optical portion of the experimental set-up used in the first balloon flight[6] (19 October 1973,

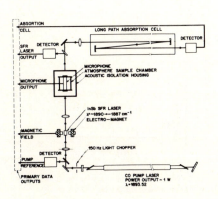

Figure 1. Schematic of the SFR laser spectroscopy set up using the opto-acoustic absorption cell and a long path absorption cell utilized for the 19 October 1973 flight. This set up was slightly modified for the 22 May 1974 flight (see text).

Figure 2. Instrumental package assembled within the flight vessel for stratospheric investigations.

launched from Palestine, Texas), and Fig. 2 shows the complete experimental set-up mounted inside the flight vessel. The general features of the experimental set-up remained unchanged for the second balloon flight[7] (22 May 1974, launched from Holloman AFB, New Mexico). The overall experimental techniques employed in these two flights are described in detail in References 6 and 7 and will not be repeated here. Flight 1 provided us with NO concentration data from sunrise up to early afternoon while the second flight provided NO concentration data from about 11:00 local time to sunset at the flight altitude. These two sets of data together provide a NO concentration information in the stratosphere through a complete diurnal cycle. Measurements of H_2O concentration are also reported from the first flight.

Figures 3(A) and (B) show opto-acoustic signals as a function of magnetic field covering the range for the observation of absorption from NO (see Table I), obtained from the first flight to illustrate the effect of sunrise. Figure 3 (A) shows the OA spectrum just prior to the visible sunrise which was expected to occur at 07:02 local time (CDT) i.e. 12:02 UT at h ~ 28 km. UV sunrise is expected to be delayed from the visible sunrise by 10 to 90 minutes depending upon the wavelength of interest[8]. The absence of the expected NO peaks at 2545 G and 2605 G denote a measured NO concentration of $\leqslant 1.5 \times 10^8$ cm^{-3}. This upper limit is set by the system noise which can be seen on Fig. 3 (A). The effect of visible and/or UV sunrise is seen on Fig. 3 (B) where the OA spectrum was taken at 12:22-12:51 CDT. The NO lines are clearly seen. The two lines seen are positively identified as those arising from NO from their absolute SFR laser magnetic field positions as well as from the relative widths of the two lines (See Refs. 6 and 7 for details). The NO concentration deduced from the OA spectrum in Fig. 3(B) is $(20 \pm 5) \times 10^8$ cm^{-3} at 12:22-12:51 CDT. Figure 4 shows a summary of the NO concentration data obtained as a function of local time. The growth of NO concentration with sunrise is clearly seen indicating a confirmation of the NO generation described by Eqs. (5) and (6).

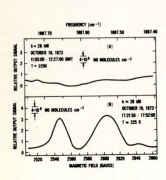

Figure 3. (A) OA signal as a function of SFR laser magnetic field obtained at 06:52-07:18 CDT on 19 October 1973. NO lines expected at magnetic fields of 2545 G and 2605 G are absent.
(B) OA signal as a function of the SFR laser magnetic field obtained at 12:28-12:55 CDT on 19 October 1974. NO absorption lines are clearly seen.

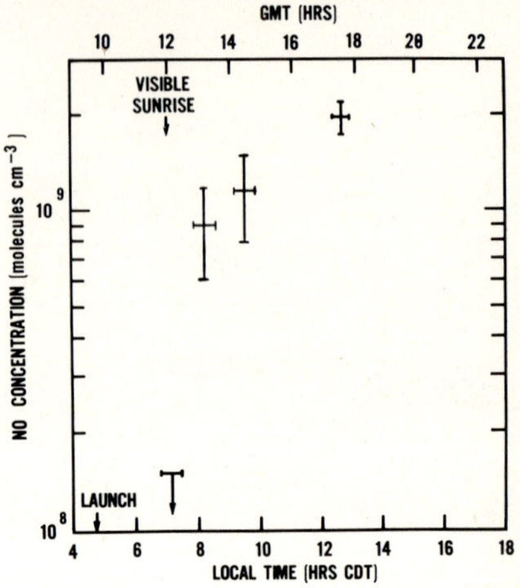

Figure 4. NO concentration as a function of the time of day (CDT) obtained on 19 October 1973.

Figures 5 (A) and (B) show the OA spectra taken at mid-morning and at sunset on 22 May 1974 obtained from the second flight. The OA signal vs the SFR laser magnetic field spectrum seen in Fig. 5(A) was obtained at 11:00-11:28 local time (MST) i.e., at 17:00-17:28 UT, and it indicates a NO concentration of $(17 \pm 5) \times 10^8$ cm^{-3}. On the other hand, the spectrum seen in Fig. 5 (B) which was taken at 20:13-20:40 MST (after the sxpected visible and UV sunrise at the balloon altitude[7]), shows no evidence of the NO lines which were clearly seen in Fig. 5 (A). We infer a NO concentration of $\leqslant 3 \times 10^8$ cm^{-3} at 20:13-20:40 MST. Again the upper limit on the NO concentration is determined by the system noise for the second flight[7]. Thus the effect of sunset is clearly seen Figure 6 (A) shows the flight altitude as a function of time for the 22 May 1974 flight and the Fig. 6 (B) shows NO concentration data as a function of time for this flight. The daytime measured NO concentration rises slowly from about $(17 \pm 5) \times 10^8$ cm^{-3} at 11:00-11:22 MST to about $(22 \pm 6) \times 10^8$ cm^{-3} at 15:00-15:28 MST. The NO data beyond about 18:00 MST are interesting in that they show the reduction of NO concentration from ~ 20×10^8 cm^{-3} to $\leqslant 3 \times 10^8$ cm^{-3} in less than 90 minutes. From Eqs. (5) - (7) the disappearance of NO is seen to be correlated with the reduction in the solar visible and/or UV radiation responsible for the NO generation (from NO$_2$). The recombination of NO and O$_3$

eventually converts all (or nearly all) of the NO into NO_2 with the sunset. A point that needs to be emphasized is that the balloon was losing altitude beyond 15:00 MST, thus the NO data seen beyond that time on Fig. 6 (B) were taken at decreasing altitudes.

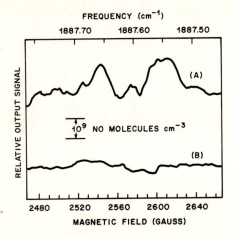
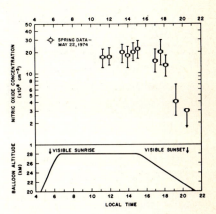

Figure 5. (A) OA signal as a function of the SFR laser magnetic field showing NO absorption lines at 2545 G and 2605 G obtained at 11:00-11:28 MST on 22 May 1974. (B) OA signal as function of the SFR laser magnetic field obtained at 20:13-20:40 MST on 22 May 1974. The NO absorption lines seen in Fig. 5(A) are absent.

Figure 6. (A) Flight profile for the spring 1974 flight. (B) NO concentration as a function of time of day (MST) measured on 22 May 1974.

Figure 7 shows the NO concentration data obtained from the fall 1973 and the spring 1974 flights plotted as a function of local time (CDT for the fall 1973 flight and MST for the spring 1974 flight). Several observations can be made:

(1) Since the NO data from the two flights were obtained at different times of the year and over different geographical regions[9] only a qualitative comparison of the absolute NO concentration should be made. Any attempt to make a one-to-one comparison could easily lead to erroneous conclusions. However, strictly on a qualitative basis, we see that for the time where data are obtained from both the flights, there appears to be a reasonable agreement in the measured NO concentration. The spring 1974 concentration is seen to be somewhat higher than the maximum NO concentration measured during the fall 1973 flight.

(2) The maximum measured NO concentration occurred at 15:00 local time and is $(22 \pm 6) \times 10^8$ cm^{-3} i.e., a VMR of $\sim$ 5 ppB at h $\sim$ 28 km, which is in reasonable agreement with other recent measurements of NO concentration obtained by a spectroscopic technique which uses the sun as a source of blackbody readiation[10] and by chemiluminescence technique[11].

(3) The presunrise concentration is $\leqslant 1.5 \times 10^8$ cm^{-3} i.e. VMR $\leqslant 0.02$ ppB at h $\sim$ 28 km while the post sunset NO concentration is $\leqslant 3 \times 10^8$ cm^{-3} i.e. VMR $\leqslant 0.02$ ppB at h $\sim$ 21.5 km. The increase in the NO concentration at sunrise appears to be delayed beyond the visible sunrise while the disappearance of NO at sunset appears to preceed the visible sunset. Thus it is tempting to correlate the observed rise and

decay in the NO concentration data with some UV wavelength which may be responsible for the NO generation reactions described in Eqs. (5) and (6). However, more work needs to be done for quantitative comparisons.

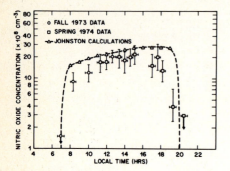

Figure 7. Combined NO concentration data from the fall 1973 and the spring 1974 flights plotted as a function of local time (CDT for the fall 1973 data and MST for the spring 1974 data). The dashed line shows results of a recent calculation by Johnston (Ref. 12).

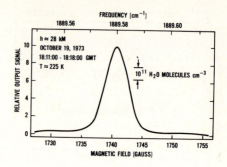

Figure 8. OA signal as a function of the SFR laser magnetic field showing H_2O absorption line at 1742 G, obtained at 13:00-13:08 CDT on 19 October 1973.

(4) The NO concentration is seen to increase slowly at times even beyond all expected visible and UV sunrise. This slow rise continues through the day and the maximum NO concentration of $(22 \pm 6) \times 10^8$ cm^{-3} is measured at 15:00 local time. This slow rise does not appear to be predicted by the reactions in Eqs. (5) and (7), if the $NO + NO_2$ concentration is assumed to remain constant. However, a recent communication from Johnston[12] suggests that throughout the daytime there is a slow dissociation of N_2O_5 which causes a small increase in the total $\cdot NO + NO_2$ concentration. This reaction reverses during the nighttime and the $NO + NO_2$ concentration remains constant in a steady state situation (i.e., averaging over the diurnal variations).

Figure 7 also shows the NO concentration as function of time obtained from a calculation by Johnston[12]. The model used by him has an instantaneous sunrise and an instantaneous sunset. The measurement appears to confirm the general shape of the NO concentration vs time curve predicted by Johnston's model[12]. The maximum calculated NO concentration value of 40×10^8 cm^{-3} is about 50% higher than the measured maximum NO concentration of $(22 \pm 6) \times 10^8$ cm^{-3} at $\sim 15:00$ local time. In addition, the increase in the calculated NO concentration at sunrise is considerably faster than the measured increase in the NO concentration. The calculated decrease in the NO concentration at sunset starts later in the day and is considerably faster than the measured variation of NO concentration at sunset. The difference in the measured and the calculated absolute concentration could be due to the higher value of $NO + NO_2$ concentration in the model calculations. The difference in the sunrise data could be due to the instantaneous sunrise of the theoretical model while the differences seen at sunset could be partly due to the instantaneous sunset of the model and partly due to the fact that the measured NO concentrations were obtained at decreasing balloon altitudes as a function of time. In any case, it is seen that the diurnal variation of the measured stratospheric NO concentration confirms the importance of the NO/NO_2 catalytic destruction cycle in the stratospheric O_3 balance as described by reactions in Eqs. (5) - (7). The NO measurements also qualitatively confirm the calculated[12,13] diurnal variation in the stratospheric NO.

A measurement of H_2O concentration obtained from the fall 1973 flight is shown on Fig. 8 in the form the OA signal as a function of the SFR laser magnetic field. A linewidth of ~ 350 MHz seen on the spectrum agrees with that predicted from Doppler broadening and pressure broadening at the stratospheric ambient pressure. Water vapor concentration is seen to be $\sim 10^{12}$ cm^{-3} i.e. a VMR ~ 1 ppM at h ~ 28 km, indicating a very dry stratosphere in reasonable agreement with other measurements[14].

The advantage of the measurement technique described here lies in the fact that this is a spectroscopic technique capable of *in situ* measurements which are independent of the solar elevation or the presence of the sun. Thus the technique is capable of measuring other minor constituents of the stratosphere by a change in the wavelength of the SFR laser. On the other hand, the spectroscopic technique that uses the solar blackbody radiation as a source[10] and measures the long path absorption of the solar radiation through the stratosphere requires deconvolution of the vertical profile of the constituent sought and is incapable of yielding data at night time. The chemiluminescence technique[11] of NO measurements is an *in situ* technique but it requires careful calibration and is incapable of measuring other constituents simply.

In conclusion, this paper has described a high resolution spectroscopy technique using a tunable SFR laser as the source of radiation which has been applied to the measurements of stratospheric NO and H_2O. The complete diurnal variation in the NO concentration is reported for the first time, and the results strongly support the theorized role of NO/NO_2 in the catalytic destruction of stratospheric O_3. This paper has also demonstrated the feasibility of using the tunable SFR laser in complex experiments outside the laboratory environment. Continuation of the present experiments in the future will be aimed towards seeking and measuring other minor constituents important in the stratospheric chemistry as well as towards making perturbation measurements. A modification of the present experimental technique could involve the replacement of the SFR laser with a high power (> 200 mW) tunable cw diode laser when it becomes available. This modification would reduce the system weight and the power requirements, which might aid towards planning flights of longer duration. Yet another laser based technique which is being implemented[15] is expected to use a low power tunable diode laser heterodyne spectrometer[16] for measurements at higher altitudes. Needless to say, tunable lasers have a very important role to play in the stratospheric investigations, and a start has been made in that direction. Future of tunable lasers in such studies is indeed very bright.

REFERENCES

1. P. Crutzen, *J. Geophys. Res.* **76**, 7311 (1971); H. Johnston, *Science* **173**, 517 (1971); E. Hesstvedt, *Can. J. Chem.* **52**, 1592 (1974).

2. M. B. McIlroy, S. C. Wofsy, J. E. Penner and J. C. McConnell, *J. Atm. Res.* **31**, 287 (1974); F. N. Alyea, D. M. Cunnold and R. C. Prinn, *Science* **188**, 177 (1975).

3. J. Molina and F. S. Rowland, *Nature* **249**, 810 (1974).

4. S. Chapman, *Mem. R. Meterol. Soc.* **3**, 103 (1930).

5. C. K. N. Patel in *Laser Spectroscopy* ed. R. G. Brewer and A. Mooradian)Plenum Press, N. Y., 1974) pp.

471-491 and references cited therein.

6. C. K. N. Patel, E. G. Burkhardt and C. A. Lambert, *Science* **174**, 1173 (1974).

7. E. G. Burkhardt, C. A. Lambert and C. K. N. Patel, *Science* **188**, 1111 (1975).

8. R. Shellenbaum, AGU meeting in San Francisco, December 1974, and private communication.

9. The flight profile for the 19 October 1973 flight can be seen in Fig. 3 of C. K. N. Patel in the *Proceedings of the expert Conference on Laser Spectroscopy of the Atmosphere*, Rjukan, Norway, 15-21 June 1975 (Optical and Quantum Electronics, to be published). The vertical profile for the 22 May 1974 flight is seen in the Fig. 6 (A) of the present paper. Due to favorable climatic conditions, the balloon for this flight stayed within 75 km of the launch site at the Holloman AFB throughout its flight.

10. M. Ackerman, J. C. Fontanella, D. Frimout, A. Girard, N. Louisnard and C. Muller, *Aeronomica Acta* **133** (1974).

11. H. F. Savage, M. Loewenstein and R. C. Whitten in the *Proceedings of the Second International Conference on the Environmental Impact of Aerospace Operations in the High Atmosphere*, San Diego, (published by Am. Meteor. Soc., Boston, Mass., 1974) pp. 5-10.

12. H. Johnston (Private communication).

13. I. A. Isaksen, *Geophysica Norvegica* **30(2)**, 1 (1973). I am indebted to Dr. E. Hesstvedt for bringing this work to my attention.

14. J. E. Harries, *Nature* **241**, 525 (1973); - *Nat. Phys. Lab. (U.K.) Reports* **DES 16** (Nov. 1972) and **DES 21** (April 1973).

15. This work is being carried out in collaboration with Dr. C. E. Hackett and Dr. D. R. Smith of Sandia Laboratories.

16. C. K. N. Patel and E. G. Burkhardt (unpublished).

SPECTROSCOPY WITH SPIN-FLIP RAMAN LASERS:
Mode Properties and External Cavity Operation.

S.D. Smith and R.B. Dennis
Heriot-Watt University
Physics Department
Edinburgh.

INTRODUCTION

The now traditional form of the spin-flip Raman laser, SFRL, uses a parallel sided crystal cavity of length about 1 cm and in c.w. operation yields resolutions varying between 600 MHz (0.02 cm^{-1})[1] and less than 100 KHz (3 x $10^{-6} cm^{-1}$)[2][3].

Particularly when operating with low resolution, this tunable source is characterised by a considerable degree of frequency instability combined with amplitude fluctuation. These problems have to some extent been obviated by the use of opto-acoustic detection[1] and double-beam spectroscopy[4]. This has enabled us to obtain good quality spectra extending over tens of wavenumbers with resolutions down to 100 MHz. Such operation is in a form particularly suitable for molecular band analysis, an example of which we present in this paper.

In this work frequency calibration using both known molecular lines and also a Fabry-Perot interferometer has been investigated to determine how continuous and linear is the frequency coverage of the SFRL. We have also studied the mode characteristics of the SFRL in some detail to give a better understanding of the laser operation.

A recent and very significant result is that we have shown oscillatory fine structure often observed on the output power modulation due to the InSb axial cavity modes to be a consequence of a coupled cavity effect between the pump laser output mirror and the spin-flip crystal. This fine structure has previously been (without much supporting evidence) attributed to transverse modes of the SFRL[1]. Such extra-cavity coupling can also be responsible for the observed amplitude instability. We interpret the behaviour of previous low resolution spin-flip spectroscopy in the light of this effect.

We also present a study of the fine tuning within a single axial mode and discuss detail of the form of the observed output power modulation. Plane wave theory is unable to explain the observed asymmetric behaviour of both these properties. Agreement is however obtained by comparison with a new theory by Firth, Wherrett and Weaire[5] which includes diffraction effects for a pump beam of finite size with a Gaussian intensity profile.

Finally, we demonstrate for the first time external cavity operation[6] of the spin-flip Raman laser (SFRL); show how this is related to input cavity modulation and indicate that the greater freedom of laser design made possible by separating the cavity optics from the active medium will facilitate the solution of many of the problems discussed.

EXPERIMENTS AND DISCUSSION

In this work we have used two c.w. SFRL systems each employing similar plane parallel InSb cavities of length 8.5 mm and electron concentration $8.5 \times 10^{14} \text{cm}^{-3}$. The first system which utilised the natural reflectivity of InSb (∼36%) had a separation between the pump laser and the InSb crystal of ∼100 cm. The second system, pumped by a stabilised Edinburgh Instruments CO pump laser (short term stability < 100 kHz), had an equivalent separation of ∼50 cm and was substantially vibration insulated. The cavity in this system was anti-reflection coated on one surface.

As the magnetic field sweeps the frequency, ν_s, of the spin-flip gain according to

$$\nu_s(B) = \nu_{co} - g^* \frac{\beta B}{h} \qquad \ldots \ldots (1)$$

through adjacent cavity mode frequencies

$$\nu_c = q \frac{c}{2nL} \qquad (q \text{ integer}) \qquad \ldots \ldots (2)$$

the output frequency is pulled from the value of ν_s. This results in reduced gain and hence periodic modulation in the SFRL output about each cavity mode frequency with period $\Delta\nu = c/2nL$. Such modulations are commonly observed provided the input pump power is sufficiently low. Typical recorder traces of the Stokes output power as a function of magnetic field are shown in Figure 1(a) for the first and in Figure 1(b) for the second experimental system. The appearance of an additional fine structure corresponding to a magnetic field period of from 2-4G is often seen. In the second system the modulation is 100% if the SFRL is operated close to threshold. A systematic alteration in the pump laser to cavity path length, d, introduced changes in this fine structure modulation period exactly corresponding to the reciprocal of this distance and absolutely corresponding in frequency to $\Delta\nu = c/2d$. We interpret these oscillations as due to feedback of Stokes radiation into the InSb cavity affecting the gain.

Molecular Spectroscopy

Most of our molecular spectroscopy has been achieved, using the first system described, with pump powers such that the SFRL operates in the semi-spin saturated regime[7]. The system was insufficiently stabilised in some respects so that the fine structure was in practice averaged out to less than 10%. The temporal frequency and amplitude instability of the SFRL in these circumstances could be explained as due to the critical nature of a 100 cm cavity in a vibrationally susceptible apparatus. A residual modulation of between 10-50% due to the InSb cavity axial modes is typically observed. The combination of these effects masks all but the strongest molecular absorption lines in a single beam spectrum and makes positive line identification and analysis difficult. Where line frequency measurement alone is vital a useful palliative is that of opto-acoustic detection as it discriminates against amplitude fluctuations of the source by responding only when the molecules absorb radiation. A complete P-branch of the $(12^00)-(00^00)$ band of OCS has been studied using various pump lines from the CO laser.

The opto-acoustic spectrum of part of this band is shown in Figure 2 for a pressure of 10 torr in a 10 cm cell. A signal/noise ratio in excess of 100:1 is achieved. Five other hot bands including one arising from the naturally occurring S^{34} isotope (relative abundance 4.22%) have been assigned.

Previously the combination of the <u>high power</u> SFRL(up to 1W) and opto-acoustic detection has been used to detect trace quantities of pollutants[8] (nitric oxide) in a relatively large buffer pressure. In our work we have also found the technique to be extremely powerful at low SFRL power ($\sim$ 10m W) combined with low gas pressure (of <100m Torr) where self-broadening is negligible. Signal to noise ratios of >10:1 are readily achieved with good opto-acoustic detector design.

For accurate spectroscopy magnetic field measurement and resettability is important due to the rapid tuning rate, $\sim$ 70MHz/G. A field sweep linear to $<1/10$G over ranges of several hundred Gauss was achieved by using subsidiary coils to the electromagnet driven by a feedback loop controlled by the output of a thermally stabilised Hall probe.

Frequency Calibration and Band Analysis

Primary frequency calibration (to accuracy 0.01 cm^{-1}) was achieved by measuring the magnetic field position of near Doppler limited known molecular absorption lines of the gases NO, DBr, H_2O and OCS using several CO pump laser frequencies. Regression fits to the data indicate a linear tuning rate of (2.35 ± 0.05) cm^{-1}/kG, corresponding to a field independent g-factor of InSb of - 48.5 ± 1.0 for all CO pump lines (1896 cm^{-1} - 1850 cm^{-1}) used in our experiments.

A band analysis of our results of OCS yields values of the band origin frequency, ν_o, and the difference between the upper and lower state rotational constants, B'- B", for each case. These results are summarised in Table 1. Comparison of measured hot band frequencies and previous infra-red data[9] shows a standard deviation of <0.01 cm^{-1} over the entire set of bands studied. This result indicates that the SFRL in the semi spin-saturated regime provides a linearly tunable source to this accuracy and no lines appear to have been missed, i.e. the frequency coverage is complete within these limits.

In the presence of a varying background the measurement of both line frequency and absolute transmitted or absorbed intensity is not obtained by opto-acoustic detection alone. It can be achieved by means of moderate speed double-beam dividing[4]. Operating with two separate room temperature pyroelectric detectors, with temporal and spatial beam splitting by a simple rotating chopper/mirror, quotient signals with a fluctuation of <2% were obtained compared to a single beam signal which exhibits modulations of >50%. We have used this technique together with an air spaced Fabry-Perot interferometer[10] to investigate the extent of continuous tunability and the fine tuning characteristics of the SFRL to a higher order than that afforded by molecular spectroscopy. Using the natural reflectivity ($\sim$ 36%) of germanium flats, anti-reflection coated on their outward facing surfaces, a simple interferometer was constructed giving a theoretical modulation depth of $\sim$ 75%. No difficulty was experienced aligning etalons of up to 60 cm spacing, but the observed modulation was somewhat degraded (typically to $\sim$ 30%) indicating the geometrical limitations. A mechanically and piezoelectrically driven length control was built into the interferometer to check alignment and also frequency content of fixed frequency sources.

Figure 3 shows a double beam SFRL spectrum where a 15 cm Fabry-Perot interferometer and a 10 cm cell of DBr (50 torr) were inserted in series in one arm of the double beam optics. A regular production of Fabry-Perot fringes is observed in the frequency range (0.57 cm^{-1}) between the R(3) isotope split doublet of DBr$^{(11)}$; these lines act as absolute frequency markers. The irregularity of spacing of the fringes (∼ 25%) indicates that the non-linearity of the SFRL tuning is a similar fraction of the inter-fringe spacing. As the length of the interferometer is increased fewer than the predicted number of orders appear within the calibration points. Assigning the missing orders on the basis of mode jumps the tuning curves shown in Figure 4 are obtained for interferometer spacings of 10, 15, 25 and 35 cm. Jumps of 0.016 cm^{-1} (500MHz) are observed. The equally spaced frequency markers obtained from such a low finesse interferometer can be readily interpolated to ∼ 10% of their frequency separation. For a 50cm interferometer this corresponds to an accuracy of 30MHz, quite suitable for Doppler limited spectroscopy. This produces a much more accurate frequency calibration technique than early work using grating spectrometers and has the advantage in simplicity over heterodyne experiments in that no secondary laser source is required and in addition the frequency range is automatically extendable over tens of wavenumbers and not limited by the i.f. bandwidth of the detector.

Mode properties

By using a 52 cm interferometer, Figure 5, it was found that the tuning within the SFRL mode is linear to ± 50MHz and the tuning rate across a mode variable from 65MHz/G (below the 70MHz/G rate of spontaneous tuning) to 40MHz/G. These results are in substantial agreement with both heterodyne[3] and spectroscopic[1] measurements.

Understanding of the tuning behaviour to date is based upon the plane-wave mode "pulling and hopping" equation for the output frequency, $\nu(B)$, as follows:

$$\nu(B) = \frac{\nu_c \Gamma_s + \nu_s(B) \Gamma_c}{\Gamma_s + \Gamma_c} \qquad \ldots \ldots (3)$$

where $\nu_s(B)$ and ν_c are defined in equations (1) and (2), Γ_s is the linewidth of the Raman gain and Γ_c that of a nearby cavity resonance. The competition between $\nu_s(B)$ and ν_c through equation (3) predicts a region of linear tuning followed by a mode jump. This is at variance with the results of Fig. 5 (and with other supporting evidence cited above).

Firth, Wherrett and Weaire[5], at our laboratory, have analysed this behaviour taking into account the finite cross section of the pump beam by obtaining solutions of the appropriate form derived from Maxwell's equations, viz:

$$\nabla^2 E + \frac{2\pi\nu^2}{c}(\varepsilon + 4\pi \chi^{NL}|E_p|^2)e = 0 \qquad \ldots\ldots (4)$$

where E is a component of Stokes field and E_p the pump field, ε the dielectric constant and χ^{NL} the resonant non-linear susceptibility.

The most novel results from the solutions are <u>frequency asymmetric</u> effects in tuning rate and power necessary to achieve threshold. The latter result also allows us to infer an asymmetry in output power as a function of magnetic field when operated at constant input power. These asymmetries are caused by focussing/defocussing effects due to a change of refractive index, proportional to pump intensity and therefore varying across the pump beam associated with the Raman gain. The fractional change of the index can be estimated from the plane-wave steady-state expression

$$\frac{\Delta n}{n} = \frac{\nu - \nu_s}{\nu} \cdot \frac{\Gamma_c}{\Gamma_s} \qquad \ldots\ldots (5)$$

and is $\sim 4 \times 10^{-5}$.

The results from this theory are summarised schematically in Figure 6. Qualitative agreement between the tuning rate curve of Figure 5c with Figure 6c can be seen. Inversion of the threshold intensity parameter, Figure 6d, also predicts the asymmetric power output form shown in Figure 1.

The asymmetry in output power is closely related, by theory, to the behaviour of the tuning curve. The tuning rate varies across a mode from around the plane wave rate

$$\frac{d\nu}{dB} = \frac{g^*\beta}{h} \cdot \frac{\Gamma_c}{\Gamma_s + \Gamma_c} \qquad \ldots\ldots (6)$$

at the high field and to approach the spontaneous value,

$$\frac{d\nu_s}{dB} = g^*\beta/h \qquad \ldots\ldots (7)$$

at low fields. As a result the tuning rate is greater and the mode hop smaller than the plane wave case.

For our experimental conditions at 5 kG, Γ_s is $\sim$ 700MHz and Γ_c $\sim$ 3GHz implying mode jumps from

$$\Delta\nu = \Delta\nu_c \frac{\Gamma_s}{\Gamma_c + \Gamma_s} \qquad \cdots\cdots\cdots \quad (8)$$

of $\sim$ 500MHz i.e. 80-90% frequency coverage.

Our conclusion from molecular band analysis that coverage is effectively complete might be explained by the inclusion of input cavity effects mentioned earlier in this paper. The frequency separation of these modulations is $\sim$ 150MHz and several such modes could be excited in the presence of a spontaneous gain linewidth of 700MHz, with perhaps hopping near the mode extremity.

External Cavity Operation of the SFRL

The mode hopping and pulling characteristics described in the previous section are inconvenient for spin-flip spectroscopy particularly if high resolution (<<100MHz) over a reasonable range is required. We report here the first observations of external cavity operation by deliberately introducing an external cavity mirror. In fact, the input cavity interaction effect, referred to earlier, and inadvertently found in many cw SFRL systems, is a form of external cavity operation.

For these experiments the second SFRL system, vibrationally stabilised and pumped by the Edinburgh Instruments CO laser was used. The InSb sample was anti-reflected on one end by a $\lambda/4$ layer of ZnS (n=2.2) which reduces the reflectivity, theoretically, to $\sim$ 1%, was held in a cryostat with anti-reflected ZnSe windows. Measurements of the effectiveness of the single surface anti-reflection coat were made by reflecting the CO pump beam of the sample at 300°K and 77°K when high absorption eliminated effects from the second surface. A surface reflectivity of $\sim$ 3% was observed.

In an initial experiment, Fig. 7(a), with the anti-reflected end facing the pump laser, the pump power was adjusted to be just above threshold giving axial modes, strongly modulated by input cavity effects as in Figure 1(b).

A plane germanium mirror, anti-reflected on its back surface was then aligned and introduced approximately 13 cm from the back surface of the InSb cavity. According to the pump power level, this mirror either induced oscillation where there was none, or Figure 7(b), increased the output power by up to a factor of $\sim$ 5, changed the form so that the oscillation occurred continuously between axial mode frequencies and introduced a third periodicity into the output power modulation. This in itself is good evidence of external cavity operation.

A final and more appropriate experiment utilised the cavity with anti-reflection side away from the laser and the "input-cavity" distance increased to 2 metres to reduce the modulation period to $\sim$ 75MHz. A coated germanium mirror of reflectivity 70% situated just behind a 10 cm focal length BaF_2 lens was then inserted to complete the cavity. The output with and without the external cavity lens and mirror is shown in Figure 8. It shows the expected very strong modulation with a period very different to both the input cavity effect and the axial mode separation. It is worthy of note that the crystal axial mode modulation is very difficult to eliminate although the InSb cavity has only $\sim$ 3% reflectivity at one end - an indication of the high gain levels in the SFRL.

An obvious advance in spectroscopic method is now to servo-control the external cavity frequency to the peak of an axial mode while sweeping the magnetic field. This should dramatically linearise the high resolution tuning behaviour.

Acknowledgements

The work reported here has been recently carried out by the Spin-Flip Group at Heriot-Watt University. We acknowledge in particular the contribution of Dr. R.J. Butcher (University of Cambridge), Dr. A.Z. Nowakowski (Gdansk Technical Institute), Dr. M.J. Colles, H.A. Mackenzie, T. Scragg and F.H. Hamza on the experimental side, and of Dr. W.J. Firth, Dr. B.S. Wherrett and Dr. D. Weaire on the theoretical side. Financial assistance from the Science Research Council and Heriot-Watt University is also gratefully acknowledged. Infra-red anti-reflection coatings were provided by Dr. J.S. Seeley and his group at University of Reading.

References

1. Butcher, R.J., Dennis, R.B., and Smith, S.D., 1975, Proc. Roy. Soc. (London) to be published.
2. Patel, C.K.N., 1974, Appl. Phys. Letts. $\underline{25}$, 112
3. Brueck, S.R.J., and Mooradian, A., 1974, I.E.E.E. $\underline{QE-10}$, 634.
4. Nowakowski, A.Z., Dennis, R.B., Hamza, F.H., and Smith, S.D., 1975, Optics Commun. to be published.
5. Firth, W.J., Wherrett, B.S., and Weaire, D., 1975, Optics Commun. to be published.
6. Scragg, T., and Smith, S.D., 1975, Optics Commun. to be published.
7. Colles, M.J., Dennis, R.B., Smith, J.W., and Webb, J.S., 1974, Optics Commun., $\underline{10}$, 145.
8. Kreuzer, L.B., and Patel, C.K.N., 1971, Science, 45.
9. Maki, A.G., Plyler, E.K., and Tidwell, E.D., 1962, J. of Res. of Nat. Bur. of Standards, $\underline{66A}$, 163.
10. Mackenzie, H.A., Smith, S.D., and Dennis, R.B., 1975, Optics Commun. to be published.
11. Tables for the Calibration of Infra-red Spectrometers, Commission on Molecular Structure and Spectroscopy. Mould, H.M., Price, W.C., and Wilkinson, G.R., London, unpublished.

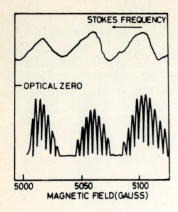
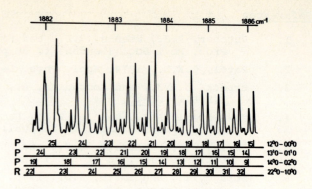

Figure 1
Stokes output power as a function of magnetic field for both experimental systems.

Figure 2
Part of the opto-acoustic OCS spectrum around 1884 cm^{-1} with assignments.

MOLECULAR CONSTANTS DERIVED FROM OCS

$$\nu = \nu_o + (B' + B'')m + (B' - B'')m^2$$

m = J for P Branch
m = J + 1 for R Branch

Transition	ν_o (cm^{-1})	$(B'-B'') \times 10^5$ cm^{-1}	Maki et. al. (1962) ν_o (cm^{-1})
$(12^0 0) - (00^0 0)$	1892.29 (.01)	6.7 (1.4)	1892.20
$(13^1 0) - (01^1 0)$	1891.685 (.002)	16.2 (.4)	1891.78
$(22^0 0) - (10^0 0)$	1872.48 (.01)	4.7 (.8)	1872.45
$(14^0 0) - (02^0 0)$	1889.640 (.004)	2.7 (1.9)	1889.66
$(14^2 0) - (02^2 0)$	1889.95 (0.08)		
$(12^0 0) - (00^0 0)$ (OCS^{34})	1879.77 (0.03)		

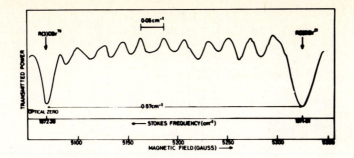

Figure 3

Double beam SFRL spectrum of 10 cm Fabry-Perot interferometer and a 10cm cell of DBr (50 torr) in series.

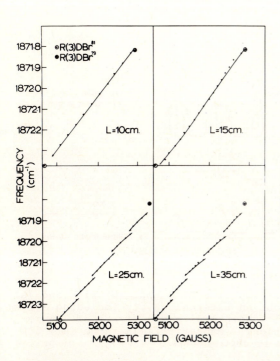

Figure 4

SFRL output frequency versus magnetic field inferred from various length Fabry-Perot etalons.

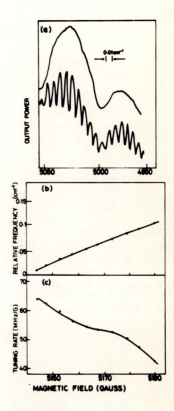

Figure 5

(a) SFRL output and its transmission through a 52cm air-spaced Fabry-Perot interferometer (lower curve); (b) inferred tuning and (c) tuning rate through a mode, all as a function of magnetic field.

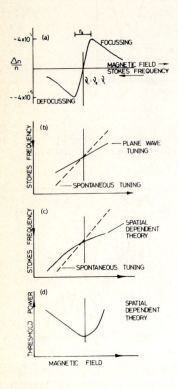

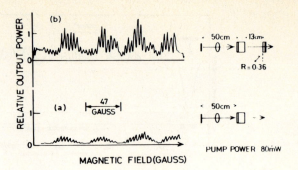

Figure 7

Comparison of SFRL output (b) with and (a) without an external cavity mirror of 36% reflection. Magnetic field range is ∿ 200 gauss at 4800 gauss.

Figure 6

Schematic representation of (a) refractive index change, (b) tuning (plane wave model) (c) tuning (FWW model) and (d) threshold factor through a mode.

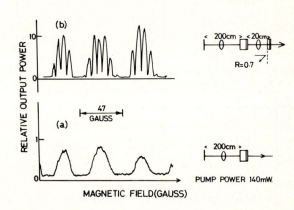

Figure 8

Comparison of SFRL output (b) with and (a) without an external cavity including a 10cm focal length lens and a mirror of 70% reflection. Magnetic field range is ∿ 150 gauss at 4800 gauss.

NEW LASER MEASUREMENT TECHNIQUES FOR EXCITED ELECTRONIC STATES OF DIATOMIC MOLECULES

R. E. DRULLINGER, M. M. HESSEL AND E. W. SMITH

Gaseous Electronics Section, 277.05
National Bureau of Standards, Boulder, Colorado 80302

Introduction

This paper will briefly outline several new laser measurement techniques which we have developed for the analysis of excited elecronic states of diatomic molecules. For molecules which have bound ground states, a visible laser is used to selectively excite a single vibration-rotation level in the electronic state of interest. We then use DC Stark effect, RF double resonance and various fluorescence techniques, as discussed in Sections I and II, to obtain excited state dipole moments, lifetimes, quenching cross sections, transition moments as a function of internuclear distance, and other molecular structure data. We have also developed new laser excitation techniques and used optical double resonance methods for excimer molecules which have repulsive ground states and are bound only in their excited states. These techniques, discussed in Sections III and IV, have been used to obtain potential energy curves, f- values, lifetimes and various kinetic rates.

I. D. C. STARK EFFECT AND R-F DOUBLE RESONANCE

To determine the dipole moment of electronically excited molecules, an argon ion laser was used to excite the $C^1\Pi$ state in the NaK molecule, RF and DC electric fields were applied and the resulting fluorescence was analyzed in terms of a fairly straightforward theoretical model.

It was found[1] that the 488 nm argon laser line would produce a fairly large population in the $(v', J') = (7, 24)$ level via the $(v', J') \leftarrow (v'', J'')$ transition $(7, 24) \leftarrow (1, 23)$ and the 496.5 nm line populates the $(7, 5)$ level via the transition $(7, 5) \leftarrow (4, 5)$. The $C^1\Pi$ state of NaK is well described by Hunds case a and each rotational level consists of two closely spaced states of opposite parity (Λ doublets). Since each rotational level in the ground state has a definite parity, the laser induced transition $(v'J') \leftarrow (v'',J'')$ will excite only one member of the Λ doublet (due to the parity selection rule).

In an unperturbed molecule, the J" = 5 state gives rise to a v" fluorescent series of Q lines (the P and R transitions being forbidden by parity) and the J" = 24 state gives rise to a series of P and R lines. If the excited molecules are perturbed by a weak electric field (10-100 volts/cm) for J' = 5 and a strong field (5000V/cm) for J'= 24, the Λ doublet states are mixed, some population is transferred to the state of opposite parity, and the parity forbidden transitions begin to appear. Figure (1) shows the fluorescence near 556 nm with the electric

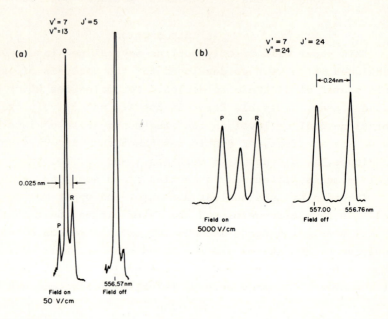

Fig. 1 Effect of Electric Field on Fluroescence Signal for NaK
(a) J' = 5 (b) J' = 24

field on and off. Figure (2) shows that the intensity of the parity forbidden lines increases with the DC electric field strength until the population in the two states of opposite parity is equalized. A theoretical analysis of this experiment[2,3] shows that this intensity is proportional to an expression of the form:

$$I = \Sigma_M F(M) G(M) X^2(M) / \left[1 + X^2(M)\right]$$
$$X(M) = 2\mu EM/\delta J(J + 1)$$

where F and G are matrix elements[2] that depend on the laser polarization and detector position, μ denotes the excited state dipole and δ the splitting of the Λ doublet (energy spacing between states of opposite parity). Since the DC electric field strength, E is known, this measurement provides (μ/δ); our statistics indicate that the accuracy of this ratio is the order of 2%. We measure δ by applying an r.f. electric field and sweeping the frequency to find the resonance.

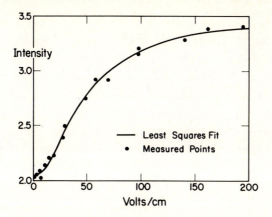

Fig. 2 Intensity of J' = 5 P-Branch of NaK as a function of electric field.

The shape of this resonance signal shown in Fig. 3 is determined by the

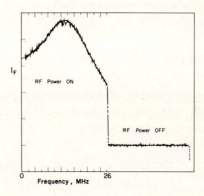

Fig. 3 Resonance signal for Λ-doublet splitting of NaK for J' = 5

usual radiative and collisional lifetimes but these data have not yet been extracted since the line is also power broadened by the r.f. electric field; an analysis of these data is currently in progress.

II. VARIATION OF TRANSITION MOMENT WITH INTERNUCLEAR DISTANCE

Another measurement technique was developed to obtain the variation of the electronic transition dipole moment as a function of internuclear distance[4]. The A value for the transition $^1\Sigma_g\ v"J" \leftarrow\ ^1\Pi_u v'J'$ is proportional to the square of the matrix element $<v'|R|v">$. We assume that $R(r)$ can be expanded in powers of the internuclear distance r,

$$R(r) = R_0 + \alpha\ r + \beta\ r^2 + \ldots,$$

the matrix elements $<v'|r^n|v">$ are calculated using the known Na_2 wavefunctions[5] and the expansion coefficients R_0, α, β, etc are determined by comparison with experiment. For example these coefficients may be determined by a least squares fit to the measured lifetimes[4] $\tau_v^{-1} = \Sigma_{v"J"} A(v'J',v"J")$ for various vibrational levels. The coefficients α, β ... may also be determined by a least squares fit to the intensities $I(v', v")$ of a $v"$ fluorescence series (since I is proportional to the A value).

To demonstrate this measurement technique, we measured the intensities of two $v"$ fluorescence series[4] excited by the 488 nm and 476.5 nm argon laser lines and used the lifetime data of Baumgartner, Demtroder, and Stock. These three independent determinations of R_0, α, β gave for the $^1\Sigma_g \leftarrow\ ^1\Pi_u$ transition in Na_2, $R_0 = 6.8 \pm 0.2D$; $\alpha = 0.4 \pm 0.1\ D/Å$; $\beta < 0.1\ \alpha$. These results are in qualitative agreement with a crude theoretical calculation by Tango and Zare; a more accurate ab initio calculation of $R(r)$ is currently in progress. Our results disagreed radically with an analysis by Callender, et. al.[6] who used a somewhat different technique, but recent changes[7] in their analysis have improved the agreement considerably.

III. EXCIMER MOLECULES - LASER PUMPING FROM THE GROUND STATE

Excimer molecules are of current interest[8] as high power laser candidates. These molecules are difficult to study by conventional spectroscopic methods because of their repulsive ground state. Actually many excimers have a small ground state well produced by Van der Waals attraction but this well is so shallow (usually a few hundred cm^{-1})

and lies at such large internuclear distances that conventional absorption spectroscopy provides very little information about the excited states. We have therefore developed new laser measurement techniques for studying this class of molecules; these techniques have been applied to Hg_2 as an example.

Previous analyses of electronically excited Hg_2 could be divided into three categories (1) low density (less than 3×10^{16} Hg atoms/cm^3) optical excitation in which the atomic resonance line is used to excite the 6 $^3P_1$ atomic state which then produces excited Hg_2 via three body recombination. (2) low pressure discharge excitation and (3) high energy (MeV) electron beam excitation at atomic densities greater than 3×10^{18} cm^{-3}. The resonance lamp excitation is restricted to low pressures because of optical depth problems and at low pressures the molecular formation rate is quite slow; this rate can be enhanced by adding N_2 but the presence of a foreign gas complicates the analysis of the Hg_2 molecular structure and kinetics. Electric discharge and electron beam excitation are inhibited by the presence of too many lines emitted by highly excited ions and neutrals. We have therefore developed a highly selective laser excitation scheme in which the 253.7 nm mercury resonance line is pumped in the line wing at 257.2 nm using the 15 mW output of a doubled Argon ion laser. There is no optical depth problem in the line wings and we have used this technique to selectively excite the 6 $^3P_1$ state at pressures up to several atmospheres where the molecular formation rate is very high. Figure 4 shows a typical laser induced fluorescence spectrum of the Hg_2 molecule. Two continuous bands are emitted, one centered at 335 nm and the other at 485 nm. For high temperatures (T > 575K) and high densities (N > 10^{17} cm^{-3}) it was found that, for any two wavelengths, λ_1 and λ_2 in these bands, the ratio of intensities $R = I(\lambda_1)/I(\lambda_2)$ is an exponential function of temperature of the form $\exp(\Delta E_{12}/kT)$. In ref. (19) it was argued that the states emitting these bands are in thermal equilibrium at these high temperatures and densities; a theoretical analysis based on this argument shows that ΔE_{12} equals the difference in energy between the states which emit at the wavelengths λ_1 and λ_2.

Thus it was possible to map out the potential curves and the f-values for these electronic states by plotting log R versus 1/kT for several values of λ_1 and λ_2. In addition, from the pressure and temperature dependence of both steady state and time dependent molecular fluorescence the basic kinetics of the pure Hg_2 system has been analyzed. For further details, see reference 9. The effect of the buffer gases He, Ar, Xe and N_2 was studied for buffer gas densities ranging from zero to

10^{20} cm^{-3} and mercury densities ranging from 3×10^{16} to 1×10^{20} cm^{-3}. Except for some minor changes in diffusion rates there were no observed effects due to the buffer gases when the mercury atom density was greater than 5×10^{17} cm^{-3}. Results such as this are very important for Hg$_2$ laser design and could not be inferred from low pressure fluorescence data.

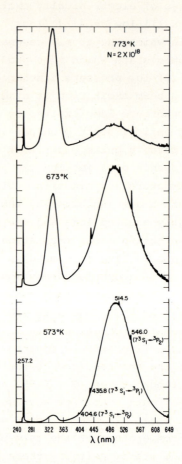

Fig. 4 Mercury fluorescence spectrum at various temperatures for a fixed atomic density of 2×10^{18} cm^{-3}.

IV. EXCIMER MOLECULES - LASER OPTICAL DOUBLE RESONANCE TECHNIQUE

The above excitation scheme can be used to create a high steady state excimer density (approximately 10^{12} cm^{-3}) in the manifold of states which arise from the 6 $^3P_0$ and 6 $^3P_1$ atomic states. This excimer population can then be probed by a second laser to look for excited state absorption or gain on transitions to the ground state. The excited state fluorescence induced by the probe laser can then be used to map out higher electronic states in a systematic manner. By chopping the probe laser and measuring the phase lag of the excited state fluorescence one can measure various inelastic rates for specific excited electronic states.

Thus far, we have made four types of measurements using this excited state fluorescence technique:

(1) The 15 mW output of the pump laser at 257.2 nm was tightly focused to 10^{-2} cm^2 and two photon pumping was observed. The second photon apparently excites only repulsive states which dissociate to Hg(7 $^3S_1$) since the only new fluorescence features observed were atomic transitions resulting from that state as seen by the spikes at 404.6, 435.8 and 546.0 nm in Fig. 4.

(2) A 1 watt 488 nm argon laser line was used as a probe laser. The probe laser was chopped and focused to 10^{-2}cm^2 colinearly with the 257.2 nm pump laser as shown in Fig. 5. The modulated fluorescence signal was measured with a lock in detector and the modulated signal

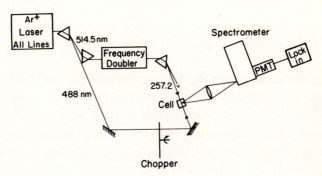

Fig. 5 Experimental setup for optical double resonance experiment.

is given by the dashed curve in Fig. 6 (the solid curve gives the unmodulated fluorescence for comparison); a positive signal corresponds to an increase in fluorescence when the probe laser is on, similarly a negative signal corresponds to a decrease in fluorescence intensity due to the probe laser. The probe laser reduced the fluorescence intensity at all wavelengths except 225 nm, 235 nm, 254 nm and 488 nm. The latter is due simply to the strong scattered light from the probe laser, but, under high resolution, the 225 nm (20 nm wide) and 254 nm (1 nm wide) features prove to be highly structured molecular bands and the narrow emission feature at 235 nm seems to be a Rydberg band. All bands will be analyzed to give molecular constants for the excited states involved.

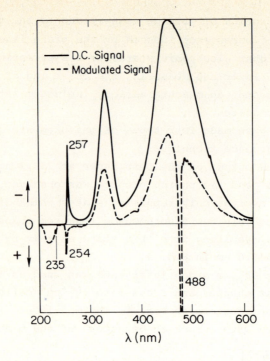

Fig.6 Hg_2 fluorescence signal with and without lock-in detection.

(3) A 1 mW HeCd laser at 325 nm focused to $10^{-2} cm^2$ gave no new fluorescence. Recent ab-initio calculations[10] in Mg_2 may be used to infer the structure of corresponding states in Hg_2 and this approach also indicates that there is no excited state absorption near 325 nm. This result is very important since the 335 nm Hg_2 fluorescence band is a serious laser candidate and the presence of excited state absorption in this vicinity would inhibit laser action.

(4) A 1 watt Nd YAG laser at 1.06 μm was focused to 0.04 cm^2 but no alteration in the steady state fluorescence was produced. This wavelength was tried because our measurements of the Hg_2 molecular structure showed that the two principal fluorescence bands at 335 nm and 485 nm are emitted by electronic states which are separated by 6500 cm^1. The presence of nonradiating (gerade) states nearby indicated that the Nd laser might be absorbed off resonance and thereby alter the populations in the states radiating at 335 nm and 485 nm. The null result obtained by this measurement probably indicates that vibrational equilibration rates exceed the off resonance pump rate.

It should be emphasized that all of the above experiments relied on the highly selective nature of laser excitation as well as the high power density which is obtainable with many visible lasers. As more lasers become available and more new fluorescence techniques are adapted to their use, it will no doubt be possible to obtain a great deal of highly specific data on molecular structure and excited state kinetics for electronically excited molecular states.

REFERENCES

1. J. Toueg, M. M. Hessel, and R. N. Zare, 29th Symposium on Molecular Structure and Spectroscopy (Ohio State University 1974).

2. S. J. Silvers, T. H. Bergman, and W. Klemperer, J. Chem. Phys. $\underline{52}$, 4385 (1970).

3. R. W. Field and T. H. Bergman, J. Chem. Phys. $\underline{54}$, 2936 (1971).

4. M. M. Hessel, E. W. Smith, and R. E. Drullinger, Phys. Rev. Lett. $\underline{33}$, 1251 (1974).

5. P. Kusch and M. M. Hessel (in preparation 1975).

6. R. H. Callender, J. I. Gersten, R. W. Leigh and J. L. Yang, Phys. Rev. Lett. $\underline{32}$, 917 (1974).

7. R. H. Callender, J. I. Gersten, R. W. Leigh and J. L. Yang, Phys. Rev. Lett. $\underline{33}$, 1312 (1974).

8. C. W. Werner, E. V. George, P. W. Hoff, and C. K. Rhodes, Appl. Phys. Lett. $\underline{25}$, 235 (1974).

9. R. E. Drullinger, M. M. Hessel, and E. W. Smith, NBS Monograph 143, US Government Printing Office, Washington, DC (1975).

10. M. Krauss and W. Stevens, private communication, 1975.

EXCIMER AND ENERGY TRANSFER LASERS

D. C. Lorents and D. L. Huestis
Stanford Research Institute

I.   Introduction to Excimers

Because of their property of having a dissociative ground state, the excimer molecules are uniquely suited as candidates for efficient high energy electronic transition lasers. For this reason considerable interest has developed recently in understanding the characteristics of these interesting molecules.

We define an excimer as a molecule that is bound only in an excited level, its ground state being unbound except perhaps for weak Van der Waals interaction. The major class of atomic species that form excimers are thus the closed shell atoms with ^1S ground states. Examples of the homonuclear diatomic excimers include all the rare gases and the Column II metals Be-Hg. Mixed excimers can be formed from diatomics of non-identical closed shell atoms such as XeHg and closed shell-open shell combinations ($^1S + {}^2S, {}^2P$ or ^3P) that yield repulsive ground states such as XeO and NaXe. Interactions with closed shell molecules are also generally repulsive and form the basis of a class of triatomic and larger excimers. $HgNH_3$ and XeN_2 are two examples of a very large class.

The basic molecular interactions of a homonuclear closed shell excimer system are illustrated in Fig. 1. In the rare gas case the bound excited states may be described as a Rydberg electron orbiting the stable molecular ion. It is the binding of the molecular ion core which produces the stability of the excimer. In such a symmetric system any ground state-excited state interaction gives rise to a pair of g-u states due to the symmetric exchange interaction, one of which is bound and the other repulsive (with the outer electron resembling the excited atomic state to which it dissociates). In heteronuclear systems such an interaction is non-existent and the excimer binding is generally weaker. Frequently the mixed excimer binding is ionic in nature deriving from an $X^+ + Y^-$ interaction. In such cases curve crossings to $X^* + Y$ states may play an important role as illustrated in Fig. 2. An example of such a class of systems is obtained in the rare gas halide interaction recently observed by

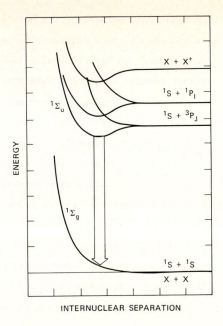

Fig. 1. Basic structure of homonuclear excimers

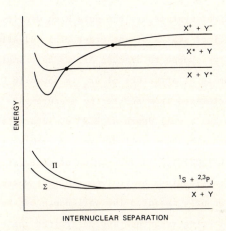

Fig. 2. Schematic state structure of mixed excimers (e.g., rare gas-halogens). With column VI elements the Σ and Π labelings are reversed.

Setser and coworkers (VS75) and by Ewing and Brau (EB75).

The class of excimer molecules have obviously important applications to the development of high energy lasers since total inversions are always easily maintained. Considerable interest has therefore developed in understanding the structural and radiative characteristics of these systems. The rare gases play a particularly important role as we shall see not only because of the multitude of excimers that can be formed with them but also because of their favorable properties for electronic excitation at high densities with high energy electron beams.

The spectroscopy of excimer systems is limited mainly to emission between the lowest excimer level and the ground state. Due to the repulsive ground state this emission is generally a single structureless continuum band when the upper state is fully relaxed. Structure in this emission has only been observed from vibrational levels near the top of the excimer well and such levels are sometimes seen also in absorption. More useful information is obtained from emission and absorption spectroscopy between excited levels of excimer systems, but very few systems have been extensively studied. For example, the work of Ginter (GB70) on He_2 excited states has provided the most detailed and complete understanding of the state structure of any excimer system to date. The early work on continuum spectra has been extensively reviewed by Finkelnberg (FP57). Recent analysis of XeO upper level spectra (HGHML75) will be discussed in this paper and work is progressing on XeHg at SRI and other laboratories.

II. Rare Gas Excimers: Ar_2

The best known excimer systems are the pure high density rare gases, all of which form strongly bonded excimer states of the order of 1 eV. Although He_2 is the best characterized of these, we choose to discuss Ar_2 as an example more typical of the heavier rare gases. The level structure of Ar_2 is shown in Fig. 3. The bound $^2\Sigma_u^+$ ion state is well characterized from scattering measurements (LOC73,MW74) and ab initio calculations (GW71) and has a well depth of 1.25 eV located at $R_e = 4.6\ a_o$. The ground state repulsive curve has also been well determined from scattering measurements and calculations (CJAM69,PSL72,GK72). In the region of interest for the excimer emission the ground state potential is given by $V(R) = A\ exp(-\alpha R)$ where $A = 5.66 \times 10^7\ cm^{-1}$ and $\alpha = 3.88\ \text{Å}^{-1}$. We have recently measured the well depth of the $^3\Sigma_u(1_u, 0_u^-)$ excimer potential by a beam scattering technique and determined the well depth to be 0.78 ± 0.03 eV (GSL75). This is consistent with a recent spectroscopic analysis of the very close lying $^1\Sigma_u(0_u^+)$ state based on the emission spectra of both the 1st and 2nd continua which arise respectively from very high lying and very low lying vibration levels of the bound excimers (MS74). The unstructured 2nd continuum band originating from the bottom two excimer wells is essentially the only emission observed from excited rare gases at densities above 10^3 torr. Based on the latest theoretical estimate of the spin-orbit effect, the $^1\Sigma_u$ excimer lies only about 800 cm^{-1} above the $^3\Sigma$, a separation that contributes less to the bandwidth of the emission than does the slope of the repulsive wall. The radiative lifetimes of the $^1\Sigma_u$ and $^3\Sigma_u$ have been determined experimentally from time decay measurements at high pressures and are 4.2 ± 0.3 nsec (KGW74) and 3.2 ± 0.3 μsec (KGW74,TH72,ORRF74) respectively. The ratio of these lifetimes is consistent with theoretical estimates based on the magnitude of the spin-orbit interactions (LEH73). The Ar_2 excimer has been observed in gas, liquid and solid phase with an emission band shape and energy essentially independent of the thermodynamic

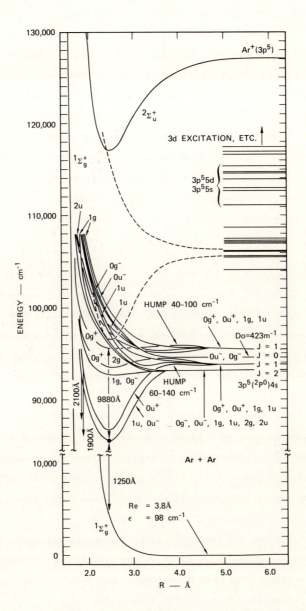

Fig. 3. Ar_2 Energy Level Diagram

tate of the medium (CRJ82). The temperature dependence of the bandwidth indicates a vibrational spacing of about 300 cm^{-1}.

Very little spectroscopic information is available on the higher excited levels of Ar_2 except for a few bands observed by Firestone et al. in the region of 1 μ (ORRF74) as indicated in Fig. 3. These states have been estimated following Mulliken's ideas on Xe_2 (RM70). There is a considerable need for extensive spectroscopic studies of the Rydberg series of the heavy rare gas excimers in order to gain a reasonable understanding of the state structure of these systems. It is of course important to note that the extensive network of states and crossings permit, under collision dominated conditions, rapid relaxation of the excited state energy to the lowest excimer levels. One of the most important properties of the high density excited rare gases is their ability to funnel ionization and excitation energy rapidly and efficiently to the lowest excited levels from which it can only be removed by radiation or collisional transfer. E-beam energy deposited in high density rare gases is delivered to the lowest excimer levels with about 50% efficiency (LO72,HEL74). The rare gases are unique in having the ability to convert such a large fraction of E-beam energy to electronic excited state energy and are therefore the best host media for energy transfer pumping of additive species.

III Mixed Excimers: XeO

The Green emission bands of the XeO excimer were first identified by Kenty et al. (KANPP46) in a high-frequency discharge in xenon-oxygen mixtures. Several investigators have studied these bands in the region of 4800-5800 Å (HH50,CCT61,Wi65,CuC74) and other emissions have been observed in the 6500-8600 Å and 2900-3200 Å regions as well (HH50, CCT61,CuC74). Each of these emission features may be correlated with the metastable levels of the oxygen atom but they are shifted, broadened and intensified by the association with an Xe atom.

We have studied the spectral and temporal behavior of XeO emissions from high density Xe-O_2 mixtures excited by a short pulse of electrons from a Febetron 706 (HGHML75). Two green band systems and a uv continuum band were observed. Densitometer traces of these emissions are shown in Fig. 4 and 5. Cooper et al. (CCT61) have provided the vibrational level assignment of the major green bands that arise from a $2^1\Sigma^+ \to 1^1\Sigma$ transition corresponding to $XeO(^1S) \to XeO(^1D)$.

Rotational analysis has not been successful because of the several isotopes of Xe. We have assigned the minor green bands to the transition $2^1\Sigma \to 1^1\Pi$ that arises from the same atomic state. The 3080 Å continuum band is a typical excimer transition from the $2^1\Sigma$ to the repulsive $^3\Pi$ state that dissociates to $Xe + O(^3P)$. Morse potentials have been used together with the spectroscopic constants to construct a set of potential

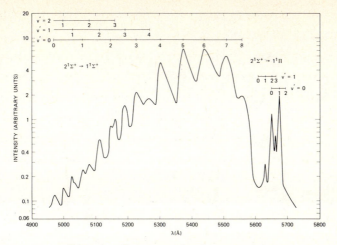

Fig. 4. XeO green band emissions

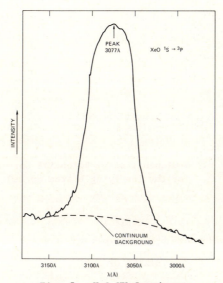

Fig. 5. XeO UV Continuum

curves shown in Fig. 6. The spectroscopic constants are given in Table I.

The ground state XeO($1^3\Pi$) curve was obtained by using Abrahamson's (Ab69) exponential potential and the $1^3\Sigma^-$ state is chosen 1.7 times as repulsive in accord with recent ab initio calculations on ArO by Stevens (St74). The 3080 Å band is consistent with the slope of the $1^3\Pi$ state and lies 800 cm^{-1} above the dissociation limit at R_e = 3.1 Å. Recent unpublished scattering measurements (FLR74) suggest that the ground state potentials should be shifted inward approximately 0.2 Å (the excited states would be shifted equally). However, Wilt (Wi65) and Tisone and Hoffman (TH74) suggest R_e = 3.2 Å. Note that a curve crossing interaction between $^3\Sigma^-$ and $1^1\Sigma^+$ provides a

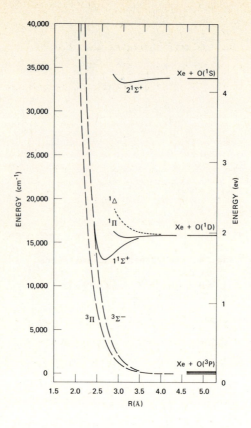

Fig. 6. Semiquantitative Potential Curves for the low-lying states of XeO

Table I

SPECTROSCOPIC CONSTANTS FOR SINGLET STATES OF XeO[a]

State	T_e[b]	ω_e	$\omega_e x_e$	R_e[c]	D_o
$2^1\Sigma^+$	33268	153	10	3.1	450
$1^1\Pi$	15600	97	8.7	3.2	222
$1^1\Sigma^+$	13068	372	12	2.65	2617

a. energies in cm^{-1}, R_e in Å

b. above $Xe + O(^3P_2)$

c. based on eyeball relative adjustment of Morse potentials referenced to Abrahamson repulsive potential (Ab69) mechanism for the rapid quenching of $O(^1D)$ by Xe that has been observed (DH70). It is also noteworthy that our analysis indicates that the $2\,^1\Sigma^+$ state has a barrier approximately 130 cm^{-1} above the dissociation limit.

The shallow well of the $2\,^1\Sigma$ state means that an association-dissociation equilibrium is easily established between the reactions

$$O(^1S) + 2Xe \rightleftarrows XeO(2\,^1\Sigma) + Xe \qquad (1)$$

at 300°K. The equilibrium constant has been calculated (HGHML75,AW75,TH74) and is given by

$$K = \frac{[XeO]}{[Xe][O]} = 6.7 \pm 0.5 \times 10^{-23} \exp[450\ cm^{-1}/RT]\ cm^3\ ,$$

thus at room temperature and 760 torr Xe

$$[XeO(2\,^1\Sigma)] \simeq .015[O(^1S)] \ .$$

To obtain information about the radiative lifetime of the $XeO(2\,^1\Sigma)$ state, G. Black et al. (BSS75) have recently measured the intensity of the green band emission as a function of Xe density. The $O(^1S)$ was produced by a pulsed source of uv that photolyzed N_2O to generate the same quantity of $O(^1S)$ in each pulse. Except for Xe densities less than 2 torr the intensity, I, is proportional to Xe density as expected for collisional equilibrium where the radiative rate exceeds the molecular formation rate. Thus

$$\frac{I}{I_o} = \frac{A_M}{A_o} K[Xe] + 1$$

where A_M and A_o are the transition probabilities of $XeO(2\,^1\Sigma)$ and $O(^1S)$ respectively. Assuming $A_o = 1.18$ sec^{-1} we find $A_M K = (2.4 \pm 0.2) \times 10^{-15}$ cm^{3}sec^{-1} which gives $A_M = 4 \pm 1 \times 10^6$ sec^{-1}. At low Xe densities the intensity of the radiation shows a quadratic dependence on Xe density from which one can extract the 3-body formation rate for $XeO(2\,^1\Sigma)$. That rate is $k_1 = (1.1 \pm 0.3) \times 10^{-31}$ cm^6/sec and since $K = k_1/k_{-1} = 6 \times 10^{-22}$ cm^3 at 300°K, $k_{-1} = (2.0 \pm 0.6) \times 10^{-10}$ cm^3/sec, the reverse rate of reaction (1).

IV. Energy Transfer Kinetics for High Density Xe-O_2 Mixtures

Lasing on the $XeO(2\,^1\Sigma \rightarrow 1\,^1\Sigma)$ transition excited in e-beam pumped Xe + O_2 mixtures has recently been demonstrated (PMR74). The energy transfer kinetics of the XeO production in such media has recently been unravelled (HGHML75) and is discussed briefly below.

From the density dependencies of the temporal behavior of the XeO and Xe_2 decay in Febetron excited $Xe-O_2$ mixtures together with the above information we have formulated a kinetic model for the formation and decay of $XeO(2^1\Sigma)$. Due to the insufficient energy available in either Xe^* metastables or Xe_2^* excimers, O_2 cannot be excited to $O(^3P) + O(^1S)$ in a single step energy transfer process. Even though $O(^1D)$ can be produced it will be rapidly deactivated to $O(^3P)$ by collision with Xe. It can be assumed therefore that the first energy transfer step results only in the production of $2O(^3P)$ atoms. The second step is an energy transfer collision between $O(^3P)$ and Xe_2^* that results in the formation of $O(^1S)$. This two-step formation process has been observed as a change in slope of the decay frequency of Xe_2^* as a function of O_2 density as shown in Fig. 7. At low O_2 densities the O_2 is mostly dissociated and the decay of Xe_2^* is dominated by $O(^3P)$ quenching and is very rapid, but at higher densities the O_2 quenching dominates.

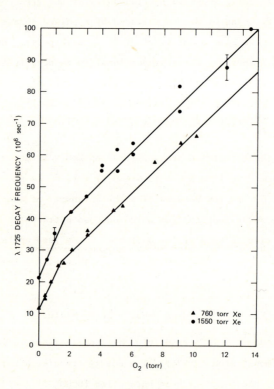

Fig. 7. Oxygen quenching of xenon excimer fluorescence

From measurements of the total $O(^3P)$ production as a function of energy deposition in Xe we were able to determine the quenching of both $O(^1S)$ and Xe_2^* by $O(^3P)$. Further, by measuring the $O(^1S)$ production (obtained from the integrated intensity of the XeO emission) we found that all of the Xe_2^* quenching by $O(^3P)$ yields $O(^1S)$. Thus even though a two-step process is required to produce the $O(^1S)$, the yield of $O(^1S)$ per Xe_2^* under optimum conditions will be 0.67. The major non-radiative quenching path for $O(^1S)$ is due to $O(^3P)$. A schematic diagram of the energy flow for the Xe-O_2 mixture is shown in Fig. 8. The rates for the reactions and their sources are given in Table II.

We can conclude that the characteristics of the XeO excimer are very consistent with the basic requirements of an efficient high power electronic transition laser. The weak binding of the XeO upper level suggests that such a laser would operate best at low temperatures and high Xe densities, suggesting liquid phase Xe/O_2 or Xe/O_3

Table II

REACTIONS IN XENON/OXYGEN MIXTURES

Reaction	Rate Coefficient (cm^3/sec, etc.)	Literature Value	Reference
$Xe^* + O_2 \to$ Products	--	2.2×10^{-10}	VS74
$Xe^* + O(^3P) \to$ Products	--	--	--
$Xe_2^* + O_2 \to 2O(^3P)$	$k_5 = 1.5 \times 10^{-10}$	--	--
$Xe_2^* + O(^3P) \to O(^1S)$	$k_6 = 6 \pm 2 \times 10^{-10}$	--	--
$Xe_2^* \to 2Xe + h\nu$	k_7 = variable to 6×10^7		LEH73
$O(^1D) + Xe \to O(^3P) + Xe$	--	$1.0 \pm 0.4 \times 10^{-10}$	DH70
$Xe + O(^1S) \to h\nu$	$k_R = 2.1 \pm 0.2 \times 10^{-15}$	$3.7 \pm .6 \times 10^{-15}$ $1.7 \pm .2 \times 10^{-15}$	WA75 CuC74
$XeO(2\,^1\Sigma^+) \to h\nu$	$k_2 = 4 \times 10^6$	5.9×10^6	WA75
$O(^1S) + O(^3P) \to$ Products	$k_3 = 2.2 \pm 0.2 \times 10^{-11}$	$1.8 \pm 0.8 \times 10^{-11}$	Sℓ75
$O(^1S) + O_2 \to$ Products	--	2.1×10^{-13}	SWB72
$O(^1S) + O_3 \to$ Products	--	$5.8 \pm 1 \times 10^{-10}$	LGS71
$O + O + Xe \to O_2 + Xe$	$2.4 \pm .5 \times 10^{-33}$		
$O + O_2 + Xe \to O_3 + Xe$	$5.1 \pm 1 \times 10^{-34}$	estimated from Jo68	
$O + O_3 \to 2O_2$	6.8×10^{-15}		

mixtures as the optimum medium. The efficient e-beam excitation of Xe_2^* together with the favorable energy transfer kinetics for producing $O(^1S)$ and removing the lower level provides for the possibility of an overall energy efficiency of $\simeq 12\%$ at 5376 Å.

This XeO system is only one example of a large class of excimers that can be produced by energy transfer in excited rare gases. Much research on the spectroscopic and kinetic properties of these interesting systems is needed to understand them and to fully realize their laser potential.

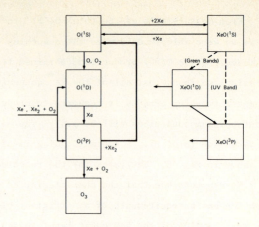

Fig. 8. Energy flow in xenon/oxygen mixtures

Acknowledgment

This research was supported by the Advanced Research Projects Agency of the Department of Defense and was monitored by ONR under Contract No. N00014-72-C-0478.

References

Ab69	A. A. Abrahamson, Phys. Rev. <u>178</u>, 76 (1969).
BSS75	G. Black, R. Sharpless, and T. Slanger, to be published.
CRJ72	O. Cheshnovsky, B. Raz, and J. Jortner, Chem. Phys. Lett. <u>15</u>, 475 (1972).
CJAM69	S. O. Colgate, J. E. Jordan, I. Amdur, and E. A. Mason, J. Chem. Phys. <u>51</u>, 968 (1969).
CCT61	C. D. Cooper, G. C. Cobb, and E. L. Tolnas, J. Mol. Spec. <u>7</u>, 223 (1961).
CuC74	D. L. Cunningham and K. C. Clark, J. Chem. Phys. <u>61</u>, 1118 (1974).
DH70	R. J. Donovan and D. Husain, Chem. Rev. <u>70</u>, 489 (1970).
EB75	J. J. Ewing and C. Brau, Phys. Rev., to be published (private communication).
FP57	W. Finkelnburg and T. Peters, Kontinuerlicke Spektren, <u>Handbuck der Physik</u> Vol. XXVIII, S. Flugge ed., Springer-Verlag Berlin (1957).
FLR74	P. B. Forman, A. B. Lees, and P. K. Rol, "Determination of Intermolecular Potentials Between Oxygen Atoms and Plume Species," part of a report (private communication), 1974.
GW71	T. L. Gilbert and A. C. Wahl, J. Chem. Phys. <u>55</u>, 5247 (1971).
GSIL75	K. Gillen, R. Saxon, G. Ice, and D. C. Lorents, private communication.
GB70	M. Ginter and R. Battino, J. Chem. Phys. <u>52</u>, 4469 (1970).
GK72	R. G. Gordon and Y. S. Kim, J. Chem. Phys. <u>56</u>, 3122 (1972).

HH50   R. Herman and L. Herman, J. Phys. Radium. $\underline{11}$ 69 (1950).

HEL74   E. Huber, D. Emmons, and R. Lerner, Opt. Comm. $\underline{11}$, 155 (1974).

HGHML75   D. L. Huestis, R. A. Gutcheck, R. M. Hill, M. V. McCusker, and D. C. Lorents, "Studies of E-beam Pumped Molecular Lasers," Technical Report No. 4, SRI No. MP 75-18, Stanford Research Institute, Menlo Park, Ca., January 1975.

Jo68   H. S. Johnson, "Gas Phase Reaction Kinetics of Neutral Oxygen Species," NSRDS-NBS20 (1968).

KANPP46   C. Kenty, J. O. Aicher, E. B. Noel, A. Poritsky, and V. Paolino, Phys. Rev. $\underline{69}$, 36 (1946).

LGSW71   G. London, R. Gilpin, H. I. Schiff, and K. H. Welge, J. Chem. Phys. $\underline{54}$, 4512 (1971).

LEH73   D. C. Lorents, D. J. Eckstrom and D. L. Huestis, "Excimer Formation and Decay Processes in Rare Gases," Final Report MP 73-2, Contract N00014-72-C-0457, SRI Project 2018, Stanford Research Institute, Menlo Park, Ca., September 1973.

LO72   D. C. Lorents and R. E. Olson, "Excimer Formation and Decay Processes in Rare Gases," Semiannual Tech. Rpt. No. 1, Contract N00014-72-C-0457, SRI Project 2018, Stanford Research Institute, Menlo Park, Ca., December 1972.

KGW74   J. W. Keto, R. E. Gleason, G. K. Walters, Phys. Rev. Lett. $\underline{33}$, 1375 (1974).

LOC73   D. C. Lorents, R. E. Olson, G. M. Conklin, Chem. Phys. Lett. $\underline{20}$, 589 (1973).

MS74   R. C. Michaelson and A. L. Smith, J. Chem. Phys. $\underline{61}$, 2566 (1974).

MW74   H. U. Mittmann and H. P. Weise, Z. Naturforsch $\underline{29a}$, 400 (1974).

RM70   R. S. Mulliken, J. Chem. Phys. $\underline{52}$, 5170 (1970).

ORRF74   T. Oka, K. Rama Rao, J. Redpath, and R. Firestone, J. Chem. Phys. $\underline{61}$, 4740 (1974).

PSL72   I. M. Parson, P. E. Siska and Y. T. Lee, J. Chem. Phys. $\underline{56}$, 1511 (1972).

PMR74   H. T. Powell, J. R. Murray, and C. K. Rhodes, Appl. Phys. Lett. $\underline{25}$, 730 (1974).

SWB72   T. G. Slanger, B. J. Wood, G. Black, Chem. Phys. Lett. $\underline{17}$, 401 (1972).

Sℓ75   T. G. Slanger, private communication (1975).

St74   W. J. Stevens, private communication (1974).

TH72   N. Thonnard and S. Hurst, Phys. Rev. $\underline{A5}$, 1110 (1972).

TH74   G. Tisone and J. Hoffman, Sandia Report SAND74-0425, Sandia Laboratories, Albuquerque, N.M. (1974).

VS74   J. E. Velazco and D. W. Setser, Chem. Phys. Lett. $\underline{25}$, 197 (1974).

VS75   J. Velazco and D. Setser, J. Chem. Phys. $\underline{62}$, 1990 (1975).

WA75   K. H. Welge and R. Atkinson, preprint. We would like to thank R. Atkinson for allowing us to see this preprint prior to its submission.

Wi65   J. R. Wilt, thesis, Dept. of Chemistry, University of California at Los Angeles (1965).

LASER FLUORIMETRY

Richard N. Zare
Department of Chemistry
Columbia University
New York, New York 10027, USA

I. INTRODUCTION

Observations of fluorescence date back before 1900 when the appearance of visible fluorescence was often noted along with color and smell in characterizing new compounds. The first primitive fluorimeters consisted of focussed sunlight as an excitation source, a glass tube (test tube) as a sample holder, and the human eye as a detector. Fluorimetric instrumentation has become vastly more sophisticated since then, but until the late 1940's it was considered unreliable by many for analytical purposes because of the difficulty of obtaining a reproducible linear dependence of fluorescent intensity on sample concentration. Modern quantitative fluorimetry may be said to have its beginnings in the development of sensitive photomultipliers whose use permit the photoelectric recording of fluorescence. With the introduction in the late 1950's of commercial spectrofluorimeters employing two monochromators, one to select the wavelength range of the excitation source, the other to analyze the wavelength dependence of the sample fluorescence, fluorimetric analysis has become perfected to the point where today complete fluorescence emission and excitation spectra can be recorded automatically for small samples of material at the flick of a switch. As fluorimetry has matured and its applications diversified, an extensive literature has grown with it (1-5).

Fluorescence, which will be taken here to mean the process whereby a material absorbs light at one wavelength and emits some fraction of the energy as light at other wavelengths, has various advantages and disadvantages compared to other analytical methods. First of all, its sensitivity is extremely high. Lower limits of detection lie in the sub-parts per million to parts per billion range for many compounds. This makes fluorimetry particularly well suited for trace analysis. On the other hand, this technique has the limitation that not every substance emits measurable luminescence for the range of excitation wavelengths conveniently available. Sometimes this fault can be turned into a virtue since a fluorescent substance can often be readily determined without preliminary separation from other substances that are either non-absorbing, non-fluorescent, or fluoresce in a different spectral region or with a different characteristic time decay so that this fluorescence can be rejected compared to the fluorescence of the substance under analysis. Moreover, it may also be possible by preparing an appropriate chemical derivative to convert a non-fluorescent substance into a fluorescent one for the purposes of analysis (fluorescence-labelling).

Despite the high sensitivity already attained in conventional fluorimetry, there are many applications where even greater analytic power is required. We address ourselves here to the potential benefits laser excitation can bring to fluorimetry. Because of the extraordinary spectral brightness of lasers, i.e. the amount of radiant energy delivered per unit wavenumber, compared to blackbody and resonance lamp sources, lasers are starting to have a strong impact on chemical analysis (6). In what follows we discuss the uses of laser fluorimetry, first in the analysis of gases and then in condensed media. This account is not meant to be comprehensive and is biased towards that work best known to the author.

II. LASER FLUORIMETRY OF GASES

A. **Elemental Analysis**. One of the most common means of analyzing a substance for its constituent elements is to use atomic fluorescence flame spectrometry (AFFS). In this field notable progress is being made by Winefordner and coworkers who are replacing the hallow-cathode or electrodeless-discharge lamp excitation source by a nitrogen-laser-pumped dye laser (7). Here the pulsed dye laser is tuned to an atomic resonance line and excites fluorescence in the flame containing metal atoms. The resultant atomic emission is dispersed by a grating monochromator and detected by a gated photomultiplier. Typical detection limits are 0.2-0.05 µg/mℓ with the most sensitive limits being for Aℓ and Ca (both 0.005 µg/mℓ). Moreover, the fluorescence signal is found to be linear in metal atom concentration over typically three orders of magnitude. These results compare favorably with the best previous results obtained from conventional AFFS. Near the detection limit, the major contribution to noise was found to be random scattering of the incident laser beam caused by refractive index inhomogeneities and particulates in the flame.

Clearly, additional progress can be made if the flame is replaced by a non-emitting homogeneous gas medium. This has been dramatically illustrated by the detection of Na atoms in an atomic vapor. Because a single atom can scatter resonant photons many times per second (in sharp contrast to most molecules), the limits of detection can be markedly reduced. For example, in 1972 Jennings and Keller (8) reported detecting 2×10^6 Na atoms/cm^3 using a cw dye laser and this year Fairbank, Hänsch, and Schawlow (9) have extended this detection limit to 1×10^2 Na atoms/cm^3, corresponding to 4×10^{-15} µg/ml. This technique is readily applied to many other elements. As ways are found to dissociate samples into their constituent atoms without background interference, tunable lasers may be expected to revolutionize elemental analysis.

B. <u>Molecular Beam Diagnostics</u>. Not all analytical problems in the gas phase require the extremely high sensitivity described above, but molecular beam experiments in which collisional processes are studied "one collision at a time" have always been limited by signal-to-noise problems. Most molecular beam studies in the past have relied on the detection of ions, either through hot-wire surface ionization, Auger ejection of electrons from surfaces by metastables, or by electron bombardment ionization followed by mass analysis. In each of these methods, the ionization process is rather insensitive to the internal state of the molecule being detected, and information on the distribution of such states has generally had to be obtained by indirect means. The use of laser-induced fluorescence as a molecular beam detector (10) overcomes this drawback and has in addition many advantages associated with its selectivity and sensitivity.

One of the first demonstrations of the power of this technique was the characterization of the molecules in a supersonic jet expansion. The mechanism of beam acceleration and dimer formation as well as the internal state distribution of the dimers has been the subject of much investigation and the processes are still not understood in detail. Sinha <u>et al</u>. (11) used various lines of an argon ion laser to excite fluorescence in a nozzle beam of Na$_2$ molecules. They found extensive cooling but disequilibrium between the rotational and vibrational degrees of freedom. The rotational distribution was described by a temperature $T_{rot} = 55°K$ and the vibrational distribution by $T_{vib} = 150°K$. Later, Sinha et al. (12) showed that the sodium dimer molecules were aligned in the hydrodynamic flow with the molecule's angular momentum vector pointing preferentially at right angles to the flow direction. This could be determined by carefully measuring the degree of polarization of the laser-induced fluorescence. Korving et al. (13) have confirmed this result using a slightly different experiment in which a magnetic field was used in determining the alignment. Recently, Bergmann et al. (14) measured the velocity distribution of individual (v,J) states of the Na_2 molecules in a nozzle beam. They used the output from a stabilized, tunable, single-mode argon ion laser which was split into two beams that cross the molecular beam perpendicular and (nearly) parallel to the beam flow direction. Much work has been done on alkali dimers because their electronic band systems occur in the visible. However, this technique is readily applied to other molecules provided that they fluoresce, their spectra is known, and lasers can be found that excite them.

Laser fluorescence detection is not limited to primary beams but can be used as well to study reaction products of crossed beam experiments. Schultz et al. (15) first applied this technique to obtain information about the BaO product formed in the reaction $Ba + O_2 \rightarrow BaO + O$. Subsequent work by Dagdigian et al. (16) showed that as few as 10^4 BaO molecules per cm^3 could be detected in a specific vibration-rotation state. Since then, several other reactions have been investigated by this technique (17-20). Two examples from current work illustrate the scope and potential of this new molecular beam detector.

In our laboratory Dr. Gregory P. Smith is carrying out an angular distribution study of the reaction $Ba + KCl \rightarrow BaCl + K$. Figure 1 shows a schematic of the apparatus. A barium beam crosses the salt beam at right angles and the BaCl product scattered through an angle θ (measured from the Ba beam) passes through a slit 22 cm distant from the reaction zone where it is detected (see Fig. 2) by laser-induced fluorescence. Figure 3 summarizes the data obtained showing that (1) in the laboratory reference frame the angular distribution exhibits forward and backward peaking with respect to the center of mass angle, suggestive of a long-lived collision complex, and (2) the angular distribution of the BaCl high vibrational levels is narrower and concentrated nearer the center of mass angle than the low vibrational levels. Smith has also been able to detect the K atoms formed in this reaction and obtain a crude angular distribution for this species. The point of this work is that laser fluorescence detection permits the measurement of angular distributions of individual internal states, an important feature of the collision dynamics, which heretofore could not be determined.

Laser fluorescence detection of reaction product internal states presents the opportunity of probing features of the reaction dynamics which previously were hidden from the experimentalist. In our laboratory, Dr. J. Gary Pruett is investigating how the internal energy of the reactants affects the internal energy of the products. Using a pulsed HF chemical laser tuned to the 1-0 P(2) line, the reaction Ba + HF → BaF + H is studied for HF(v=0) and HF(v=1). Figure 4 shows the BaF excitation spectrum for HF laser "off" and "on." Figure 5 shows an expanded region of the high vibrational levels in Fig. 4, clearly illustrating the production of new vibrationally excited BaF reaction products. It is too soon to say what fraction of the HF excitation appears as BaF vibration, but a preliminary analysis indicates that this fraction is less than 0.8. In any case, these types of experiments are a large step towards realizing the measurement of state-to-state reaction rates.

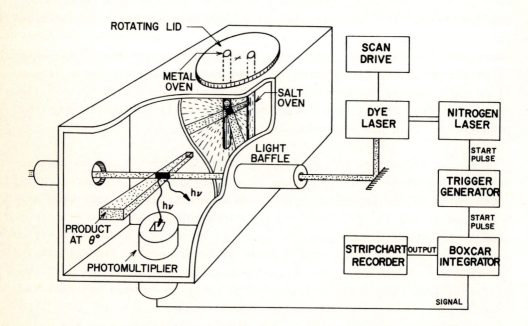

Fig. 1. Schematic diagram of the laser fluorescence detection of the angular distribution of products formed by a crossed beam reaction.

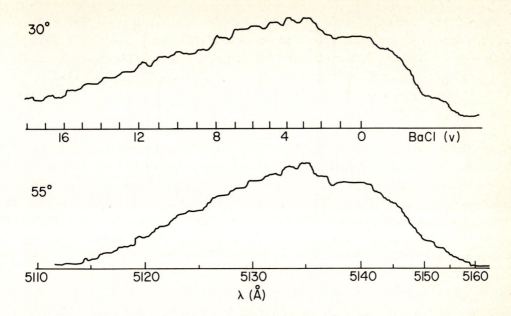

Fig. 2. Excitation spectra of the BaCℓ reaction product at θ = 30° and 55° with respect to the barium beam. The bandheads are labelled but not pronounced because of the high rotational excitation of the product.

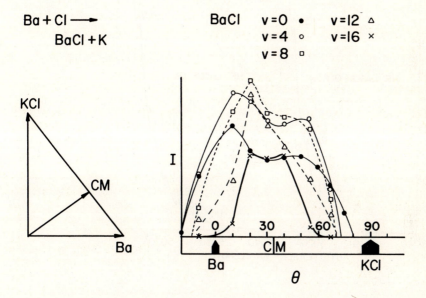

Fig. 3. Summary of angular distribution data for individual BaCℓ(v) product states taken from bandhead measurements (G. Smith, preliminary analysis).

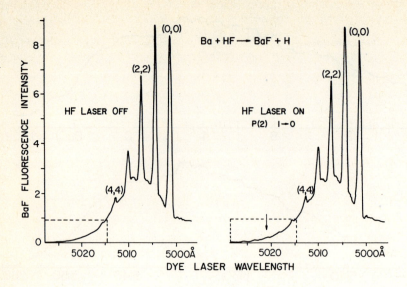

Fig. 4. Excitation spectra of BaF formed in the reaction Ba + HF with HF(v=0), "laser off" and a fraction of HF excited to v=1, "laser on."

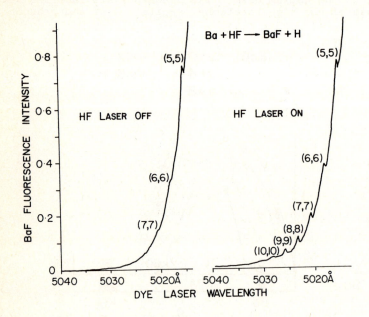

Fig. 5. An expanded portion of Fig. 4.

III. LASER FLUORIMETRY OF CONDENSED MEDIA

Applications of lasers to liquids and solids has concentrated on spectroscopic and kinetic studies, as well as nonlinear phenomena, rather than trace analysis. Nevertheless, laser fluorimetry in condensed media may turn out to be one of the most important applications of lasers. Most analytical problems involve matter at least initially in the condensed phase. One obvious possibility is to combine laser fluorescence detection with chromatographic separation. Berman et al. (21) have demonstrated this by the detection of aflatoxins (carcinogenic metabolites) in thin-layer chromatograms at the sub-nanogram level. Applications to liquid chromatography are also being explored. In this regard, A. Beatrice Bradley in our laboratory has investigated the detection of rhodamine 6G dissolved in water or in ethanol. Using a pulsed nitrogen laser, Bradley finds linearity of fluorescence signal with concentration from 0.5 µg/mℓ, to 6×10^{-7} µg/mℓ. This lower limit represents less than 1 part per trillion and suggests that laser fluorescence will become a powerful analytic tool in those applications requiring the utmost in sensitivity. A further indication of this is the recent work of Harrington and Malmstadt (22) who incorporated a pulsed dye laser into a spectrofluorimeter and obtained substantial improvement in the instrument's sensitivity. Because of the universality of condensed media and the sensitivity inherent in laser fluorimetry, this technique can be expected to enjoy rapid growth as an analytical tool.

IV. REFERENCES

1. S. Udenfriend, Fluorescence Assay in Biology and Medicine (Academic Press, New York, 1962).
2. D. M. Hercules, Fluorescence and Phosphorescence Analysis (Wiley-Interscience, New York, 1966).
3. C. A. Parker, Photoluminescence of Solutions (Elsevier Publishing Co., Amsterdam, 1968).
4. G. G. Guilbault, Fluorescence (M. Dekker, New York, 1967); Practical Fluorescence (M. Dekker, New York, 1973).
5. J. D. Winefordner, S. G. Schulman, and T. C. O'Haver, Luminescence Spectrometry in Analytical Chemistry (Wiley-Interscience, New York, 1972).
6. J. I. Steinfeld, "Tunable Lasers and Their Application in Analytical Chemistry," to appear in CRC Critical Reviews in Analytical Chemistry.
7. N. Omenetto, N. N. Hatch, L. M. Fraser and J. D. Winefordner, Anal. Chem. 45, 195 (1973); L. M. Fraser and J. D. Winefordner, Anal. Chem. 44, 1444 (1972); 43, 1693 (1971).
8. D. A. Jennings and R. A. Keller, J. Am. Chem. Soc. 94, 9249 (1972); R. A. Keller, Chemtech, p. 626 (1973).
9. W. M. Fairbandk, Jr., T. W. Hänsch, and A. L. Schawlow, J. O. S. A. 65, 199 (1975).

10. R. N. Zare and P. J. Dagdigian, Science <u>185</u>, 739 (1974).
11. M. P. Sinha, A. Schultz and R. N. Zare, J. Chem. Phys. <u>58</u>, 549 (1973).
12. M. P. Sinha, C. D. Caldwell and R. N. Zare, J. Chem. Phys. <u>61</u>, 491 (1974).
13. J. Korving, A. G. Visser, B. S. Douma, G. W. 't Hooft, and J. J. M. Beenakker, Ninth International Symposium on Rarefied Gas Dynamics C. 3-1 (July, 1974).
14. K. Bergmann, W. Demtröder, and P. Hering, Applied Physics (submitted for publication).
15. A. Schultz, H. W. Cruse and R. N. Zare, J. Chem. Phys. <u>57</u>, 1354 (1972).
16. P. J. Dagdigian, H. W. Cruse, A. Schultz, and R. N. Zare, J. Chem. Phys. <u>61</u>, 4450 (1974).
17. H. W. Cruse, P. J. Dagdigian, and R. N. Zare, Faraday Disc. Chem. Soc. <u>55</u>, 277 (1973).
18. P. J. Dagdigian and R. N. Zare, J. Chem. Phys. <u>61</u>, 2464 (1974).
19. P. J. Dagdigian, H. W. Cruse, and R. N. Zare, J. Chem. Phys. <u>62</u>, 1824 (1975).
20. G. P. Smith and R. N. Zare, JACS <u>97</u>, 1985 (1975).
21. M. R. Berman and R. N. Zare, Anal. Chem. <u>47</u>, 1200 (1975).
22. D. C. Harrington and H. V. Malmstadt, Anal. Chem. <u>47</u>, 271 (1975).

V. ACKNOWLEDGMENTS

This work was supported in part by the Air Force Office of Scientific Research and by the National Science Foundation.

SELECTIVE PHOTOCHEMISTRY IN AN INTENSE INFRARED FIELD

R.V.Ambartzumian, N.V.Chekalin, Yu.A.Gorokhov, V.S.Letokhov,
G.N.Makarov, E.A.Ryabov
Institute of Spectroscopy, Academy of Sciences
142092, Moscow, USSR

1. Since the publication last year of our work clearly showing the feasibility of infrared-laser-radiation induced chemical reactions /1/ /2/, at least two more groups in the United States have achieved macroscopic isotope separation effect as we had done by pumping the molecular vibrations by intense resonant infrared radiation /3/, /4/.

The first attempt to separate isotopes by intense infrared laser radiation was made by us as early as 1972. At that time we tried to dissociate selectively ammonia molecules of one isotopic species by focused tea laser radiation, as we do it now, but failed to achieve separation of nitrogen isotopes because of reasons that are more or less obvious for us now.

More careful investigation of the processes which take place during the interaction of a strong infrared laser field with resonantly absorbing molecules gave the following general picture /5/. In the region where the intensity reaches 0.6 GW and higher, a part of the molecules dissociate instantaneously, the others remaining excited in very high-lying vibrational levels. Very few collisions one needed for the excited molecules to reach the dissociation barrier by V-V exchange processes.

The high selectivity of collisionless dissociation is obvious. The collisional dissociation of highly excited molecules is selective due to the very small number of collisions necessary to dissociate.

In this report we shall give some evidence supporting the above picture. The main items of the report will be: selective infrared photochemistry and some questions of interaction of polyatomic molecules with a resonant intense infrared field.

2. <u>Experiment</u>. The experimental set-up consisted of a tea CO_2 laser with frequency-selective cavity. The output was from 1.5 to 2.5 joules in 85 nsec, depending on oscillation frequency. The linewidth was 0.035 cm^{-1}.

The gas in the cell was irradiated at the repetition rate of
1.6 cycles per second, and after irradiation the residual gas in the
cell was examined spectroscopically and, in the case of SF_6, mass
spectra were usually taken. In the cases when spectra of dissociated
products were studied, we used a multichannel registration system
connected to a scanning monochromator (MDR-2). A very fast gas flowing system was used to avoid contamination of the initial sample by
chemical reactions with the dissociated products.

3. Dissociation of BCl_3 and boron isotope enrichment.

The first successful experiments on boron enrichment were made
in /1/, where selectively dissociated isotopic $^iBCl_3$ reacted with
O_2, giving first ^iBO of single isotopic species and subsequently
$^iB_2O_3$ solid product which deposited on the walls of the cell. The
dissociated BCl_3 was studied in /5/ and we shall only briefly comment on this work. In contrast to /6/, a strong dipole transition was
observed in the region of 2600 Å, which belongs to the BCl radical.
An interesting feature of this transition is that it is present only
in the delayed phase of luminescence. This fact strongly suggests
that the BCl radical is formed in successive chemical reactions of
the dissociated products of BCl_3.

The experiments on macroscopic enrichament of boron isotopes
were made in a glass cell 12 cm long with i.d. 20 mm. The radiation
was focused by lens with f = 12 cm. The beam cross-section in the
focal plane was 2x1.5 mm^2 giving a power density of $\sim 10^9$ W/cm^2. The
dissociation of BCl_3 without radical acceptor is reversable, and we
could not find any decomposition of BCl_3 after several hours of irradiation. Most of the enrichment studies were made with O_2 radical acceptor, which formed the stable product B_2O_3. When HBr was used as
a radical acceptor a peak enrichment factor of K $\simeq$ 18-20 was reached
where K is defined as

$$K(^{10}B/^{11}B) = \frac{[^{10}BCl_3]_* [^{11}BCl_3]_0}{[^{11}BCl_3]_* [^{10}BCl_3]_0}$$

with $[BCl_3]_*$ and $[BCl_3]_0$ the final and initial concentrations of
BCl_3. It should be noted here that the reaction with HBr is reversible, and the BCl_3 isotopes return to their initial concentrations
in 30 minutes.

Fig. 1. showes (a) the changes in the i.r. absorption spectrum
of BCl_3 after irradiation of various isotopic species of BCl_3 and
(b) the dependance of K on the initial pressure of BCl_3, the par-

tial pressure of O_2 being held constant (20 torr).

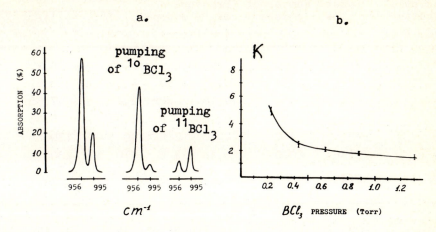

Fig. 1.

The experiments also show that though the amount of reacted BCl_3 decreases with increase of O_2 the K increases monotonicaly, giving the evidence of importance of vibrational deactivation and thermal heating processes /7/.

4. Enrichment of sulphur isotopes by selective dissociation of SF_6.

In contrast to BCl_3, the dissociation of SF_6 by infrared radiation in a glass cell is irreversible, and no additional gas was needed for chemical binding of the dissociated products. This may be connected with fluorine atom reactions at the walls of the cell.

The Table 1 lists the enrichment factors obtained in various experiments, mostly by dissociating the $^{32}SF_6$ isotope. The enrichment was determined by analyzing the mass spectrum of residual SF_6 in the cell /2/, /7/.

From the table it is seen that the highest enrichment factor $K(^{34}S/^{32}S)$, defined as in section 3, reached 2800. The main products of chemical reactions are sulphur oxifluorides, presumably SOF_2, determined from mass-spectrometer and infrared spectra measurements.

Table 1

Irradiated molecules	Laser line	Number of pulses	Pressure SF_6 + acceptor (torr)	Enrichment factors		
				$K(33/32)$	$K(34/32)$	$K(32/34)$
$^{32}SF_6$	P(12)	100	0,18	5,35	16,1	-
-"-	-"-	-"-	0,18 + 2NO	4,55	8,65	-
-"-	-"-	-"-	0,18 + 2HBr	2,8	7,55	-
-"-	-"-	-"-	0,18 + 2 H_2	2,36	5,53	-
-"-	-"-	400	0,18 + 2 H_2	40	1200	-
-"-	P(16)	2000	0,18 + 2 H_2	270	2800	-
$^{34}SF_6$	P(40)	500	0,18 + 2 H_2	-	-	18

Fig. 2. shows the infrared spectra of SF_6 before (a) and after (b) irradiation. The number of shots was chosen so as to equalize the concentrations of $^{34}SF_6$ and $^{32}SF_6$

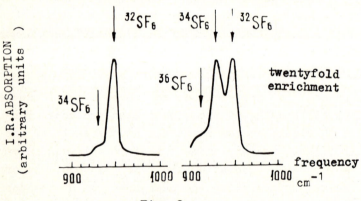

Fig. 2.

Below we shall discuss the restrictions on the highest achievable selectivity of dissociation processes.

Fig. 3. shows the dependence of the enrichment factor $K(^{34}S/^{32}S)$ as a function of the initial pressure of SF_6. The various curves correspond to different number of irradiation pulses.

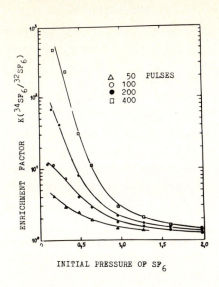
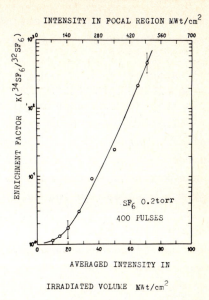

Fig. 3.   Fig. 4.

It is seen that the enrichment drops with increasing pressure. A possible explanation is that at the lowest pressure (0.18 torr of SF_6) the molecules which cannot reach the dissociation barrier mostly deactivate on the walls of the cell (1.2 cm i.d.) because of the large V-T relaxation time (τ_{V-T} = 150 μsec.torr and the diffusion time at 0.2 torr is 450 μsec). At higher pressures the excited molecules deactivate in the irradiated volume giving rise the thermal heating which leads to nonselective dissociation of molecules, and therefore decreasing the enrichment factor. It was also discovered that the enrichment increases exponentialy with incident power (Fig.4). It is difficult to explain such a dependence, but this data gives an indication of the number of absorbed infrared photons, necessary to dissociate the molecule. At the lowest power densities,presumably, the radiation is not sufficiently intense to directly dissociate the molecules, but after the pulse they remain in highly exited vibrational states and can reach the dissosiation threshold by means of V-V transfer in collisions, partially deactivating and causing thermal dissociation. At the intensity of $\sim$ 70 MW/cm^2 the dissociation becomes very effective. This can occur only in the case when most of the molecules are near the dissociation barrier. Combining the dependence of $\langle n \rangle$, the average number of absorbed photons per molecule given in the next section, on incident power, one can estimate that it is necessary to absorb approximately 200-250 quanta for the dissociation of one SF_6 molecule.

5. Absorption of intense laser pulses in SF_6.

In this section some results are presented on absorption of CO_2 laser pulses in gaseous SF_6. The absorption in SF_6 was studied by direct measurements of the energy absorbed in the cell at various pressures and incident intensities. The length of the absorption cell was varied with pressure so that approximately only 10% of the energy was absorbed in the cell, and this permitted us to consider the power in the irradiated volume of the cell to be nearly constant. The measurements were made using both focused and unfocused beams. In the case of focused beam power density was computed taking the average cross-section of the beam. Then the absorbed energy, measured in CO_2 quanta energy units, was devided by the number of molecules in the irradiated volume. The number of absorbed quanta per molecule in irradiated volume, denoted as $\langle n \rangle$, is plotted in Fig. 5 (unfocused beam).

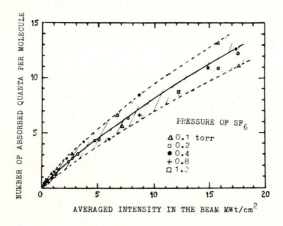

Fig. 5.

The pressure at which the measurements were made was low enough to avoid rotational relaxation (the lowest pressure was 0.1 torr), and only a small fraction (Q) absorbed the laser radiation. The collision time in SF_6 is 80 nsec at one torr, the laser pulse 85 nsec. The value of Q is not known exactly but it is mainly determined by the ratio of laser linewidth (0.035 cm^{-1}) and the width of the Q-branch in the absorption spectrum of SF_6 (~2 cm^{-1}), and also can be estimated from the dependance of enrichment on number of laser pulses at a given pressure (Fig.3). The estimates give that Q lies between 10^{-1} and $\sim 3 \cdot 10^{-2}$. This means that in reality the interacting molecules absorb at least 10 times more than indicated by $\langle n \rangle$. One can see that at very moderate intensities the molecules gain several eV of energy from the radiation field. To obtain the value of $\langle n \rangle$ at higher power levels, the laser beam was focused and the average intensity was calculated. These measurements of $\langle n \rangle$ were put on the same (Fig.5) plot and the measured $\langle n \rangle$ in unfocused beam and in focused beam coincide in the overlapping region (Fig.6, O - unfocused beam, Δ - focused beam)

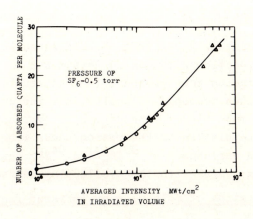

Fig. 6.

Addition of 16 torr of H_2, as a buffer gas, doubles $\langle n \rangle$ supporting the statement of low value of Q.

The measurements of $\langle n \rangle$ at given intensity (12 MW/cm^2) and various pumping frequencies are shown in Fig.7 together with the absorption band contour of SF_6. One can see that $\langle n \rangle$ strongly correlates with the absorption contour of the band. This fact indicates that only small portion of rotational levels are pumped by laser radiation.

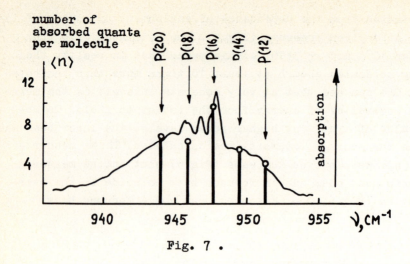

Fig. 7.

This also immediately gives that the higest selectivity which can be achieved in reaction products is determined by the degree of overlapping of absorption band contours (normalised to equal concentrations) of various isotopic molecules, but the enrichment of residual gas may be much higher. The degree of overlapping in our case was obtained by extrapolation of absorption bands wings by exponent. The results on enrichment of $^{33}SF_6$ and $^{34}SF_6$ against $^{32}SF_6$ are in quite well agreement with the overlapping degree.

6. Photodissociation of molecules by infrared radiation.

The fact that the molecule which interacts with intense laser field absorbs much more energy than the dissociation energy, stimulates interest on what are the products of dissociation, i.e. what is the result of primary photochemical act? The radicals formed in such dissociation process were studied by examining the time resolved spectra of instantanoues luminescence, which accompanied the dissociation process. To be sure that the detected radicals are formed in the dissiciation only, the spectra were taken in first 150 nsec, the time which coincided with the laser pulse. Also with the same purpose the pressures of the dissociated gases were less than one torr i.e. the detected radicals were formed by collisionless phase. The results /8/ immideately indicated that the dissociation by infrared differ strongly from photolysis produced by ultraviolet. For example: ethelene dissociates loosing four hydrogen atoms simultaneously and the C_2 radical is seen; the same is with $C_2F_2Cl_2$ which looses four halogens. The CF_3Cl gives the CCl radical. The SiF_4 molecule dissociates forming SiF radicals /9/. These data indicate that the molecule being in the intense infrared field absorbs certain amount of energy and then explodes breaking several bonds simultaneously.

7. <u>Discussion</u>. It is still puzzling how can the molecule absorb so much energy ($\sim 20-30$ eV). From the radiation field taking in account anharmonicity, but several conclusions might be made at now.

The compensation of anharmonicity due to dynamic Stark-broadenning in the field 10^9 W/cm^2 can explain only excitation to $v \simeq 10$, not more /5/. This, of course, cannot explain the fast dissociation of molecules with $D_o \geqslant 5-10$ eV. Our results on measuring $\langle n \rangle$ show that dynamic Stark-broadenning can be neglected in our case at least.

Another possibility was discussed in /9/. It was based that the transitions spectra for the highly excited levels become continious, and therefore no anharmonicity problem arises. But it should be taken in account that the oscillator strength between such transitions is very low and it is difficult to explain the dissociation in this way though the consideration of this possibility must be examined quantitavely.

Another opportunity to get rid of the excitation of extremely high lying vibrational levels is the concept of fast intramolecular collisionless relaxation. But the experimental data which are discussed below evidence that the excitation (energy) is stored in the v_i-th mode which absorbs the laser radiation. It is true in collisionless stage, untill the collissions distribute the stored energy to the vibrational manyfold. This is supported by various observations. For example the measurements of $\langle n \rangle$ in similiar experiments with CH_3NO_2 molecule show that $\langle n \rangle$ is different for various vibrational modes which absorb laser radiation, and this difference reaches the factor 1.6. The dissociation products are the same but with different vibrational temperatures /8/. This is impossible if there is very high rate intramolecular collisionless V-V relaxation process, which equalibrates the vibrational temperatures and does not allow to have more energy in any mode than the energy of dissociation D_o. The character of dissociation of other molecules gives the same; the molecules do not dissociate by the lowest D_o dissociation bond. More direct evidences on absence of fast intramolecular V-V relaxation were obtained by studing the changes in electronic absorption spectrum during vibrational pumping in the OsO_4 molecule. This method is carefully described in /10/. The electronic absorption spectrum of OsO_4 molecule represents an electronic-vibrational structure with ν_1 mode, which is only Raman active. The active mode in the infrared is ν_3 which can be pumped by CO_2 laser. The time resolved measurements show that the time of intramolecular V-V energy exchange is not less than 1,5 μsec. The shift to the red of the absorption boundary in these experiments reached more than 10000 cm^{-1}, indicating that at least 12-th vibrational level of ν_1 mode was populated significantly, after V-V relaxation of the excitation energy present in the mode ν_3. As it is near impossible to transfer all the energy from ν_3 mode to ν_1 mode in V-V exchange processes, we can conclude that initial population in ν_3 mode was at the vibrational levels much higher than the V=12.

On the other hand we see that the dissociation goes by various bonds. It seems that the molecule accumulates energy in the mode which is under pumping to the extent untill the mode coupling begins to play a role and the ammount of energy it can absorb is enormously high. After absorption of such energy the molecule becomes unstable and explodes breaking more than one bond.

The way how the molecule overcomes the difficulties connected with anharmonicity in vibrational frequencies during the absorption is not clear. One way is that the absorption spectrum in high intencity resonant field changes drasticaly. This must be investigated both experimentaly and theoretically.

Concerning other molecules as SiF_4, CCl_2F_2 for isotope separation technique one should analyze the band structure: there must be no accidental resonances of composite vibrations with principal vibrations. Our experiments showed that the CCl_4 molecule dissociate effectively when pumped through the $\nu_3 + \nu_2$ vibration. This observation broadens the number of molecules for this technique: WF_6, MoF_6 and many others have composite vibrations in the operation region of CO_2 laser. To eliminate overlapping effect in molecules with small isotopic shift dynamic cooling of the gases might be used /11/. The molecules containing C, H, O atoms are less suitable for isotope separation because of their tendency to form dimers and eximers. However, the effect of fast dissociation in intense infrared field through multiphoton absorption prooved to be extremely usefull and efficient method in obtaining isotope selective chemical reactions induced by infrared radiation.

References

1. R.V.Ambartzumian, V.S.Letokhov, N.V.Chekalin, E.A.Ryabov
   Lett to JET Ph. 20, 597 (1974)

2. R.V.Ambartzumian, Yu.A.Gorokhov, V.S.Letokhov, G.N.Makarov
   Lett to JET Ph. 21, 375 (1975)

3. S.M.Freund, J.J.Ritter. Chem.Phys. Lett. 32, 255 (1975)

4. J.L.Lyman, R.J.Jensen, J.Rink, C.P.Robinson, S.D.Rockwood.
   Preprint LA-UR 75707

5. R.V.Ambartzumian, V.S.Doljikov, N.V.Chekalin, V.S.Letokhov,
   E.A.Ryabov. Chem.Phys. Lett. 25, 515 (1974), JET Ph. 68,
   N7 (1975)

6. S.D.Rockwood. Preprint LA-UR 75-684

7. R.V.Ambartzumian, Yu.A.Gorokhov, V.S.Letokhov, G.N.Makarov
   JET Ph. (to be published)

8. R.V.Ambartzumian, N.V.Chekalin, V.S.Letokhov, E.A.Ryabov.
   Chem. Phys. Lett.(to be published)

9. N.R.Isenor, V.Merchant, R.F.Hallsworth, M.C.Richardson
   J.Canad.Phys. 51, 1281 (1973)

10. R.V.Ambartzumian, V.S.Letokhov, G.N.Makarov, A.A.Puretzki.
    JET Ph, 68, 1736 (1975)

11. O.Hagens, W.Henkes, Z.Naturforshc., 15a, 851 (1960)

LASER MAGNETIC RESONANCE (LMR) SPECTROSCOPY OF GASEOUS FREE RADICALS

P. B. Davies
Department of Physical Chemistry,
University of Cambridge,
Cambridge, CB2 1EP, England.

and

K. M. Evenson
National Bureau of Standards
Boulder, Colorado, U. S. A.

INTRODUCTION

Recently a great deal of interest in free radicals has arisen due to the discovery by radio astronomy of a number of free radicals in inter-stellar space and due to the importance of these very reactive molecules in determining the structure of the upper atmosphere. Especially important is the role they play in the formation of the ozone layer.

Four of the eight diatomic interstellar molecules discovered so far and one of the triatomic molecules are free radicals. The number might be greater if it were not so difficult to obtain sufficient concentration of these reactive molecules in the laboratory in order to determine their radio spectra. In spite of this difficulty, considerable progress has been made during the past decade in elucidating the structure of diatomic and triatomic free radicals in the gas phase from their microwave and electron paramagnetic resonance spectra. These spectra yield accurate rotational, fine structure and hyperfine parameters Only molecules with electronic spin and/or orbital magnetic momenta have electron resonance spectra while microwave spectroscopy is generally applicable to any small transient species which can be generated in sufficiently high concentration to detect microwave absorption. The limiting sensitivity of electron paramagnetic resonance spectroscopy and microwave spectroscopy is about 5×10^{10} radicals/cm^3 and this is insufficient to detect the more reactive diatomic and triatomic radicals, which are of particular interest in astrophysics and free radical chemistry, such as CH, NH, NH$_2$.

Compared with microwave spectroscopy, an increase in sensitivity by a factor of more than a hundred has been achieved using absorbtion spectroscopy in the far infrared, due to the increase of the absorption coefficient with frequency. Although there are no tunable, coherent radiation sources in the far infrared there are several cw far infrared gas lasers which are suitable oscillators with accurately measured line frequencies. In place of a tunable oscillator, rotational transitions of free radicals in the 70 to 1000 μm region have been tuned into coincidence with these far infrared laser lines with a magnetic field. At wavelengths of 5 and 10 μm vibrational transitions can be Zeeman tuned into coincidence with an infrared laser. We have called this technique laser magnetic resonance (LMR) because of its many analogies with EPR. The range over which molecules can be tuned depends on the transition involved and in favorable cases can be as large as 2 cm^{-1} at magnetic fields of 20 kG. The sensitivity is further enhanced by using an intracavity absorption cell. It has been estimated, by generating measured concentrations of hydroxyl radicals by the gas phase titration of hydrogen atoms with NO_2, that a density as small as 2 x $10^8/cm^3$ can be detected by LMR with a one second time constant.[1]

Suitable near coincidences with laser lines are necessary, and often LMR spectra of several rotational transitions can be obtained with sufficient sensitivity to detect transitions at high rotational energies. The requirement of a near coincidence of a rotational transition and laser has become less of a problem with the discovery of many optically pumped lasers which are suitable as sources and which extend the range of transitions to energies below 20 cm^{-1}. These low energy sources permit spectroscopy of non-hydride radicals which have much smaller rotational constants. Although most LMR spectra arise from rotational transitions within an electronic state, weak electric and/or magnetic dipole spectra in OH and NO originating from transitions between the spin-orbit components of the $^2\pi$ ground state have been measured.[2]

Experimental Details

An LMR spectrometer is shown in Figure 1. The laser oscillates between mirrors A and B in a single longitudinal mode. The cavity can be tuned to the center of the gain curve by cavity length adjustment of mirror B. The polyethylene or polypropylene membrane beam splitter separates the intracavity sample cell on the left from the lasing medium on the right. Earlier designs incorporated discharge excitation in H_2O, D_2O etc. to generate cw lines in the 30 to 150 cm^{-1} range. Many more lines are now available by pumping CH_3OH, CH_3F and other gases with CO_2 laser radiation which is coupled in through a small hole in mirror B. The recent development of CO_2 pumped far infrared waveguide

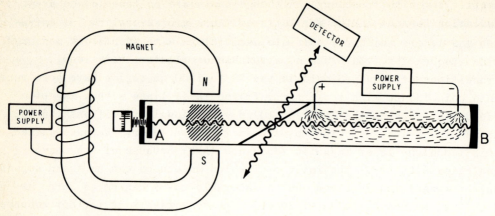

FIG. 1.   Schematic diagram of laser magnetic resonance spectrometer.

lasers has provided more new frequencies for LMR spectroscopy.[3] The intracavity absorption cell is located between the pole caps of a large electromagnet on which modulation coils have been wound. The cell is part of a fast flow system in which radicals are generated either directly in a 2400 MHz microwave discharge or by adding reactants to the products of a discharge close to the laser cavity.

The beam splitter is adjustable near the Brewster angle and polarises the radiation. By rotation about the laser axis, transitions with either $\Delta M = 0$ or ± 1 can be selected. A small fraction of the radiation is coupled out by reflection to the detectors which are Golay cells or helium cooled bolometers. One detector continuously monitors laser output and the other is synchronized with the modulation frequency applied to the magnetic field.

SENSITIVITY

We have recently performed several experiments which yield some important parameters on intracavity laser absorption. One would like to know: 1) what is the absolute sensitivity of the spectrometer ? 2) how does the signal depend on the various parameters of the laser ? and 3) how much is gained by operating inside the laser cavity compared with outside ?

Third order laser theory from Sargent, Scully and Lamb,[4] pp.152 and 153, gives the dimensionless intensity, I_n, at line center:

$$I_n = \frac{F_1 - \frac{\nu}{2Q} - L}{\frac{1}{2} \frac{\gamma ab}{\gamma} F_1} \qquad (1),$$

where F_1 is the first order factor, ν is the laser frequency, Q is the quality factor of the cavity, L is the loss due to the intracavity absorbtion, $\gamma_{ab} = \frac{1}{2}(\gamma_a + \gamma_b)$ is the spontaneous emission and inelastic collision decay rate, and γ is total decay constant. The LMR signal is proportional to $\frac{dI_n}{dH}$ therefore

$$\frac{dI_n}{dH} = - \frac{\frac{dL}{dH}}{\frac{1}{2}\frac{\gamma_{ab}}{\gamma} F_1} \qquad (2),$$

Now, solving for F_1 from equation (1) and substituting into equation (2):

$$\frac{dI_n}{dH} = - \frac{dL}{dH}\left(\frac{4}{\nu}\frac{Q\gamma}{\gamma_{ab}}\right)\left(1 - \frac{I_n \gamma_{ab}}{2\gamma}\right) \qquad (3),$$

but

$$P_\ell = \frac{2\hbar^2 \gamma_a \gamma_b}{\delta^2} I_n ,$$

where P_ℓ is the laser power, and δ is the electric dipole matrix element. Therefore:

$$\frac{dP_\ell}{dH} = - \frac{16\hbar^2 \gamma_a \gamma_b Q}{\delta^2 \nu}\left\{\frac{\frac{1}{2}(\gamma_a + \gamma_b) + \gamma_{ph}}{\frac{1}{2}(\gamma_a + \gamma_b)}\right\}\left\{1 - I_n \frac{\gamma_{ab}}{2\gamma}\right\}\frac{dL}{dH} \qquad (4),$$

where γ_{ph} is the elastic collision contribution to the dipole decay. The terms in the second bracket indicate an intensity dependence; however, the laser output power was varied by a factor of 23, and $\frac{dP_\ell}{dH}$ varied by only 30%. We therefore conclude that $I_n \frac{\gamma_{ab}}{2\gamma}$ is much less than 1 and can be ignored.

γ_a and γ_b, the upper and lower level spontaneous emission and inelastic collision decay constants, would be expected to exhibit a pressure dependence. However, the pressure dependence in the first bracket is in both the numerator and denominator and hence cancels.

A LMR signal from NO_2 was found to be proportional to the pressure of NO_2 over nearly three decade in pressure. Thus, the response of the spectrometer is accurately linear.

An experiment was performed in which $\frac{dP_\ell}{dH}$ for an oxygen sample was monitored as the D_2 added to the D_2O in the laser discharge was varied while the laser was oscillating at 108 μm. The LMR signal was found to be approximately proportional to the pressure of D_2 in the range of 0.08 to 0.34 Torr of D_2. The LMR signal increased by a factor of 6!

Next, a hole coupled end mirror was used and the signal monitored as the brewster membrane was rotated to spoil the Q of the laser. In this case, the signal was found to be proportional to r, the reflectivity of the polyethylene, and hence $\frac{dP_\ell}{dH} \propto Q$.

The overall sensitivity was also measured by calculating the reflectivity of the brewster membrane as a function of angle and measuring the change in the laser voltage. The laser was found to exhibit a signal to noise of 1 with a one second time constant with an absorbtion of 10^{-9} at 84 μm. Since the sample cell is approximately 2 cm long, we can therefore detect an absorbtion coefficient of about 5×10^{-10} cm^{-1}. This number is about 5 times less than that observed in the most sensitive microwave systems; however, because of the ν^2 dependence in the absorbtion coefficient, the spectrometer is more than a hundred times more sensitive per molecule than a microwave spectrometer.

This same sensitivity (10^{-9}) corresponds to a 10^{-5} change in the laser output power; hence intracavity absorption is about 10,000 times more sensitive than single pass external absorption.

In summary:

$$\frac{dP_\ell}{dH} \propto P_{res} \, Q \, \frac{dL}{dH} ,$$

where P_{res} is the D_2 pressure in the laser cavity, Q is the Q of the cavity, and $\frac{dL}{dH}$ is the absorption.

SPECTRA

The LMR spectra of a number of diatomic radicals have been measured, and assigned to $^2\pi$, $^3\Sigma$ or $^1\Delta$ electronic states; for NH ($X^3\Sigma^-$) the spectrum of the radical in the v = 1 excited state has also been detected.[5] In most cases the rotational and fine structure parameters from electronic spectroscopy permit the Zeeman pattern to be calculated and an assignment made. Accurate measurement of the magnetic field positions of the LMR lines and of the laser frequency yields the rotational and fine structure parameters with higher accuracy. In addition hyperfine structure can usually be resolved, leading to new structural information. In this short review we have selected "representative" examples from the diatomic and triatomic radicals that have been detected by LMR, listed in Table 1, that illustrate different features of LMR spectroscopy.

TABLE 1

Free Radicals detected by LMR

O_2	$(X^3\Sigma^-)$	NO_2	$(^2A_1)$
NO	$(X^2\pi)$	HCO	$(^2A')$
OH	$(X^2\pi)$	HO_2	$(^2A'', \, ^2A')$
CH	$(X^2\pi)$	NH_2	$(^2B_1)$
^{14,15}NH	$(X^3\Sigma^-)$	PH_2	$(^2B_1)$
^{14}NH	$(X^3\Sigma^-, v = 1)$		
PH	$(X^3\Sigma^-, a^1\Delta)$		

The CH radical was one of the first species detected by LMR.[6] It had eluded detection by microwave and electron resonance techniques for several years and only recently has its radiofrequency spectrum been discovered in the interstellar medium.[7] The LMR spectrum was detected in an oxy-acetylene flame burning inside the laser cavity, using the 118.6 μm line of the water vapour laser as source oscillator. Fig. 2 shows the spectrum with the electric vector of the laser radiation polarised parallel (π) and perpendicular (σ) to the magnetic field.

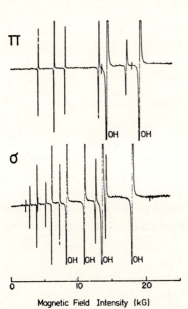
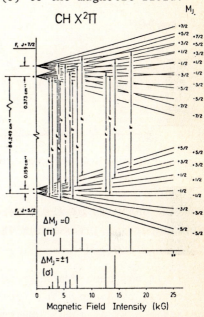

Fig. 2. LMR spectrum of $CH(X^2\Pi)$ in a low pressure oxyacetylene flame, with the 118.6 μm water vapour laser. In addition to absorption lines of CH there are several intense lines of OH.

Fig. 3. Magnetic field dependence of the Zeeman components of the N, J = (2, 5/2) and (3, 7/2) rotational levels of $CH(X^2\Pi)$ and assignment of the 118.6 μm LMR spectrum.

The rotational energies of CH have been calculated by Douglas and Elliott[8] from analysis of the $^2\Delta - ^2\pi$ electronic spectrum of CH. The transition N = 2, J = 5/2 → N = 3, J = 7/2 (in Hund's case (b) nomenclature) is nearly coincident with the 118.6 μm laser line of H_2O and the energies of the Zeeman sublevels of this rotational combination are shown in Figure 3 with the assignment denoted by vertical lines. A least squares fit of the spectrum leads to a redetermination of the rotational constant (B = 14.162 cm^{-1}), rotational distortion constant (A/B = 1.99 $\pm$ 0.2) and differences in the lambda doublet separations of the two rotational levels (ν_Λ (7/2) - ν_Λ (5/2) = 0.213 $\pm$ 0.002 cm^{-1}). Hyperfine structure arising from magnetic hyperfine interaction of the proton spin with the unpaired electron was not resolved in the spectra in Fig. 3, which were recorded at relatively high pressures (several Torr). However, spectra of CH can also be obtained at much lower pressures by reacting OH with carbon suboxide, C_3O_2, and the transitions then exhibit saturated absorption (i.e. Lamb dips) and the hyperfine splitting is resolved as shown in Fig. 4.

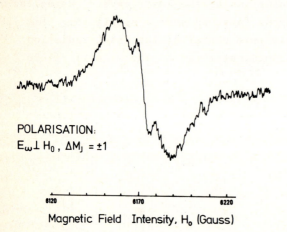

FIG. 4.
Saturated absorption in $CH(X^2\Pi)$ at 118.6 μm.

When the analysis of the hyperfine splittings is completed it should yield important information about the electronic structure of the $^2\pi$ ground state of CH.

It has been demonstrated by work on HO_2 that the absence of structural parameters for free radicals in the gas phase does not prevent the assignment of their LMR spectra. Spectra of HO_2 were first reported by Radford, Evenson and Howard[9] and assigned to three transitions of the type $N_2 \rightarrow (N + 1)_3$ in symmetric top notation, (which is a good approximation for this near prolate molecule). Several more transitions have now been reported including spectra with optically pumped far infrared lasers.[10] Hougen[11] has developed general procedures for the analysis of this type of LMR spectrum arising from triatomic radicals in orbitally non degenerate states and doublet spin states where each rotational level is split by the spin-rotation interaction. Spectra can be assigned to P, Q and R branches, where $\Delta M_J = -1, 0, 1$, which have patterns dependent on the size of the spin-splitting and the degree of mismatch of laser and rotational transition.

The zero field rotational transition energies obtained from HO_2 LMR spectra were fitted to asymmetric rotor formula to yield the rotational and distortion constants of the radical. (The complete geometry of HO_2 can only be determined if spectra of other isotopic variations can be measured.) The constants are the first structural parameters obtained for HO_2 in the gas phase and are in good agreement with calculated rotational constants (Table 2). The rotational and distortion parameters can be used to predict microwave transition frequencies and a microwave spectrum at 65 GHz recently detected by Beers et al[12] and then by Saito has now been assigned to the $0_{00} \rightarrow 1_{01}$ transition in HO_2.

TABLE 2

Rotational constants and symmetric top centrifugal distortion constants for HO_2 obtained from LMR spectra[10], and rotational constants from <u>ab initio</u> calculations[b]

	LMR[a]	SCF[b]	CI[b]
A	20.358(3)	21.368	20.577
B	1.1179(8)	1.035	0.938
C	1.0567(5)	0.987	0.897
D_K	0.0041(3)		
D_{NK}	0.00012(1)		
D_N	0.0000042(8)		

a: one standard deviation given in brackets.

b: D.H. Liskow, H.F. Schaefer and C.F. Bender, J. Amer. Chem. Soc., <u>93</u>, 6734 (1971).

Part of the 84 μm spectrum could not be assigned to transitions in the ground electronic state of HO_2. The intensity of this spectrum optimised under different experimental conditions and increased at higher concentrations of metastable $O_2 a^1\Delta$. It has been suggested that this spectrum arises from HO_2 in the low lying electronically excited $^2A'$ state and a possible excitation process could involve energy transfer from $O_2{}^1\Delta$: $HO_2\,(^2A'') + O_2\,(^1\Delta) \rightarrow HO_2\,(^2A') + O_2\,(^3\Sigma^-)$.

LMR absorption by more than one species with the same laser oscillator is often observed. Figure 5 is a recording of the 118.6 μm LMR spectrum between 8 and 9 kG of the gaseous products produced by reacting hydrogen atoms with solid phosphorus.[13] The spectrum contains absorption lines from four radical species; ●, PH_2 in the ground vibronic state; □ , PH in the $^3\Sigma^-$ ground state; ■ , PH in the low lying $a^1\Delta$ excited state; ○ , OH radicals which are present as an impurity.

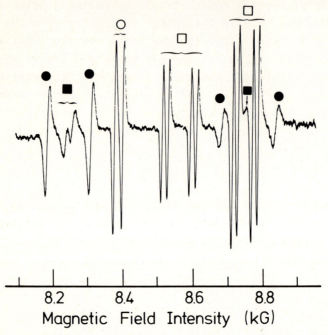

Fig.5. LMR spectrum with the 118.6 μm water vapour laser ($E_\omega \perp H_o$) of the gaseous products produced by passing hydrogen atoms over red phosphorus powder.

PH_2 is a bent triatomic radical in a doublet spin state, and the assignment of the spectrum was assisted by accurate rotational, distortion, spin-rotation constants and term values up to N = 10 determined from the electronic spectrum. The 118.6 μm LMR spectrum was identified with the $5_{32} \rightarrow 6_{42}$ rotational transition in PH_2.[14] These levels have antisymmetric nuclear spin functions i.e. the effective proton spin is zero. If the rotational assignment is correct then each transition should show a doublet hyperfine splitting only, due to interaction of the unpaired electron with the ^{31}P (I = $\tfrac{1}{2}$) nucleus. The spectrum of PH_2 at 118.6 μm consists of an extensive series of doublets between 7 and 14 kG. Two of these can be identified in Fig. 5, with splittings of about 150 gauss.

The four line hyperfine patterns in Fig. 5 cannot arise from PH_2 in either ortho or para modification. The most likely source is PH which would be expected to show a doublet splitting from the protons and a doublet from the ^{31}P nucleus. The rotational constant, spin-spin and spin-rotation parameters for PH, $X^3\Sigma^-$, have been determined from analysis of the $^3\Pi - {}^3\Sigma^-$ band system, and were used to calculate the lower rotational and fine structure levels at zero field using standard formulae. The $N = 4 \rightarrow 5$ transition was identified as the likely origin of the 118.6 μm spectrum. The magnetic sublevels as a function of field were calculated by diagonolisation of the Hamiltonian matrix set up in Hund's case b and the assignment of the LMR transitions was made using the optically determined zero field parameters to predict the spectrum.

PH $a^1\Delta$ has the same electronic configuration as the $^3\Sigma^-$ state and the rotational constants are similar; the LMR spectrum also arises from the $4 \rightarrow 5$ rotational transition.[13] The electronic Zeeman effect is particularly simple to calculate in this case and arises from interaction of the orbital magnetic moment with the field. To second order the Zeeman energies are $W = \frac{\Lambda^2}{J(J+1)} M_J H + A H^2$ where A is a small M_J^2 dependent term and Λ is the projection quantum number along the internuclear axis, (2 for Δ states). The calculated LMR spectrum can be brought into agreement with measurements within experimental error if the optically determined rotational constant is changed from 8.440 to 8.4427 cm^{-1}. The lines marked ■ in Fig. 5 are hyperfine components of the $M_J = -2 \rightarrow -1$ transition. The hyperfine structure of PH $^1\Delta$ differs considerably from the $^3\Sigma^-$ state and is characterized throughout the spectrum by a large phosphorus splitting of about 500 gauss and a much smaller proton splitting of about 10 gauss which is not always resolved.

Although most examples of LMR involve tuning rotational transitions into coincidence with far infrared lasers, it is also possible to tune vibration-rotation transitions into coincidence with an infrared laser. LMR spectra at 5 μm using the CO laser as source have been measured for NO [15] and NO_2.[16]

SUMMARY

Both OH and CH have been detected in the interstellar medium. Although the 118.6 μm CH spectrum is not of direct interest in radio-astronomy it is likely that the lower energy rotational transitions should be accessible with one of the optically pumped lasers. Several r.f. transition frequencies have been calculated for HO_2 from the constants determined by LMR and are listed by Hougen et al.[10] The most accurate predictions are for transitions involving $K_a = 0$ states where the spin splitting should be vanishingly small. The more accurate rotational energies available from analysis of NH[5] and NH_2[17] LMR spectra, for example, should greatly assist in searching for these species in the interstellar medium.

Kaldor et al[18] have used LMR at 5 μm with a CO laser to measure small amounts of nitric oxide in polluted atmospheres. Another important application of LMR is the measurement of concentrations in free radical reactions, in particular those of importance in atmospheric and combustion chemistry. The method has recently been used to measure the rates of reaction of OH with NO, NO_2 and CO in a discharge flow system[1] and the results are in satisfactory agreement with measurements by other techniques. Measurement of OH reaction rates with CH_3F, CH_3Cl and other fluorocarbons are planned. Much attention has recently focussed on the role of ClO in atmospheric chemistry and LMR detection of the radical using the 700 μm line of ethanol would be a convenient way of monitoring its reactions. Much less is known about NH_2 reactions and LMR detection has provided a new way for following its reactions in flow systems.

The measurement of relative radical concentration by LMR is analagous to gas phase e.p.r. under conditions of constant line width i.e. pressure, modulation amplitude and laser power, the concentration is proportional to the peak to peak height of the first derivative signal. For reactions which cannot be treated by first order kinetics or are second order in radical concentration, measurement of absolute concentrations may be necessary to derive rate constants. In this case calibration against a measured density of other species is necessary. Filling factor differences could be eliminated by using the same laser line for transient and calibrant species.

Acknowledgement

We thank Dr. H.E. Radford for many stimulating discussions particularly with respect to the sensitivity of the apparatus.

REFERENCES

1. K.M. Evenson and C.J. Howard, J. Chem. Phys., $\underline{61}$, 1943 (1974).

2. K.M. Evenson, J.S. Wells and H.E. Radford, Phys. Rev. Lett., $\underline{25}$, 199 (1970); M. Mizushima, K.M. Evenson and J.S. Wells, Phys. Rev., $\underline{A5}$, 2276 (1972).

3. H.E. Radford, Appl. Phys. Lett., to be published.

4. M. Sargent, M.O. Scully and W.E. Lamb, Laser Physics, published by Addison - Wesley 1974.

5. H.E. Radford and M.M. Litvak, Chemical Physics Lett., in press.

6. K.M. Evenson, H.E. Radford and J.M. Moran, Appl. Phys. Lett., $\underline{18}$, 426 (1971).

7. O.E.H. Rydbeck, J. Ellder and W.M. Irvine, Nature, $\underline{246}$, 468 (1973).

8. A.E. Douglas and G.A. Elliot, Can. J. Phys., $\underline{43}$, 496 (1965).

9. H.E. Radford, K.M. Evenson and C.J. Howard, J. Chem. Phys., $\underline{60}$, 3178 (1974).

10. J.T. Hougen, H.E. Radford, K.M. Evenson and C.J. Howard, J. Mol. Spect., $\underline{56}$, 210 (1975).

11. J.T. Hougen, J. Mol. Spect., $\underline{54}$, 447 (1975).

12. Y. Beers, C.J. Howard and J.T. Hougen, J. Mol. Spect., to be published.

13. P.B. Davies, D.K. Russell and B.A. Thrush, Chem. Phys. Lett., to be published.

14. P.B. Davies, D.K. Russell and B.A. Thrush, Chem. Phys. Lett., to be published.

15. P.A. Bonczyk, Chem. Phys. Lett., $\underline{18}$, 147 (1973).

16. S.M. Freund, J.T. Hougen and W.J. Lafferty, J. Mol. Spect., to be published.

17. P.B. Davies, D.K. Russell, B.A. Thrush and F.D. Wayne, J. Chem. Phys., $\underline{62}$, 3739 (1975).

18. A. Kaldor, W.B. Olson and A.G. Maki, Science, $\underline{176}$, 508 (1972).

High resolution laser spectroscopy of the D-lines of on-line produced radioactive sodium isotopes

by

H. T. Duong, P. Jacquinot, P. Juncar, S. Liberman, J. Pinard, J. L. Vialle

Laboratoire Aimé Cotton, C.N.R.S. II, Bât. 505, 91405-Orsay, France

and

G. Huber, R. Klapisch, C. Thibault

Laboratoire René Bernas, C.N.R.S., 91406-Orsay, France.

New methods of laser spectroscopy using atomic beams enable ones to obtain resolution beyond the Doppler limit with atoms in quantity much smaller than with conventional spectroscopic methods. This makes possible the study of hyperfine structures and isotope shifts in long series of unstable isotopes of the same elements produced by nuclear reaction, some of them far from stability This can extend our knowledge about shapes and magnetic and electric moments of nuclei as a function of the number of neutrons.

Numerous isotopes of sodium, even short lived ones have been found recently to be produced more or less easily, and studied using a mass spectrometer on line with an accerator [1] In addition, the fact that the rhodamine 6G C.W. dye laser is particularly well adapted to investigate the spectral properties of the sodium D lines, made this element an especially good candidate for this type of experiment.

More precisely in the case we are dealing with, one was interested inasmuch as it seemed to be possible, in measuring nuclear magnetic and quadrupole moments and of course isotope shifts.

In the experiments which we have done, the sodium isotopes were produced in the spallation of aluminium by 150 MeV protons from the Orsay synchrocyclotron [2]. A molten aluminium target heated up to 900°C has been used. The spallation reaction can be written as : $^{27}A\ (p, 3pxn)^{25-x}Na$ (x taking values between 0 and 5).

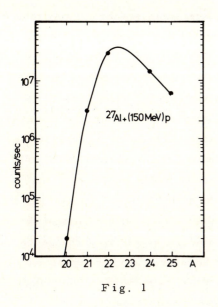
Fig. 1

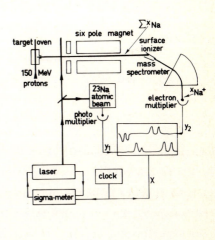
Fig. 2

This means that all the isotopes having their mass number between 20 and 25 are simultaneously produced, although with different yields, as can be seen on Figure 1. But as they are all present in the source, it is necessary to use a mass spectrometer in order to separate them. A schematic view of the experimental set-up is shown in Fig. 2. The principle of the experiment consists in selectively exciting the atoms leaving the target in the form of an atomic beam with a tunable single-mode dye laser which induces an optical pumping in the hyperfine sublevels of the ground state (see Figure 2). The consequent change in the populations of these sublevels is detected using a six-pole magnet, but this point will be described below with a little more detail. After the six-pole magnet the atoms are ionized by impinging on a hot rhenium surface, in order to be separated by the mass spectrometer. Separated ions are then counted by an electron multiplier. Of course, as we were interested in measuring isotope shifts, we found it easier to compare the recorded spectrum of a given isotope to the corresponding spectrum of the stable one provided by an auxiliary atomic beam (the signal in this latter case was monitored by the fluorescence light). The laser and the device (called "sigmameter") which we used to servocontrol and to tune its frequency have already been described elsewhere [3][4]. Several points are to be emphasized : first, although the total amount of ^{25}Na atoms in the beam leaving the oven approaches about 10^8 atoms/sec., because of the different losses the counting rate at the detector lies between 1 500 c/sec. and at the best 3 000 c/s. . Secondly, one will notice that the diffusion delay time in the molten metal of the target depends very strongly on the temperature and decreases for instance from 1 min. to about 10 sec. as the temperature varies from 850°C to 900°C.

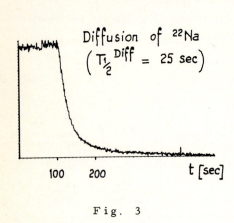

Fig. 3

An example of recorded diffusion delay time is given in Figure 3, with ^{22}Na, whose half-life is 2.6 years, very long compared to the diffusion time. Actually this diffusion time appears to be the most important limitation preventing studies of shorter lived isotopes. To overcome this difficulty it would be necessary to change the target in our experimental set-up. It has been shown however in another experiment [1] that the diffusion properties of heated graphite allow the study of sodium isotopes of millisecond half-life and the use of this technique is being considered.

Concerning the magnetic detection [5] it is well known since the hydrogen maser, that a six-pole magnet has the property to focus atoms in the sublevel $m_J = +1/2$ of the ground state, whereas on the reverse, it defocuses atoms in the sublevel $m_J = -1/2$. In the case of Na atoms (see Figure 4) the ground state is split into two hyperfine sublevels which correspond for example, to F=2 and F=1 for the stable isotope (because of I=3/2). Therefore, in a strong magnetic field, these two sublevels are split again according to the diagram into two equally populated sets of magnetic sublevels corresponding to $m_J = \pm 1/2$. After resonant optical interaction, it is easy to see that an optical pumping will occur which will modify the relative populations of these magnetic sublevels and will lead consequently to a change in the counting rate of the detector. In the case of the D_1 line for instance, the

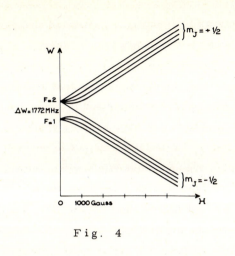

Fig. 4

observed signal could be described as follows : since with transitions starting from F=1 the optical pumping can only increase the population of F=2 , these transitions therefore increase the number of atoms in $m_J = +1/2$, which corresponds to an increasing of the signal ; the reverse being true with transitions starting from F=2 . In the case of the D_2 line the situation is a little more complicated due to the fact that the F'=3 hyperfine sublevel of the upper level ($3\,^2P_{3/2}$) can only be reached starting from the F=2 sublevel of the ground state and can only decay on the same F=2 ground state sublevel, so that in an isotropic arrangement it is not possible to get any optical pumping using this transition. But fortunately, breaking up the isotropy of the system by using a small magnetic field (about 2 gauss) combined with a suitably polarized exciting light, the optical pumping works again on the Zeeman hyperfine sublevels of the ground state, taking advantage of the fact that the one with $m_F = -2$ corresponds to $m_J = -1/2$. It is then easy to see that a σ^+ pumping will clear out this $m_F = -2$ sublevel and will therefore lead to an increasing of the signal, whereas a σ^- pumping will fill the $m_F = -2$ sublevel and will lead to decreasing of the signal.

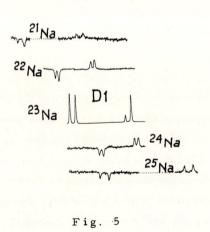

Fig. 5

Figure 5 shows a display of the recordings that we obtained on the D_1 line for all the isotopes studied according to their position in the frequency scale with respect to the stable one. From these recordings one can get both isotope shifts and magnetic hyperfine constants which give the nuclear magnetic moments. Since the D_1 line is a $J=1/2 \to J=1/2$ line, one cannot determine from it the nuclear quadrupole moment. In order to get it one has to study the hyperfine structure of the D_2 line. An example of the recordings which have been obtained for ^{25}Na is given in Figure 6. The two traces correspond to σ^+ or σ^- case and, as it has been explained above, one can observe in particular the expected change in sign of the signal for the component corresponding to the highest F values.

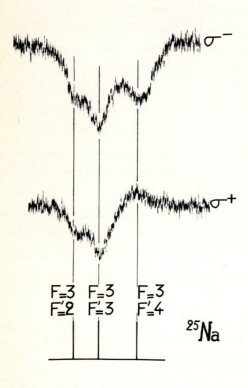

The numerical results that we have obtained are summarized in Table 1. One will notice that the measured values of isotope shifts follow the law of normal plus specific mass effect within the experimental errors as is better seen for the corresponding values obtained for the reduced mass effect. All nuclear magnetic moments of these isotopes, except for ^{25}Na, were already known with a very high accuracy and our values are in good agreement with them. The value that we found for ^{25}Na also agrees very well with the one recently measured by another group using a different technique 6. Concerning the nuclear quadrupole moments we can only give estimates and only for the odd isotopes : for ^{21}Na even the sign is not quite sure.

Fig. 6

Further experiments would be needed of course in order to increase the accuracy of nuclear quadrupole moment measurements on the isotopes that we have already studied. On the other hand, numerous heavier isotopes with mass numbers up to 33, have their fundamental nuclear properties quite unknown at the present time, except for their masses and their lifetimes which lie in the few- milliseconds range. This would require using quite different nuclear reactions (with a high energy proton accelerator) but the principle of the spectroscopic experiments would remain essentially the same.

Table 1

Mass number	21	22	23	24	25
Isotope shift (in 10^{-3}cm^{-1})	53.5 (2)	25.3 (2)	0	23.1 (4)	44.6 (2)
Reduced Isotope shift	2.010 (17)	0.990 (13)	0	0.990 (22)	2.00
μ/μ_N	*	*	*	*	+3.685 (022) [+3.683 (004)**]
b_{3p} (in MHz)	-1.5 (2.0)				+7 (3).

* See text
** Deimling et al. [6]

References

[1] R. Klapisch C. Thibault A. M. Poskanser R. Prieels C. Rigaud E. Roeckl
Phys. Rev. Lett. 29, 1254 (1972)
R. Klapisch R. Prieels C. Thibault A. M. Poskanser C. Rigaud E. Roeckl
Phys. Rev. Lett. 31, 118 (1973)
[2] G. Huber C. Thibault R. Klapisch H. T. Duong J. L. Vialle J. Pinard
P. Juncar P. Jacquinot
Phys. Rev. Lett. 34, 1209 (1975)
[3] S. Liberman J. Pinard
Appl. Phys. Lett. 24, 142 (1974)
[4] P. Juncar J. Pinard
To be published in Optics Commun.
[5] H. T. Duong J. L. Vialle
Opt. Commun. 12, 71 (1974)
[6] M. Deimling R. Neugart H. Schweickert
Z. Physik A 273, 15 (1975).

COMPARISON OF SATURATION AND TWO-PHOTON RESONANCES

V.P. Chebotayev
Institute of Semiconductor Physics,
Novosibirsk, 630090, USSR

The problem of obtaining narrow radiation resonances in any wavelength range remains one of the most actual problems in physics. It is successfully solved in the optical region at interaction of two or several waves with the same or different frequencies in a gas. By the present time a lot of the methods of obtaining narrow resonances are predicted and realized: Lamb dip, saturation absorption resonances, two-photon resonances, resonances at interaction of unidirectional waves, non-linear optical Ramsey resonance and so on. The recent progress in this field is described in /1/.

Each of these resonances has some advantages and its own field of application: spectroscopy inside a Doppler line, optical standards, carrying out fundamental experiments, precise measurements of frequencies of forbidden transitions, selective excitation of levels and laser photochemistry.

The most important results have been obtained by using the method of saturation absorption (SA). The phenomenon of formation of a Lamb dip /2/ is the basis of this method. The absorption of a standing wave in the centre of a line undergoes the resonance dip whose width is equal to a homogeneous width. Every travelling component of a standing wave "burns" a dip in the population difference in the atoms whose velocities satisfy the resonance conditions $v = \pm \frac{\Omega}{K}$ (K is a wavenumber, Ω is frequency detuning from the centre of a line). In the centre the dip lines become one and the saturation is amplified since both waves interact with the same atoms. The resonance width is determined by the homogeneous width of the transition. The use of the long-lived states of the vibrational-rotational transitions permits to to obtain extremely narrow resonances.

The principal cause of the resonance broadening of the vibrational-rotational transitions of molecules is collisions. When the Doppler frequency shift at scattering by a characteristic angle θ is more than a homogeneous linewidth the broadening is determined by an elastic scattering cross-section σ /3/ ($\sigma = 10^{-14}$ to 15cm^2). The resonances with the width of 10^3 Hz can therefore be obtained at pressures of about 10^{-5} torr. Under these conditions a free path can be much more than the transverse sizes of beams in ordinary lasers and the transit effects influence upon a shape of the narrow resonances. The telescopic expanders of the laser radiation are necessary for obtaining the intense narrow resonances. The telescopic expander inside a cavity permits to obtain more intense supernarrow resonances suitable for laser frequency stabilization as compared with the resonances in the telescopic system outside a cavity. The scheme of the laser with the telescopic cavity is given in Fig. 1. The investigation of the narrow resonances requires the use of stable lasers with high long- and short-term stability and the special methods of recording. The recording of the oscillation power resonance in He-Ne laser at $\lambda =$ = 3.39 u with a methane absorber is shown in Fig. 2 /4/.

One can see a hyperfine structure of a methane line and crossing resonance. The observation of many similar curves has shown that the change of profile (see Fig. 2) is connected with a recoil doublet. The halfwidth of the resonance was about 4 KHz. Its intensity is enough for frequency stabilization. Recently, Hall and Borde have obtained the direct resolution of the recoil doublet using the telescopic expander outside a cavity./5/

Resonance narrowing may be obtained in the transit system due to specific influence of slow atoms. The atoms having small transverse velocities interact with the field for a long time. Owing to this the saturation for such atoms is more and the Lamb dip width is narrower than for the atoms having the most probable thermal velocity. We have carried out the special investigations of this phenomenon. The pressure dependence of the resonance width /6/ is shown in Fig. 3*. At the value

---

* Recently, the similar results have been obtained in /5/.

of the parameter $\Gamma\tau_o \cong 0.1$ (2Γ is a homogeneous width, $2\tau_o$ is the transit time of a particle with the most probable velocity) it is possible to reduce the width by a factor of about two as compared with the transit one. The experiments made with using the laser beam of 6 mm in diameter at $p \sim 10^{-5}$ torr permit to resolve the hyperfine structure of methane. Of course, the resonance intensity was extremely small in this experiment (Fig. 4). The reduction of the resonance width in comparison with $1/\tau_o$ is observed in the pressure range of 10^{-5} torr; it is connected with the influence of slow atoms. It has been noted even in 1969 /7/ but still the phenomenon has not been discovered due to small resonance intensity. Thus, the considerable resonance narrowing may be obtained in the transit system without using the telescopic cavities.

Further progress in obtaining supernarrow resonances can be connected with the type of a resonance in transit systems new for the optical region; we call it the non-linear optical Ramsey resonance /8/. Note that for one atom with a fixed velocity v Ramsey resonance arises at interaction with two widely separated beams in field linear approximation. A quantity and a sign of an absorbed energy depend on field phases of the first and the second beams in z_1 and z_2 points, respectively. As applied to the optical region an energy is a rapidly oscillating function v with zero mean value. The absence of the transit Ramsey resonance due to polarization transfer is connected with the velocity distribution of particles and the small wavelength in comparison with the cavity sizes. We have taken notice of that a structure with a period of $\Delta v = \frac{\lambda}{2\tau_o}$ arises after the non-linear interaction of particles with two widely separated fields in the velocity distribution of particles (τ_o is the transit time between the beams). It results in the following: in the third beam the contribution of these atoms to absorption and amplification at averaging over v proves to be non-compensated that leads to resonance formation. The resonance width depends on the transit time T. The Ramsey resonance shape in three separated beams is shown in Fig. 5.

Obtaining the supernarrow resonances permits to turn to the investigations of the influence of the recoil effect, the Doppler quadratic shift which are important for standards. The results of the investigations of the resonance centre shift due to the Doppler quadratic effect are summarized in Fig. 6. At $\Gamma\tau_o > 1$ and large fields the shift approaches to its "mean" value equal to $\frac{KT}{mc^2}\omega$. In the transit region $\Gamma\tau_0 \ll 1$ the resonance shift depends on both the field and the gas pressure. It is connected with the fact that the velocity of atoms effectively interacting with the field changes with the change of these parameters. At $\Gamma\tau_0 > 1$ practically all the atoms with the velocity projection $v = 0$ effectively interact with the field. Very stable lasers have allowed to measure the red temperature shift. For the methane line the shift is equal to 0.5 Hz/°K.

The resonance width reduction involves the sharp reduction of its intensity. The most intense resonance is observed at the saturation parameter $G = 1$, as $G \sim 1/\Gamma^2$ the optimal power is in inverse proportion to Γ^2. The gas density reduction is accompanied by the linear absorption reduction. The absolute saturation resonance intensity is

$$\Delta I [w] = 6 \cdot 10^3 \frac{P^3 [torr] \frac{\partial \Gamma}{\partial P} [\frac{MHz}{torr}] L [cm] S [cm^2]}{\sqrt{\frac{T^\circ [°K]}{m [a.t.w]}}} q \qquad (1)$$

where L is an absorption length, S is a beam cross-section, P is a gas pressure, q is a fraction of particles on the level, $\partial\Gamma/\partial P$ is collisional broadening. In methane at $P \sim 10^{-5}$ torr, $\partial\Gamma/\partial P \sim 10$ MHz/torr $T = 300°K$, $q = 0.1$, $L = 10^2$ cm, $S = 1$ cm^2 the absolute resonance intensity ($\Delta I \sim 10^{-10}$ w) is on the limit of recording with the available radiation receivers. The use of the telescopic expanders permits to increase the resonance intensity by 10^2 to 10^4 times. However, the supernarrow resonance recording and especially their use for the optical standards causes the serious difficulties.

The new type of resonances has no similar limitations at extremely low gas pressures. The main idea of the method consists in recording the resonance of a refractive index not in the direct measurement of a phase of the light that has passed, but in the observation of the change of the laser frequency spectrum caused by the resonance change of the refractive index. The small changes of the refractive index of a medium in the cavity results in the small relative changes of oscillation frequency. However, the absolute changes of the frequency can be appreciable. This feature is used in laser gyroscopes in which a very small phase change at rotation is expressed in the appreciable frequency difference of beatings of two waves. Let the medium with the resonance absorption be inside the cavity. The oscillation frequency ω_{os} near the centre of the absorption line is given by the expression (the influence of the active medium and linear absorption is neglected)/1/

$$\omega_{os} - \omega_c = = (\omega_{os} - \omega_0) \frac{1}{1 + \frac{(\omega_{os} - \omega_0)^2}{\Gamma^2}} \beta \; ; \quad \beta = \frac{2\sqrt{\pi} \, A \, C}{\Gamma} \; ;$$

ω_c is cavity frequency, ω_0 is frequency of absorption line; (2) where C is the light speed, A is the absolute magnitude of the absorption resonance per unit length. For SA $A = G \mathcal{æ}_0/\Gamma$ where $\mathcal{æ}_0$ is linear absorption per unit length, $G \ll 1$. The sign (-) corresponds to the resonance absorption reduction (the case of frequency pulling), the sign (+) to the absorption increase (the case of frequency repulsion). When recording the absorption resonances it is necessary that the laser radiation spectrum should be much narrower than the resonance width. In this case the situation is quite different. Let the laser resonator undergo perturbations due to which the density of the frequency distribution $P(\omega_c)$ is the uniform function. With no absorber the radiation spectrum is given by the expression:

$$P(\omega_{os}) = P(\omega_c) \frac{\partial \omega_c}{\partial \omega_{os}} = P(\omega_c) \left[1 + \beta \frac{1 - \delta^2}{(1 + \delta^2)^2} \right]$$

$$\delta = \frac{\omega_{os} - \omega_0}{\Gamma} \tag{3}$$

The radiation spectrum of He-Ne laser with CH_4 cell is shown in Fig. 7 /9/. The spectrum resonance possesses the following features:
1. Its relative intensity is independent of a gas density and determined by the absorption-to-collision broadening ratio. For methane the spectrum resonance contrast has the magnitude of about 1.
2. The spectrum recording is made by the heterodyne method. The recording sensitivity increases by several orders in comparison with the power resonance recording.
3. The possibility of obtaining the supernarrow resonances with the help of lasers with bad frequency stability.
The resonance widths of about 1 KHz obtained at present by using the methods described are limiting. The obtaining of the resonance widths of 10^2 Hz and less is real.

The new possibilities for the obtaining of narrow resonances are opened by two-photon resonances. Their formation cannot be associated with the effects of population change. The conditions of the resonance at absorption of two opposite moving atoms allowing for the Doppler shift is $\omega \to \omega - k v$, $\omega' \to \omega + k v$. In the resonance $\omega_{12} = \omega + \omega' = 2\omega$

inspite of the velocity all the atoms take part in absorption. As a result, the sharp peak with a homogeneous width of the forbidden transition arises in the centre of the line /10/.

Let us consider the three-level system in Fig. 8. The transition probability $W_{1\to 2}$ of the particle excited at the moment $t = 0$ from the level 1 to the level 2 under the action of the field E and E' is:

$$W_{1\to 2} = \gamma_2 \int_0^\infty |a_2|^2 dt = -2Re\left\{\frac{iP_{02}E'}{\hbar} \int_0^\infty e^{-i\Omega' t} a_0 a_2^* dt\right\} \quad (4)$$

γ_i is a rate of decay of i level, P_{ik} are matrix elements of the dipole transition $i\to k$, a_i is probability amplitude, $\Omega = \omega - \omega_{01}$, $\Omega' = \omega' - \omega_{02}$. $W_{1\to 2}$ is determined by Fourier component from the dipole moment of an atom. If there is the population of the level 0 the field produces the dipole moment on frequencies ω' and ω_{02}. However due to the non-linear, coherent effects the dipole moment of the transition $0\to 2$ occurs on the other frequencies depended on detunings Ω and Ω'. Fourier components determine the interesting features of the line of the Raman scattering type and so on. Inspite of the simplicity of the scheme the expression for $W_{1\to 2}$ is unwieldy. The simplest and most convenient expression for analysing $W_{1\to 2}$ is

$$W_{1\to 2} = \frac{2(P_{01}E)^2(P_{02}E')^2}{\hbar^4\left[\left(\frac{\gamma_1+\gamma_2}{2}\right)^2+\Omega^2\right]} Re\left\{\frac{1}{\gamma_1\left[\frac{\gamma_1+\gamma_2}{2}-i(\Omega'+\Omega)\right]} + \frac{1}{\gamma_0\left(\frac{\gamma_0+\gamma_2}{2}-i\Omega'\right)} + \frac{1}{\left[\frac{\gamma_1+\gamma_2}{2}-i(\Omega'+\Omega)\right]\left[\frac{\gamma_0+\gamma_2}{2}-i\Omega'\right]}\right\} \quad (5)$$

To obtain the probability for the scattering scheme it is sufficient to replace Ω with $-\Omega'$. The first term corresponds to the two-quantum process, the second one to the stepwise excitation, and the third one is related to the interference of these processes. To obtain the line shape in a gas it is sufficient to go on to the Doppler shifted frequencies $\omega \to \omega \pm kv$, $\omega' \to \omega \pm k'v^*$ and integrate over velocities. Here two cases arise:

The first one $\Omega \gg k\bar{v}$, $\bar{v}$ is a medium-thermal velocity. It has been considered in /10/ and yields the known two-photon absorption resonance in a standing wave field ($\omega = \omega'$, $E = E'$, $kv = -k'v$)

$$W_{1\to 2} = \frac{4 P_{01}^2 P_{02}^2 |E|^4}{\hbar^4 \Omega^2} Re\left\{\frac{1}{\gamma_1}\left[\frac{\gamma_1+\gamma_2}{2}-i(2\omega-\omega_{12})\right]\right\} + \text{Dopp width term}$$

---
* The signs ($\pm$) depend on the wave propagation direction in relation to the observation axis.

This resonance possesses a number of the features important for the spectroscopy: it does not undergo Doppler broadening, no recoil effect, small collisional broadening, and it permits to obtain supernarrow resonances in the transit systems. Its applications for the spectroscopic and other problems are described in /11/. Our interest to TPR is connected with the possibility of its use on the 1S - 2S transition of H in the experiments on the precise measurement of Rydberg constant /12/.

We shall consider the influence of the recoil effect, collisions, transit effects, and, at least, the second order Doppler effect on the TPR width. In some cases the influence of the enumerated factors on the width of saturation resonances (Lamb dip) is estimated for comparison.

1. The recoil effect leads to a well-known shift of the resonance line of absorption and radiation by values of $+\delta$ and $-\delta$, respectively, where $\delta = \frac{k^2 \hbar}{2m}$. Therefore the Lamb dip conditioned by the transitions from the upper and lower levels splits up into two due to the recoil effect. They are $\pm \delta$ away from each other /13/. Their amplitudes depend on the ratio of the level relaxation constants. At the different relaxation constants and great homogeneous width the Lamb dip is asymmetrical and its maximum does not coincide with the centre of the line. As it has been shown in /14/ the supplementary shift of the dip connected with the coherent processes at the field-particles interaction occurs together with the field growth. These features can greatly influence upon the precision of the experiments especially in the optical spectral region. For instance, the resonance shift due to the recoil effect on the 1S - 2S transition of hydrogen has the value of about 1 MHz.

When using TPR a quite different situation arises. At the absorption of two opposite moving photons a pulse amounts to zero. Therefore in contrast to the saturation resonances the centre of the two-photon resonance does not shift due to the recoil effect. It is an advantage of the two-photon resonance. In future it can prove to be very important for high resolution spectroscopy in X-ray and γ-ray regions.

2. The problem of the collisions influencing upon the two-photon absorption is little studied. Therefore we limit ourselves to qualitative

considerations of the collisions influence upon the TPR width. It follows from the very nature of the resonance (all the particles to equal extent and irrespective to their velocity take part in the production of the resonance) that the change of particles velocity when coming into collision will not influence on the interaction and, hence, lead to line broadening. It is the distinction of the collision influence on the form of the narrow saturation resonances and the TPR form. We would remind that at a very low gas pressure the elastic particle scattering plays a determinative role in the Lamb dip broadening: after the collision the particle does not interact with the field any more* and the broadening is determined by a total elastic scattering cross-section even in the case when the processes connected with the phase change do not play an appreciable role. From the above-mentioned it follows that the TPR broadening at collisions may be caused only by the processes connected with the oscillator phase change. It is possible at the different amplitudes of the particles scattering on the upper and lower levels. As the lower and upper levels at the two-photon absorption usually belong to the same configuration the perturbation of neibouring levels at collision is of the same character and the cross-section of the phase change can be far less than the total elastic scattering cross-section. The circumstance that the upper level at the two-photon absorption is not optically connected with the ground state and the resonance character of the collisions does not become apparent can also promote the small cross-section in comparison with the collisional broadening of the saturation resonances. In accordance with the data of the collisional broadening of the lines which are not optically connected with the ground state the upper limit for the collisional TPR broadening is considered to be 10 MHz/torr. Thus, at the pressure of about 10^{-6} torr the collisional broadening is about 10 MHz.

3. At the great excited state natural lifetime and the small value of the collisional broadening the TPR width can be determined by the transit effects. Though this broadening cause is not principal the real technical means limit the TPR line width because of the final transit time. We would note that the Lamb dip broadening due to the transit effects is determined by the cross dimensions of a laser beam (the

---

* The condition under which it occurs is evident: Doppler frequency shift at scattering by a typical angle θ must be more than the homogeneous linewidth. For instance, θ has an order of magnitude of 10^{-2}, for particles with the mass $m = 14$ and velocity of $5 \cdot 10^4$ cm/sec we have $\sim 10^5$ Hz.

particle whose velocity projections onto the beam axis are equal to zero interact with the field). The advantage of the TPR lies in the fact that they permit to use the longitudinal beams and without great efforts to increase the field interaction time by several orders. For the beam length of 5 m, the particles with atomic mass of 100 and the temperature of 300°K we obtain the linewidth of 20 Hz. We would note that the linewidth is realized in drift beam tubes with a beam of atoms of caesium. Respectively, the relative width of the TPR line in the optical spectral region is 5 to 6 orders less than the resonance width in SHF range.

4. In contrast to the Lamb dip the TPR is not connected with the change of level populations (saturation effect). Its contrast is about 100% at any field intensity. The field intensity necessary for the TPR observation is determined by the means of the experimental tool. The simple estimations indicate that the fields at which the TPR is observed in sodium produce the saturation of about 10^{-6-7}. At such fields it is possible not to take into account the absorption saturation and the field broadening. The saturation effects may be expected in very strong fields. The two-photon absorption of the opposite moving photons produces the homogeneous saturation and the saturation at the absorption of the photons moving in the same direction is inhomogeneous. Therefore when both waves interact with the same particles the dip similar to the Lamb dip may arise in the centre of the absorption line in the strong fields. Allowing for collisions the widths of the TPR and of the two-photon absorption dip are different. In addition, this dip is shifted due to the recoil effect.

5. The above enumerated factors do not play a determinative role in achieving the widths from 10^{-13} to 10^{-15} under certain conditions. Now we dwell on the influence of the second order Doppler effect which is, to our mind, principal and the most important limitation in achieving the extremely small widths of the TPR. The relativistic effect of time reduction in moving coordinate system leads to a well-known red line shift quadratically dependent on the particle velocities. Due to velocity distribution of particles every atom has the resonance frequency dependent on its absolute velocity. Therefore the lines of the ensemble of particles undergo the supplementary broadening. The TPR line form with allowing for the second order Doppler effect has been analysed in

/15/. By an order of magnitude the supplementary broadening of the TPR line is equal to the line shift for the particles possessing the medium thermal velocity.

If the homogeneous width Γ is less than the magnitude the TPR line is inhomogeneously broadened due to the motion of atoms. Therefore the magnitude $\frac{v^2}{c^2}\omega$ may be taken for the inhomogeneous Doppler width of the TPR line due to the second order Doppler effect. It is important that at the inhomogeneous broadening of the two-photon absorption line the field interacts with the particles whose absolute velocities meet the conditions of the resonance. Therefore at the two-photon absorption on the upper level the particles are excited in the narrow interval of the absolute velocities. This phenomenon affords the chances for producing monokinetic beams.

For the hydrogen 1S - 2S transition the linewidth Γ is about 10^4 Hz at T = 300°K (the relative width is 10^{-11}). When transiting to heavier atoms and lower temperature the width of about 10 Hz and less can be obtained. The same can be obtained when using the low-kinetic beams of atoms or molecules.

Thus, this method permits to obtain the resonances with the relative widths of about 10^{-11} to 10^{-13}.

REFERENCES

1. V.S.Letokhov, V.P.Chebotayev."The Principles of Non-linear Laser Spectroscopy". Nauka, 1975.

2. W.E.Lamb, Jr., Phys.Rev., 134A, 1429 (1964).

3. S.N.Bagayev, Ye.V.Baklanov, V.P.Chebotayev. JETP Lett., 16, 15, 1972.

4. S.N.Bagayev, L.S.Vasilenko, V.G.Gol'dort, A.K.Dmitriyev, N.M.Dyuba, M.N.Skvortsov, V.P.Chebotayev. 2nd Physics Gas Laser Symposium, Novosibirsk, June, 1975.

5. J.L.Hall, C.Bordé. 2nd Physics Gas Laser Symposium. Novosibirsk, June, 1975.

6. Ye.V.Baklanov, B.Ya.Dubetsky, Ye.V.Titov, V.M.Semibalamut. Sov. J. of Quant. El., in press.

7. S.G.Rautian, A.M.Shalagin. JETP Lett., 9, 636, 1969.

8. Ye.V.Baklanov, B.Ya.Dubetsky, V.P.Chebotayev. Preprint No. 27, IPS, Novosibirsk, 1975.

9. S.N.Bagayev, L.S.Vasilenko, V.G.Gol'dort, A.K.Dmitriyev, M.N.Skvortsov, V.P.Chebotayev, in press.

10. L.S.Vasilenko, V.P.Chebotayev, A.V.Shishayev. JETP Lett., 12, 161, 1970.

11. Physics Today, 17 July, 1974. B.Cagnac, G.Grynberg, and F.Firaben. J.Phys., 34, 645 (1973).

12. Ye.V.Baklanov, V.P.Chebotayev. Proceedings of III Vavilov Nonlinear Conf., June, 1973. Novosibirsk.
Opt.Spectr. 38, 384, 1975.

13. A.P.Kol'chenko, S.G.Rautian, R.I.Sokolovsky. JETP, 55, 1864(1968).

14. Ye.V.Baklanov, Opt. Comm., 13, 54, 1975.

15. Ye.V.Baklanov, V.P.Chebotayev . Sov. J. of Quant. El. 2, 606, 1975.

16. J.L.Hall, C.Borde. Phys.Rev.Lett., 30, 1101, 1973.

17. S.N.Bagayev, V.P.Chebotayev. JETP Lett., 16, 614, 1972.

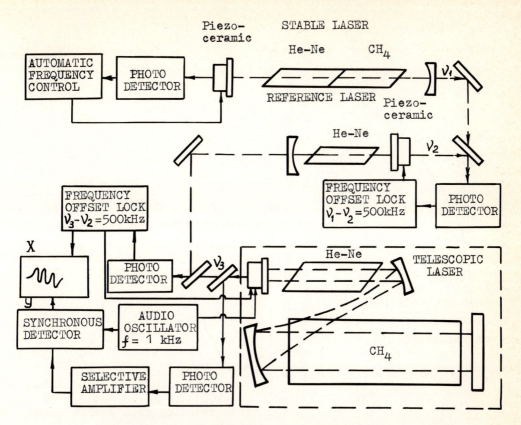

Fig.1. Experimental arrangement for observation of power resonances in telescopic laser with methane cell.

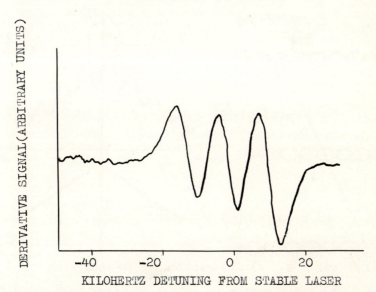

Fig.2. Recording of hyperfine structure of CH_4 line with the help of telescopic cavity it have been observed. For the first time with help of telescopic expander outside cavity have observed in /16/.

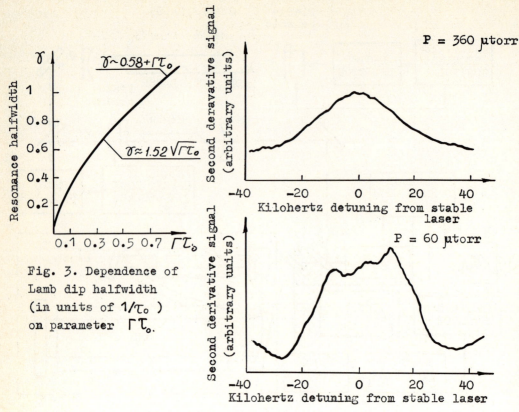

Fig. 3. Dependence of Lamb dip halfwidth (in units of $1/\tau_0$) on parameter $\Gamma\tau_0$.

Fig. 4. Shape of the second derivative function of power from frequency in He-Ne laser with CH_4 absorber at two methane pressures.

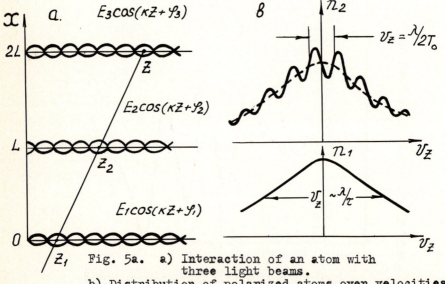

Fig. 5a. a) Interaction of an atom with three light beams.
b) Distribution of polarized atoms over velocities:
n_1 - after transit through the first beam;
n_2 - after transit through the second beam.

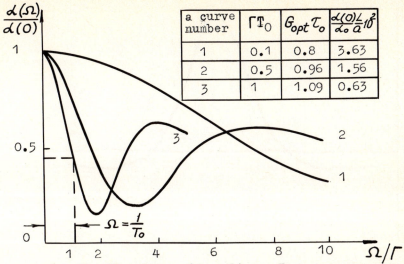

Fig. 5b. The shape of non-linear Ramsey resonance.
$\alpha(\Omega)$ is addition to absorption coefficient due to interaction of atoms with three beams; α_0 is unsaturated absorption coefficient $\alpha(0)/\alpha_0$ is relative resonance amplitude; Γ is homogeneous halfwidth of transition line.

Quadratic Doppler shift $\quad \omega' = \omega - \frac{1}{2}\frac{v^2}{c^2}\omega$

Red temperature shift $\quad \Delta_t = -\frac{\kappa T}{mc^2}\omega_0$

Shift in transit system $\quad \Delta = F \cdot \Delta_t$

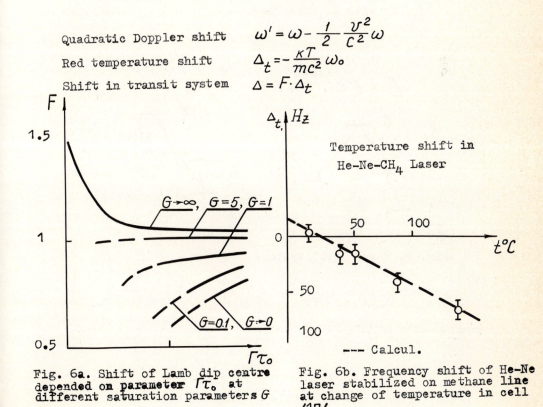

Fig. 6a. Shift of Lamb dip centre depended on parameter $\Gamma\tau_0$ at different saturation parameters G

Fig. 6b. Frequency shift of He-Ne laser stabilized on methane line at change of temperature in cell /17/.

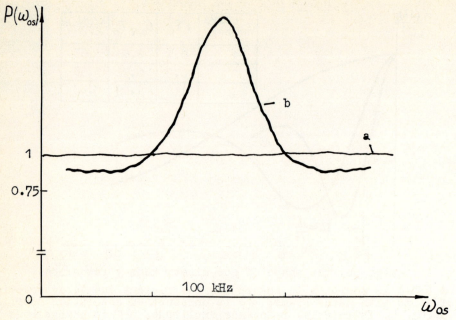

Fig. 7. Radiation spectrum of He-Ne laser at $\lambda = 3.39 \mu$
a) without absorber,
b) with CH_4 absorber.

$$\mathcal{E} = E e^{i\omega t} + E' e^{i\omega' t} + C.C.$$

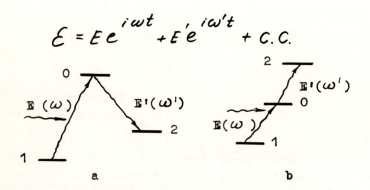

Fig. 8. Three level systems

HIGH RESOLUTION TWO-PHOTON SPECTROSCOPY

B.Cagnac
Laboratoire de Spectroscopie Hertzienne de l'ENS
Université Pierre et Marie Curie - 75230 PARIS CEDEX 05

Many two-photon spectroscopy experiments have been made since the advent of the LASER (cf references [1] and references herein). But we shall restrict our interest in this paper to high resolution spectroscopy, i.e. to experiments in vapors without Doppler broadening.

The Doppler broadening in atomic transitions is explained by the momentum conservation in the whole system "atom and photons": the energy and the momentum an atom can absorb, are bound together by a precise relation depending on the atomic velocity. In a multiphotonic transition experiment, we can choose the directions of the LASER beams in such a manner that the vectorial sum of their wave vectors $\vec{k}_i$ is zero, i.e. the whole momentum of the photons $\Sigma \hbar k_i$ is also zero. In this condition all the atoms undergo the multiphotonic transition for the same values of the photon energies, whatever their velocities may be [2].

In the case of two-photon transitions, the two vectors $\vec{k}_i$ must be equal and opposite; this can be obtained by reflecting one LASER beam back on itself with a mirror, as it was noticed by Vasilenko, Chebotaev and Shishaev [3].

With Biraben and Grynberg we did precise calculations in order to apply this phenomenon to high resolution spectroscopy [2]. In addition to polarization effects and selection rules, we showed two important points: 1°) The light power which is necessary to produce a visible effect is not too high, if the whole power is distributed on a frequency interval narrower than the natural width Γ_e corresponding to the excited level E_e; some experiments can be made with power much smaller than one Watt - 2°) In many cases the light shifts (or a.c. Stark effect) will be negligible.

These light shifts have been exhaustively studied by Cohen-Tannoudji [4]; and we can calculate, for example, the displacement δE_g of the ground state energy level E_g in terms of the matrix elements of the interaction hamiltonian $\mathcal{H}$ between the atom and each of the two oppositely travelling waves (which we assume to be identical: same intensity, polarization and circular frequency ω):

$$\delta E_g = \sum_i \frac{2 \langle g|\mathcal{H}|i \rangle \langle i|\mathcal{H}|g \rangle}{\Delta E_i}$$

$\Delta E_i = E_i - E_g - \hbar\omega$ is the energy defect of each intermediate state involved in the perturbation calculation.

On the other hand we suppose that the two-photon transition is far below the saturation, i.e. its probability $\mathcal{P}_{ge}$, at the center of the resonance ($E_e-E_g=2\hbar\omega$), is much smaller than its natural width Γ_e:

$$\mathcal{P}_{ge} = \frac{4}{\Gamma_e} \left| \sum_i \frac{2<g|\mathcal{H}|i><i|\mathcal{H}|e>}{\hbar \Delta E_i} \right|^2 << \Gamma_e$$

Taking the square root we obtain:

$$\sum_i \frac{2<g|\mathcal{H}|i><i|\mathcal{H}|e>}{\Delta E_i} << \hbar \frac{\Gamma_e}{2}$$

The first term of this inequality is very similar to the formula giving δE_g; if a particular level E_i is predominant in the calculation, and if the two matrix elements $<i|\mathcal{H}|e>$ and $<i|\mathcal{H}|g>$ are of the same order of magnitude (i.e. in other words one should compare the oscillator strengths f_{ie} and f_{gi} of the two one photon steps), this inequality leads to $\delta E_g << \hbar\Gamma_e$. These assumptions will be true in many cases (not in all cases) which allow high resolution spectroscopy.

First experiments with sodium.

The first demonstrating experiments[5] have been made in Paris and in Harvard by Levenson and Bloëmbergen on the case of sodium, because the energy levels of sodium (cf Figure 1) are in the spectral range of Rhodamin 6G, which remains until now the most efficient dye. With the Rhodamin 6G we can investigate two transitions starting from the 3S ground state: one to the 5S excited state, and another one to the 4D excited state. In the two cases, the presence of the 3P levels close to the half energy gap magnifies the two photon probability. (This magnification can be increased by using two LASERS at different wavelengths, if one accepts some residual Doppler broadening[6]).

All these levels are splitted in two sublevels, by hyperfine interaction for the S levels, by fine interaction for the 4D level (its hyperfine splitting is negligible). In consequence the transition to 4D level is splitted in 4 components. But the transition to 5S level is only splitted in 2 components because of the special selection rule $\Delta F=0$ between S levels (the spins can be affected only by the coupling with orbital momentum; and we have no coupling when the orbital momentum is null).

The detection of the two photon transitions is made by collecting photons spontaneously reemitted from the excited level: either ultraviolet photons emitted in cascade after the infra-red transition, or visible photons directly reemitted, with an interference filter or a monochromator in order to eliminate the stray light of the LASER.

The first experiments[5] used pulse-LASER; but the precision is increased by the use of a CW LASER[7] [8] [9]. Figure 2 shows a typical experimental set-up: the beam of the tunable LASER is focused into the sodium cell in order to increase the energy density, and reflected back by a concave mirror, whose center coïncides with the focus of the lens. (Remark: we do not need to use a well collimated sodium beam as in reference (10). An optical isolator is needed, between the LASER and the lens, in order to prevent interference effects of the return beam inside the LASER cavity; this isolator can use the Faraday effect (α Bromonaphtalene solution, or flint glass) or the birefringence of a quarterwave plate. A photomultiplier collects the spontaneously reemitted photons at the choosed wavelength; and the current of the photomultiplier is recorded versus the frequency of the tunable LASER; a Fabry-Pérot interferometer is used in order to scale the frequency axis.

Figure 3 shows the curves we obtained in these conditions for the transition 3S-5S in sodium: the lower curve (b) is the scaling of the Fabry-Pérot. The upper curve (a) corresponds to the photomultiplier current; it shows two narrow peaks corresponding respectively to F=1 and F=2 transitions. The width of the peaks, less than 10MHz, is mostly due to the frequency jitter of the LASER: the total natural width due to the lifetime and to the transit time through the focus is less than 3 MHz: but the Doppler width would be of the order of 2000MHz. (The LASER power is about 30mW, and the atomic density of the order of 10^{13} atoms/cm^3; this figure 3 is an improvement of the results of reference (9)).

Similar studies have been made for the transition 3S-4D[7] [8]. The upper curve of figure 4 shows the four narrow peaks corresponding to the transitions:
- from 3S,F=2 to 4D,J=5/2 (a) and to 4D,J=3/2 (b)
- from 3S,F=1 to 4D,J=5/2 (c) and to 4D,J=3/2 (d)

On this figure the frequency scale is more confined; and that permits to see the small Doppler back-ground due to the absorption of two photons of the same travelling wave. In agreement with theory[2] the total area under the 4 peaks is almost twice the area under the Doppler background.

Applications of the method

The applications of this new spectroscopic method are numerous:

1/ Precise measurements of <u>fine and hyperfine splittings</u>. The previously reported experiments have permitted preliminary measurements:

of the fine splitting of 4D level: 1028.5 ± 3 MHz

of the hyperfine splitting of the 5S level: 150 ± 10 MHz.

The first result is ten times more precise than precedent measurement by conventional method; the second result was unknown. Both measurements agree with recent measurements on atomic beams[10] [11] and level crossing[12]; they can be still improved by one order of magnitude. Other higher excited levels 6S and 5D have been investigated by the Harvard team[13].

2/ Study of the splitting of the levels in low external fields: <u>Zeeman effect</u> in magnetic fields[8] [14] or <u>Stark effect</u> in electric fields[15].

The second curve of figure 4 shows the Zeeman pattern obtained for the 4D transition in a weak magnetic field, 170 Gauss, with the same circular polarization σ^+

for the two oppositely travelling wave; i.e. we select the transitions δm=2. This curve fit very well with the theoretical pattern (positions and intensities) represented on the lower diagram of the figure 4.

The upper curve of figure 5 shows the Stark pattern obtained in Stanford for the same transition in a electric field of 2.5kV/cm. It can be compared with the four normal transitions recorded without external field on the lower curve of figure 5: this experiment gives not only the Stark splitting, but also the absolute Stark shift. A theoretical interpretation permits to obtain the polarizabilities of the two 4D levels.

3/ Study of <u>collisional broadening and shift at low pressure</u>[16] is illustrated on figures 6 and 7. The upper curve of figure 6 shows the four transitions to the 4D level recorded with a sodium cell enclosing a foreign gas (Neon at the pressure of 1 Torr); you can compare it with the lower curve recorded in a cell of pure sodium and you note the broadening of the optical transitions due to collisions.

In this experiment the LASER beam is splitted and irradiate simultaneously two sodium cells, one with foreign gas and the other without foreign gas. The signals obtained from the two cells are simultaneously recorded; and the relative positions of the peaks of the two curves have an actual signification. The figure 7 shows another recording of the first component only (labelled a) with higher gas pressure (5 Torr) and with a larger frequency scale; the pressure shift obviously appears. These preliminary experiments with the Neon give for the broadening the value 32±5 MHz/Torr and for the shift the value -(7±1)MHz/Torr.

4/ <u>The selective population of one single level</u>, even if other levels are close to it, at distance smaller than the Doppler width, opens other possibilities. One example is given by the investigation of collisional transfer between the two fine sublevels (J=3/2 and 5/2) of 4D level[17]. For the detection of the two photon transitions we can select with a monochromator the line 5682Å, which is emitted only from the sublevel $4D_{3/2}$ (the selection rules forbid the transition from $4D_{5/2}$ to $3P_{1/2}$); in this condition we can observe only the transitions (b) and (d) as you see on the figure 8A recorded with a cell of pure sodium. But the introduction of low pressure of foreign gas (0.2 Torr Neon for figure 8B; 1 Torr Neon for figure 8C) induces the appearance of the transitions (a) and (c). The intensities of the transitions (a) and (c) are proportional to the number of sodium atoms transferred from the $4D_{5/2}$ level to the $4D_{3/2}$ level; one deduces the cross section of this collisional transfer $2.7±0.9.10^{-14} cm^2$.

5/ Precise measurements of <u>isotopic shifts</u> as we will see further with two photon transitions on Neon atoms.

6/ <u>Precise measurements of atomic energy levels</u> and all the metrologic applications, as it will be illustrated by the experiment on hydrogen. In the future the stabilization of dye LASER on two photon transitions will furnish many secondary standards of wavelength, and may be the primary standard.

Experiments with Neon

Neon is also a favourable case for two photon spectroscopy, because Neon, as well as other rare gas atoms, has many levels; and often one of them can be used as as intermediate level, very closed to the half energy gap. Figure 9 shows the energy levels involved in the experiments made in our group with the collaboration of Mrs Giacobino[18].

The initial state for the two photon transition is the metastable 3s,J=2 state. The atom is excited from the ground state to this metastable state by a weak discharge (one obtains about 10^{11} metastable /cm^3 in a cell where the pressure is about 1 Torr). The atom is excited to the final state 4d',J=3 by absorption of two photons at 5923Å. Two energy levels lie close to the half energy gap and permit to observe the two photon transitions with a C.W. LASER, whose power is about 50mW. The resonance is detected by collecting photons emitted from the excited state to the 3p',J=2 state at 5902Å.

The signal observed from the natural Neon is shown on the lower curve of the figure 10: we see the two peaks corresponding to the even isotopes ^{20}Ne and ^{22}Ne. The scaling of the Fabry-Pérot interferometer (upper curve) permits to measure the isotopic shift: 2780.0±2.5 MHz (twice the LASER frequency splitting). Figure 11 shows the signal obtained from a cell filled with the odd isotope ^{21}Ne. The hyperfine components due to the nuclear spin can be compared with the theoretical intensies and fitted positions plotted below. Other levels can be studied in Neon.

Experiments with hydrogen

Several authors had noticed the interest to study the transition from the 1S ground state to the 2S metastable state in the hydrogen atom, with two photons at 2430Å (twice the wavelength of the Lyman α line)[2][7][19]. This experiment has been realized recently by Hänsch and co-workers in Stanford[20]; it needs much more power than the Sodium or Neon experiments. A coumarine dye LASER at 4860Å is pumped by a nitrogen LASER; the frequency is doubled with a Lithium formate crystal. The metastable atoms are detected by collisional transfer from 2S metastable state to 2P state, from which they emit the Lyman α line. Figure 12 shows the preliminary results obtained in Stanford recording the intensity of the Lyman α fluorescence versus the dye LASER frequency. The widths of the two hyperfine components are limited only by the dye LASER line width of about 120 MHz (mostly due to the pulse duration). By direct comparison with the Balmer β line, the Lamb shift of the 1S ground state has been measured.

Many other applications are possible even in molecular spectroscopy, although the oscillator strengths are small in the molecules. Two photon transitions have been observed for example in NO and CH₃F molecules[21][22]. The variety of the applications will depend of the technical improvements of the tunable LASER.

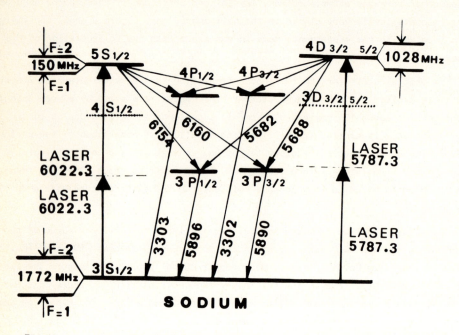

FIGURE 1: Lower energy levels of sodium. The wavelengths of the indicated transitions are given in Angström units.

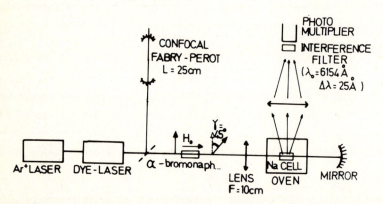

FIGURE 2: Experimental set up for the study of Doppler free two photon transitions in sodium using a CW dye LASER.

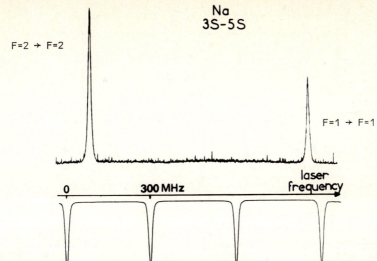

FIGURE 3: Two photon transition 3S-5S in sodium. The current of the photomultiplier is plotted versus the LASER frequency. The signal of a Fabry Pérot interferometer (lower curve) permits to scale the frequency axis.

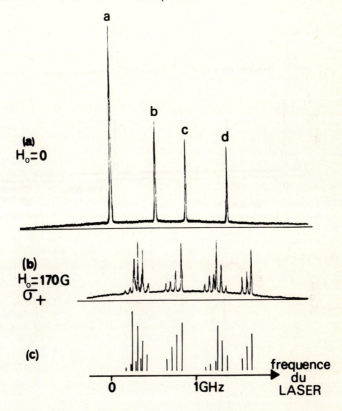

FIGURE 4: Zeeman effect on the two photon transition 3S-4D in sodium - (a) without magnetic field - (b) in a low magnetic field - (c) theoretical pattern (positions and intensities) in the same magnetic field.

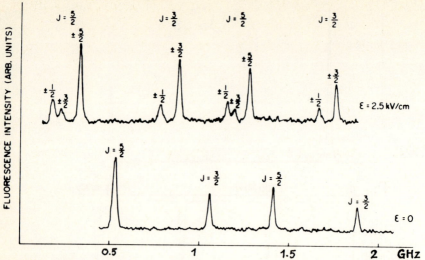

FIGURE 5: Stark effect on the transition 3S-4D in sodium: lower curve without electrical field - upper curve with an electrical field 2.5kV/cm (reference 15)

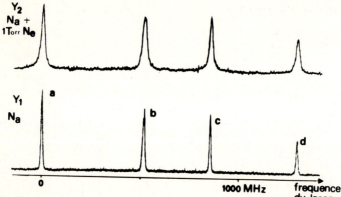

FIGURE 6: Collisional broadening of the two photon transitions 3S-4D in sodium: lower curve Y_1 without foreign gas - upper curve Y_2 with a sodium cell containing 1 Torr of Neon.

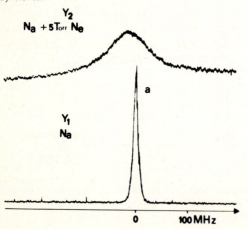

FIGURE 7: Collisional shift of the same transition as in figure 6. The component (a) only is shown with enlarged scale. Y_1 cell of pure sodium - Y_2 sodium cell containing 5 Torr of Neon.

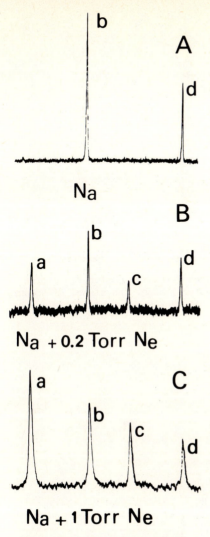

FIGURE 8: Collisional transfer between the two levels 4D. Same experiment as in figure 6, but one selects the wavelength emitted only by the level J=3/2.
A - cell of pure sodium
B - sodium cell containing 0.2 Torr of Neon
C - sodium cell containing 1 Torr of Neon

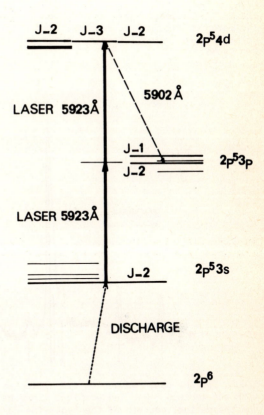

FIGURE 9 : some energy levels of the neon atom, showing the two photon transition studied.

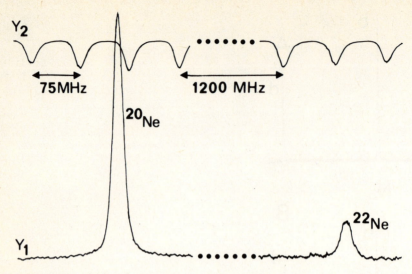

FIGURE 10: Isotopic shift of the even isotopes of Neon. The two photon transition indicated on figure 9 is recorded with a cell of natural Neon. The signal of a Fabry Pérot interferometer (upper curve Y_2) permits to measure the exact splitting of the two peaks.

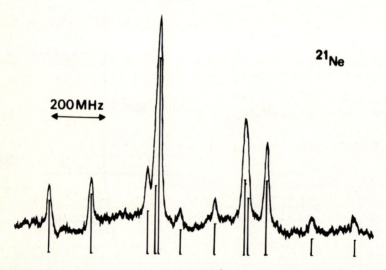

FIGURE 11: The same two photon transition as in figure 10 observed on a cell filled with odd isotope ^{21}Ne. The vertical lines below the curve indicate the best fit of the theoretical pattern of the hyperfine components.

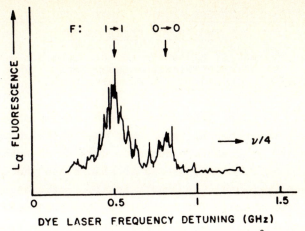

FIGURE 12: Two photon transition 1S-2S in Hydrogen (λ = 2430.7 Å) (reference 20)

REFERENCES

(1) I.D.ABELLA - Phys.Rev.Lett. 9 (1962) 453
 A.M.BONCH-BRUEVICH and V.A.KHODOVOI - Sov.Phys.Usp. 85 (1965) 3
 A.GOLD - "Two Photons Spectroscopy" in the course of the Enrico Formi School (Varenna 1967).
 J.M.WORLOCK - in Laser Handbook (North Holland) 1972, tome II, p.1323
(2) B.CAGNAC, G.GRYNBERG and F.BIRABEN - J.Physique 34 (1973) 845.
(3) L.S.VASILENKO, V.P.CHEBOTAEV and A.V.SHISHAEV - JETP Lett. 12(1970) 161
(4) C.COHEN-TANNOUDJI - Thèse, Annls de Physique 7 (1962) 423, 469
 E.B.ALEXANDROV, A.M.BONCH-BRUEVICH, N.N.KOSTIN and V.A.KHODOVOI - JETP Lett. 3 (1966) 53
 P.PLATZ - Appl.Phys.Lett. 14 (1969) 168
 C.COHEN TANNOUDJI and J.DUPONT-ROC - Phys.Rev. A5 (1972) 968
(5) F.BIRABEN, B.CAGNAC and G.GRYNBERG - Phys.Rev.Lett. 32 (1974) 643
 M.D.LEVENSON and N.BLOEMBERGEN - Phys.Rev.Lett. 32 (1974) 645
(6) J.E.BJORKHOLM and P.F.LIAO - Phys.Rev.Lett. 33 (1974) 128
(7) T.W.HANSCH, K.HARVEY, G.MEISEL and A.L.SCHAWLOW - Opt.Comm. 11 (1974) 50
(8) F.BIRABEN, B.CAGNAC and G.GRYNBERG - C.R.Acad.Sc. Paris 279 (1974) B 51 et Phys. Lett. 48A (1974) 469
(9) F.BIRABEN, B.CAGNAC and G.GRYNBERG - Phys.Lett. 49A (1974) 71
(10) D.PRITCHARD, J.APT and T.W.DUCAS - Phys.Rev.Lett. 32 (1974) 641
(11) H.T.DUONG, S.LIBERMAN, J.PINARD and J.L.VIALLE - Phys.Rev.Lett. 33 (1974) 339
(12) K.FREDRIKSSON and S.SVANBERG - Phys.Lett. 53A (1975) 61
(13) M.D.LEVENSON and M.M.SALOUR - Phys.Lett. 48A (1974) 331
(14) N.BLOEMBERGEN, M.D.LEVENSON and M.M.SALOUR - Phys.Rev.Lett. 32 (1974) 867
(15) K.C.HARVEY, R.T.HAWKINS, G.MEISEL and A.L.SCHAWLOW - Phys.Rev.Lett. 34 (1975) 1073
(16) F.BIRABEN, B.CAGNAC and G.GRYNBERG - J. de Phys. Lettres 36 (1975) 41
(17) F.BIRABEN, B.CAGNAC and G.GRYNBERG - C.R.Acad.Sc. Paris 280 (1975) B 235
(18) F.BIRABEN, E.GIACOBINO and G.GRYNBERG - Phys.Rev.Lett. (to be published)
(19) E.V.BAKLANOV and V.P.CHEBOTAEV - Optics Comm. 12 (1974) 312.
(20) T.W.HANSCH, S.A.LEE, R.WALLENSTEIN and C.WIEMAN - Phys.Rev.Lett. 34 (1975) 307
(21) R.G.BRAY, R.M.HOCHSTRASSER and J.E.WESSEL - Chem.Phys.Lett. 27 (1974) 167
(22) W.K.BISCHEL, P.J.KELLY and C.K.RHODES - Phys.Rev.Lett. 34 (1975) 300

OPTICALLY INDUCED ATOMIC ENERGY LEVEL SHIFTS AND TWO-PHOTON SPECTROSCOPY

J. E. Bjorkholm and P. F. Liao
Bell Telephone Laboratories
Holmdel, New Jersey   07733

During the last year and one-half there has been a great deal of interest in experimental techniques which allow Doppler broadening to be eliminated or greatly reduced in two-photon spectroscopy. In these techniques, atoms are caused to absorb one photon from each of two laser beams of equal, or nearly equal, frequencies which propagate in opposite directions. The basic ideas involved were pointed out in 1970[1], however it wasn't until early 1974 that the first experimental demonstrations were carried out.[2,3,4] Since then, the counter-propagating beams techniques have been used to carry out a wide variety of high resolution measurements.

Long-lived excited states are accessible via two-photon absorption making it possible, in principle, to achieve extremely narrow linewidths using Doppler-free two-photon spectroscopy. Thus, ultra-precise measurements are an attractive possibility. Caution must be used in carrying out such measurements, however, since the light used to induce the two-photon transitions can also cause the atomic energy levels to be shifted.[*] These shifts are known as the ac Stark effect[5] and they are due to virtual transitions between atomic levels caused by nonresonant light. Because two-photon transitions generally involve virtual transitions to an intermediate atomic state, such level shifts are <u>intrinsic</u> to two-photon processes (and to other multiphoton processes also).

Previous observations of the ac Stark effect in optical transitions[6] have required high-power lasers since Doppler broadening limited the available resolution. Recently, however, we have observed ac Stark shifts of several hundred MHz[7] using low-power, cw dye lasers and two-photon spectroscopy with greatly reduced Doppler broadening. Thus, when carrying out high resolution two-photon spectroscopy (even with low-power lasers), it is important to account for energy level shifts caused by the ac Stark effect. In many instances such level shifts simply cannot be neglected.

1. The ac Stark Effect

Optically induced shifts of atomic energy levels is a subject about which there has been considerable interest. These energy level shifts have probably been most extensively studied in optical double resonance experiments in which the radiation that optically pumped the atomic vapor also produced "light shifts".[8] Such shifts were observed as small changes (several hundred Hz) in the microwave or rf transition frequencies between various hyperfine components of the atomic ground state. Earlier studies of level shifts produced by electromagnetic radiation considered the shifts produced by radio frequency[9] and microwave radiation[10].

The shift of an atomic level n, induced by the optical field $\vec{E} = \mathrm{Re}\vec{E}_o e^{-i(kz-\omega t)}$ is given by perturbation theory as[5,11]

$$\Delta E_n = h\delta\nu_n = \frac{1}{4} \sum_m \left\{ \frac{|\vec{P}_{mn} \cdot \vec{E}_o|^2}{E_n - E_m - \hbar\omega} + \frac{|\vec{P}_{mn} \cdot \vec{E}_o|^2}{E_n - E_m + \hbar\omega} \right\} . \qquad (1)$$

The summation is taken over all unperturbed atomic states $|m\rangle$ having the unperturbed energy E_m, and $\vec{P}_{mn}$ is the electric-dipole matrix element between states m and n. Natural damping of the states has been neglected so that Eq. 1 is applicable only as long as 1/h times the energy denominator is large compared to the natural linewidth of the levels.

For simplicity we now restrict our attention to the two-level atom shown in Fig. 1. Equation (1) then reduces to

$$\delta\nu = \frac{1}{4} \frac{|\vec{P}_{12} \cdot \vec{E}_o|^2}{h^2(\nu_o - \nu)} , \qquad (2)$$

where $h\nu_o = E_2 - E_1$ and ν is the frequency of the applied light. The two levels experience equal, but opposite, shifts. For $\nu < \nu_o$, the

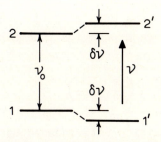

Fig. 1. Energy level shift for the case $\nu_o > \nu$, where $h\nu_o$ is the energy separation of the unperturbed energy levels 1 and 2 and ν is the frequency of the applied light. The levels 1' and 2' are the perturbed levels which have been shifted by $h\delta\nu$, where $\delta\nu$ is given by Eq. (2).

energy separation between the levels is increased by the perturbation; for $\nu > \nu_o$, it is decreased.

2. A Simple Model

In order to gain physical insight into the nature of optically induced energy level shifts, we use the following simple model to obtain Eq. (2) from considerations based on energy conservation.

Consider two systems. One is the unperturbed atom having energy levels 1 and 2 separated in energy by $h\nu_o$. The other system is a linearly polarized radiation field composed of N photons of energy $h\nu$. In the absence of an interaction between the two systems the atom is in its ground state 1 and the total energy of the two systems is $E_1 + Nh\nu$. Now let the two systems interact. The light can now induce an atomic transition from level 1 to level 2 with the absorption of a photon. Because $h\nu \neq h\nu_o$, energy is not conserved and the transition must be a virtual transition; i.e., the atom can only stay in level 2 for a time duration on the order of $1/(\nu_o-\nu)$, as allowed by the uncertainty principle. Nonetheless, the atom can spend the fraction f of its time in level 2, during which time (N-1) photons are present in the combined system. We now assume symmetric energy level shifts as shown in Fig. 1; we further assume that the sum of the energies of the two systems when interacting is equal to the sum of their energies when not interacting. We obtain

$$E_1 + Nh\nu = (1-f)[E_1-h\delta\nu+Nh\nu] + f[E_1+h\nu_o+h\delta\nu+(N-1)h\nu] ,$$

which yields

$$\delta\nu = \frac{f}{1-2f}(\nu_o-\nu) . \qquad (3)$$

Clearly the atomic energy levels must shift in order to conserve energy on average. Using a simple rate equation approach, one obtains the following expression for f;

$$f = \frac{1}{2}\left(\frac{1}{1+1/2W\tau}\right) . \qquad (4)$$

In the above, W is the transition rate induced by the radiation field of intensity I and τ is the spontaneous lifetime of the atomic system. These quantities are given by[12]

$$W = \frac{\lambda^2 I}{8\pi h \nu \tau} \frac{1}{2\pi^2 \tau \left[(\nu_o - \nu)^2 + \left(\frac{1}{2\pi\tau}\right)^2\right]} , \qquad (5)$$

$$I = \frac{c}{8\pi} E_o^2 , \qquad (6)$$

and

$$\frac{1}{\tau} = \frac{16\pi^3 \nu^3 P_{12}^2}{\hbar c^3} . \qquad (7)$$

Substitution of equations 4-7 into Eq. 3 yields

$$\delta\nu = \frac{P_{12}^2 E_o^2}{4h^2} \frac{(\nu_o - \nu)}{\left[(\nu_o - \nu)^2 + \left(\frac{1}{2\pi\tau}\right)^2\right]} , \qquad (8)$$

which is identical to Eq. (2) for $|\nu_o - \nu| \gg 1/2\pi\tau$. This analysis clearly indicates the source of the level shifts to be the non-energy-conserving virtual transitions.

3. Energy Level Shifts and Two-Photon Absorption

The previous section has demonstrated the intimate connection between optically induced energy level shifts and virtual transitions. Two-photon absorption is also intimately connected with virtual transitions. One can consider the two-photon absorption process as being two sequential single-photon virtual transitions. First there is a virtual transition from the ground state g to an intermediate state; this is followed by a second virtual transition from the intermediate state to the final state f. Associated with each virtual transition are energy level shifts; these result in a shift of the two-photon g → f transition frequency. These energy level shifts are therefore <u>intrinsic</u> to the two-photon absorption process.

In the experiments which will be described in the next section, the energy level shifts were made large by decreasing the size of one of the energy denominators in Eq. (1). In doing so, we also resonantly enhanced[13] the two-photon transition rate W_2, which for linearly polarized light, is given by

$$W_2 = A I_1 I_2 \left| \sum_m \langle f|z|m\rangle\langle m|z|g\rangle \left\{ \frac{1}{E_m - h\nu_1} + \frac{1}{E_m - h\nu_2} \right\} \right|^2 . \qquad (9)$$

In this equation, the summation is carried out over all intermediate states of energy E_m, A is a constant, I_1 and I_2 are the light intensities at the frequencies ν_1 and ν_2, respectively, and $h(\nu_1+\nu_2) = E_f - E_g$. Equations (9) and (1) are of a similar form, and the two-photon g → f transition rate is related to the energy level shifts $\delta\nu_g$ and $\delta\nu_f$. For the special case in which only one intermediate state is of importance (applicable to our experiments), we find

$$W_2(g \to f) \propto - (\delta\nu_g)(\delta\nu_f) \ . \tag{10}$$

This shows that the energy level shifts will be large whenever the two-photon transition rate is large, regardless of the amount of resonant enhancement. Also notice that since W_2 is proportional to the product of $\delta\nu_g$ and $\delta\nu_f$, if the two level shifts are significantly different, it is possible to have a large optically induced shift of the two-photon transition frequency even though W_2 is small. Thus, large shifts of the two-photon resonance frequency can occur in the absence of saturation of the two-photon transition.

3. The Experiment

We have made measurements of atomic energy level shifts associated with two-photon absorption in a vapor of atomic sodium.[7] The counter-propagating beams from two cw dye lasers operating at different frequencies induced two-photon transitions from the 3S ground state to the 4D excited state. The lasers were tuned such that either the $3P_{3/2}$ or the $3P_{1/2}$ intermediate state was nearly resonant ($\lambda_2 \sim 589$ nm and $\lambda_1 \sim 569$ nm) as shown in Fig. 2. The transitions were monitored by detecting the 330 nm fluorescence (4P-3S transition) resulting from the decay from the 4D levels. In making an experimental run the frequency ν_2 was held constant while ν_1 was swept repetitively through the various two-photon resonances. Energy level shifts of the 3S and 4D states were unambiguously observed as shifts in the resonant frequencies of the two-photon transitions as the laser intensities were changed. The resonance denominators in Eq. (1) insured that the shifts of the 3S levels were primarily induced by the light at 589 nm, while the 4D levels were shifted by the 569 nm light. An example of our data is shown in Fig. 3a where the 3S level is shifted 745 MHz by 26 mW of 589 nm light. Since the level shift is proportional to light intensity, spatial nonuniformities of the intensity within the region of observation will cause the observed two-photon absorption lines to be broadened and distorted. An example is shown in Fig. 3b where the 4D level is shifted

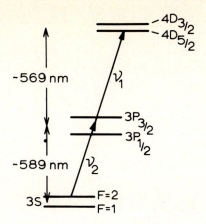

Fig. 2. The pertinent energy levels of atomic sodium. The optical frequency ν_2 is near the 3S → 3P transition frequency while ν_1 is near the 3P → 4D transition frequency.

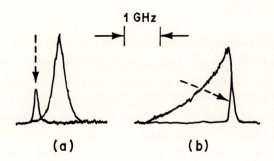

Fig. 3. Oscillographs of the 3S(F=2) → $4D_{5/2}$ two-photon absorption line. The frequency ν_1 increases to the right. Each contains two lines, one obtained with both lasers at low (<3 mW) power (shown by arrows) and the other obtained by increasing (a) only the 589-nm power to 26 mW, and (b) only the 569-nm power to 30 mW.

using 30 mW of sharply focused 569 nm light. The shifts caused by the 3S and 4D level shifts are in opposite directions, as expected.

In Fig. 3a it is apparent that, in addition to the level shift, there is line broadening. This broadening is not due to spatial nonuniformities. Instead it occurs because of Doppler effects; different atomic velocity groups experience different level shifts since the applied light has a different apparent frequency for each velocity group. To understand this, we calculate the two-photon resonance condition for the group of atoms moving with speed v along the direction of propagation for light at ν_1. Once again, specialize to the case where only a single intermediate state, n, need be considered. We follow the procedure used in Ref. 14, but include the level shifts and assume that $|\delta\nu/(\nu_o-\nu)| \ll 1$. The resonance condition is found to be

$$\nu_1 + \nu_2 = \frac{1}{h}(E_f - E_g) + \frac{|\vec{P}_{ng}\cdot\vec{E}_2|^2 - |\vec{P}_{nf}\cdot\vec{E}_1|^2}{4h^2(\nu_{ng}-\nu_2)}$$

$$- \frac{v}{c}\left[(\nu_2-\nu_1) + \frac{\nu_1|\vec{P}_{nf}\cdot\vec{E}_1|^2 - \nu_2|\vec{P}_{ng}\cdot\vec{E}_2|^2}{4h^2(\nu_{ng}-\nu_2)^2}\right] . \quad (11)$$

The shift of the two-photon resonance frequency is given by the second term and its size is $|\delta\nu_{go}| - |\delta\nu_{fo}|$, the difference between the shifts for the zero-velocity group. Clearly the shift of the two-photon resonance frequency is zero if $|\delta\nu_{go}| = |\delta\nu_{fo}|$. The width of the two-photon line is determined by the term in brackets and, in most cases, is proportional to it.[14] The quantity $(\nu_2-\nu_1)$ is proportional to the residual Doppler broadening (which exists because $\nu_1 \neq \nu_2$); the rest of the term gives the broadening due to the level shifts. Notice that the level shifts can produce line narrowing as well as line broadening. For the conditions of Fig. 3a only broadening was possible.

The intensity and frequency dependence of the 3S level shift, for near resonance with the $3P_{3/2}$ level, was measured and found to be $\delta\nu \approx 1.4 \times 10^{15} I_2/(\nu_o-\nu_2)$, where I_2 is in units of W/cm^2. The functional form is as given by Eq. (2) and the numerical factor agrees well with theory.

4. Conclusions

Because two-photon transitions involve virtual transitions, optically induced energy level shifts are always present in two-photon

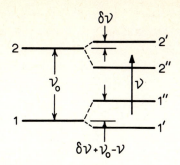

Fig. 4. The unperturbed energy levels 1 and 2, separated in energy by $h\nu_o$, are ac Stark shifted and split into the primed and double-primed levels by the application of light at the frequency ν, where $\nu < \nu_o$. The shifts and splitting are symmetric and $\delta\nu$ is given by Eq. (2).

absorption spectra. Whether the shift is significant depends upon its magnitude relative to the experimental resolution required. Large energy level shifts are not necessarily associated with resonant enhancement; they are, however, directly connected with high two-photon transition rates. In our experiments, resonant enhancement was used to obtain large two-photon transition rates, and measurable energy level shifts, at low laser power levels.

Energy level shifts may also be put to good use. For instance, the intensity dependence of the shifts provides an accurate method for measuring transition dipole moments. In fact, such a measurement has very recently been carried out in high resolution two-photon absorption spectroscopy of NH_3 using two infra-red lasers, resonant enhancement, and opposed-beam techniques.[15]

In high-resolution experiments where optically induced energy level shifts are significant, it may be necessary to take spectra at several intensity levels and then extrapolate to the case of zero shift. In this situation it is important to point out that at high light intensities Eq. (2) does not adequately describe the level shift. A more general expression for the shift is given by[5]

$$\frac{\delta\nu}{(\nu_o-\nu)} = \frac{1}{2}\left(1 + \frac{|\vec{P}_{12}\cdot\vec{E}_o|^2}{h^2(\nu_o-\nu)^2}\right)^{1/2} - \frac{1}{2}. \qquad (12)$$

In the limit $|\vec{P}_{12}\cdot\vec{E}_o|^2/h^2(\nu_o-\nu)^2 \ll 1$, this expression reduces to that given by Eq. (2); in the opposite limit the shift becomes linearly proportional to E_o and independent of $(\nu_o-\nu)$.

In actuality, the ac Stark effect splits an atomic level into two levels having mixed character. With reference to Fig. 4, the location of level 1' is given by Eq. (12); the location of the additional level 1" is given by[5]

$$\frac{\delta v''}{(v_o - v)} = -\frac{1}{2}\left(1 + \frac{|\vec{P}_{12} \cdot \vec{E}_o|^2}{h^2(v_o - v)^2}\right)^{1/2} - \frac{1}{2}. \tag{13}$$

This level can be detected via two-photon spectroscopy. However, even if Doppler-free techniques are employed, the two-photon line for the level 1" will remain Doppler broadened. This may possibly be the explanation for the broad, weak feature observed in our earlier experiments[13] (which were tentatively attributed to a two-step absorption process involving collisions).

Presently, we are making measurements of optically induced energy level shifts in the high intensity regime described by Eq. (12) and we are investigating the splitting of the levels indicated by Eq. (13).

Acknowledgment:

The authors are indebted to J. P. Gordon of Bell Laboratories for many stimulating discussions and to W. Happer of Columbia University and P. Berman of New York University for helpful discussions with regards to the ac Stark splitting.

References

1. L. S. Vasilenko, V. P. Chebotaev, and A. V. Shishaev, Pis'ma Zh. Eksp. Teor. Fiz. 12, 161 (1970) [JETP Lett. 12, 113 (1970)].

2. F. Biraben, B. Cagnac, and G. Grynberg, Phys. Rev. Lett. 32, 643 (1974).

3. M. D. Levenson and N. Bloembergen, Phys. Rev. Lett. 32, 645 (1974).

4. T. W. Hänsch, K. C. Harvey, G. Meisel, and A. L. Schawlow, Opt. Commun. 11, 50 (1974).

5. For a comprehensive review of ac Stark effect theory and early experiments see A. M. Bonch-Bruevich and V. A. Khodovoi, Usp. Fiz. Nauk. 93, 71 (1967) [Sov. Phys. Usp. 10, 637 (1968)].

6. A. M. Bonch-Bruevich, N. N. Kostin, V. A. Khodovoi, and V. V. Khromov, Zh. Eksp. Teor. Fiz. 56, 144 (1969) [Sov. Phys. JETP 29, 82 (1969)]; Peter Platz, Appl. Phys. Lett. 14, 168 (1969); B. Dubreuil, P. Ranson, and J. Chapelle, Phys. Lett. 42A, 323 (1972).

7. P. F. Liao and J. E. Bjorkholm, Phys. Rev. Lett. $\underline{34}$, 1 (1975).

8. For a recent review see W. Happer, "Progress in Quantum Electronics" (Pergamon Press, Oxford, 1971), Vol. 1, Pt. 2, pg. 51.

9. F. Bloch and A. Siegert, Phys. Rev. $\underline{57}$, 522 (1940).

10. S. H. Autler and C. H. Townes, Phys. Rev. $\underline{100}$, 703 (1955).

11. Masataka Mizushima, Phys. Rev. $\underline{133}$, A414 (1964).

12. For example, see Amnon Yariv, "Quantum Electronics" (John Wiley and Sons, New York, 1967). Our expression for $1/\tau$ differs by a factor of 2 from that given on pg. 206 of the above. This factor results because we consider an atom which is driven by a linearly polarized field. Such an atom will emit light of only one polarization; hence the density of states is reduced by a factor of 2.

13. J. E. Bjorkholm and P. F. Liao, Phys. Rev. Lett. $\underline{33}$, 128 (1974).

14. J. E. Bjorkholm and P. F. Liao, IEEE J. Quantum Electron. $\underline{QE-10}$, 906 (1974).

15. W. K. Bischel, P. J. Kelly, and C. K. Rhodes, Talk 7.2 presented at the Conference on Laser Engineering and Applications, May 28-30, 1975, Washington, D.C., U.S.A.

* Theoretical aspects of level shifts in two-photon absorption have been considered by B. Cagnac, G. Grynberg, and F. Biraben, J. Phys. (Paris) $\underline{34}$, 845 (1973), and P.L. Kelley, H. Kildal, and H.R. Schlossberg, Chem.Phys.Lett. $\underline{27}$, 62 (1947).

INFRARED LASER STARK SPECTROSCOPY

Yoshifumi UEDA and Koichi SHIMODA
Department of Physics
University of Tokyo
Bunkyo-ku, Tokyo 113, JAPAN

§1 Introduction

Two salient features of Stark spectroscopy using CO_2 and N_2O lasers are its high accuracy in frequency determination of absorption liens and extremely high sensitivety in detection of very weak transitions. Its deficiency is that the operating frequency is restricted in the region from 900 cm^{-1} to 1100 cm^{-1}.

In order to clarify its applicability or non-applicability to a systematic study of infrared spectra of molecules, let us start with an assessment that what molecules might profitably be studied by using the Stark spectroscopy with CO_2 and N_2O lasers. The results are summarized in Table I. Although it was not intended to be complete in the preparation of the table, an extensive search was made in "Tables of Molecular Vibrational Frequencies" edited by Shimanouchi[1], which includes 336 entries of molecules, their fundamental vibrational frequencies, mode symmetries, and other information.

When a molecule satisfies the following conditions, it should be a good candidate for a systematic study of Stark spectroscopy using CO_2 and N_2O lasers:

The first condition is that it must be a light molecule having a molecular weight of less than about 60. If the molecular weight is greater than this, its rotational transitions in the vibrationally excited states of as high as 1000 cm^{-1} above the ground state can usually be observed to an accuracy of ±100 kHz by a modern microwave spectrometer of high sensitivety. When the molecular weight is less than about 30 and consequently the rotational constants are larger than about 30 GHz, rotational transitions of the molecule will appear in the far-infrared, where precision spectroscopy has not yet been well developed. Rotational analysis of the spectrum of light molecules can be performed favorably by Stark spectroscopy using CO_2 and N_2O lasers; it spans a wide frequency range of 6 THz and has a sensitivity higher than microwave spectrometers.

The second condition is that it must have a significant Stark shift. For a symmetric top molecule, this means that it must have at least an appreciable electric dipole moment. Since the CO_2 and N_2O lasers oscillate at an interval of 1 cm^{-1} to 2 cm^{-1} on the average in the range from 900 cm^{-1} to 1100 cm^{-1}, the spectral lines of molecules which have an electric dipole moment of about 1 debye can somehow be observed

Table I Molecules on which Strak spectroscopic study is feasible using CO_2 and N_2O lasers

Molecule	Mol. wt.	Mol. sym.	Infrared active bands and their symmetry	A_0	B_0	C_0	Dipole moment (Debye)
NT_3	23	C_{3v}	$\nu_4(e)$ 996				
H_2CO	30	C_{2v}	$\nu_6(b_2)$ 1167	9.4	1.3	1.1	μ_a=2.34
HDCO	31	C_s	$\nu_5(a')$ 1041; ν_6 1074	6.6	1.16	0.99	
D_2CO	32	C_{2v}	$\nu_3(a_1)$ 1106; $\nu_5(b_1)$ 990; $\nu_6(b_2)$ 938	4.7	1.08	0.87	
CH_3NH_2	31	C_s	$\nu_8(a')$ 1044	3.53	0.75	0.72	μ_a=0.304; μ_c=1.297
CH_3ND_2	33	C_s	$\nu_8(a')$ 997; $\nu_7(a')$ 1117	1.75	0.55	0.52	μ_a=0.265; μ_c=1.299
CD_3NH_2	34	C_s	$\nu_8(a')$ 973; $\nu_7(a')$ 913; $\nu_{12}(a'')$ 1077				
CD_3ND_2	36	C_s	$\nu_6(a')$ 1123; $\nu_7(a')$ 880; $\nu_8(a')$ 942; $\nu_{12}(a'')$ 1077; $\nu_{14}(a'')$ 910	1.74	0.55	0.52	μ_a=0.265; μ_c=1.299
CH_3OH	32	C_s	$\nu_7(a')$ 1060; $\nu_8(a')$ 1033	4.25	0.82	0.79	μ_b=1.440; μ_c=0.885
CH_3OD	33	C_s	$\nu_8(a')$ 1040	3.65	0.78	0.73	
CD_3OH	34	C_s	$\nu_4(a')$ 1024; $\nu_8(a')$ 988; $\nu_{11}(a'')$ 877	2.35	0.66	0.64	
CD_3OD	36	C_s	$\nu_4(a')$ 1024; $\nu_6(a')$ 1060; $\nu_8(a')$ 938; $\nu_{10}(a'')$ 1080	2.15	0.63	0.60	
H_2S	34	C_{2v}	$\nu_2(a_1)$ 1183	10.3	9.03	4.72	μ_a=0.897
HDS	35	C_s	988; ν_2 1090	9.67	4.84	3.14	μ_a=1.02
D_2S	36	C_{2v}	$\nu_2(a_1)$ 855	5.47	4.69	2.46	
CH_3CCH	40	C_{3v}	$\nu_5(a_1)$ 931; $\nu_8(e)$ 1053		0.285		μ_a=0.7835
CH_3CCD	41	C_{3v}	$\nu_8(e)$ 1051		0.26		μ_a=0.769
CD_3CCH	43	C_{3v}	$\nu_7(e)$ 1048		0.245		μ_a=0.784
CD_3CCD	44	C_{3v}	$\nu_7(e)$ 1048		0.225		μ_a=0.772
$^{10}BH_3CO$	41	C_{3v}	$\nu_3(a_1)$ 1083		0.30		μ_a=1.794
$^{10}BD_3CO$	44	C_{3v}	$\nu_3(a_1)$ 888		0.25		
$^{11}BH_3CO$	42	C_{3v}	$\nu_3(a_1)$ 1073; $\nu_6(e)$ 1106		0.29		
CH_3CN	41	C_{3v}	$\nu_4(a_1)$ 920; $\nu_7(e)$ 1041		0.307		μ_a=3.92
CD_3CN	44	C_{3v}	$\nu_3(a_1)$ 1110; $\nu_6(e)$ 1046		0.26		μ_a=3.92
CH_3NC	41	C_{3v}	$\nu_4(a_1)$ 945; $\nu_7(e)$ 1129	5.36	0.335		μ_a=3.83
CD_3NC	44	C_{3v}	$\nu_3(a_1)$ 1117; $\nu_4(a_1)$ 877; $\nu_6(e)$ 1058; $\nu_7(e)$ 900		0.286		
DN_3	44	C_s	$\nu_3(a')$ 954	20.6	0.40	0.39	μ_a=0.847
CH_3CHO	44	C_s	$\nu_8(a')$ 1113; $\nu_9(a')$ 919	1.90	0.34	0.30	μ_a=2.55; μ_b=0.87

Table I. continued

Molecule	Mol. wt.	Mol. sym.	Infrared active bands and their symmetry	Rot. const. (1/cm) A_0 B_0 C_0	Dipole moment (Debye)
HCOOH	46	C_s	$\nu_8(a'')$ 1033	2.58 0.40 0.35	$\mu_a=1.391; \mu_b=0.26$
DCOOD	48	C_s	$\nu_4(a')$ 945; $\nu_5(a')$ 1040	1.69 0.39 0.32	
AsH_3	78	C_{3v}	$\nu_2(a_1)$ 906; $\nu_4(e)$ 1003	3.75	$\mu_c=0.22$

Table II. Molecules studied by Stark spectroscopy with CO_2 and N_2O lasers. (May 1975)

Molecule	Reference	Institute	Researcher
$^{14}NH_3$ (ν_2)	J.chem.Phys. 52(1970)3592	Harvard Univ. U.S.A.	F.Shimizu
$^{14}NH_3$ (ν_2)	to be published in Japan.J.appl.Phys.	Univ. of Tokyo Japan	K.Shimoda et al.
$^{14}NH_3$ ($2\nu_2,\nu_4$) *	J.mol.Spectrosc. 55(1975)131	N.R.C. Canada	J.W.C.Johns et al.
$^{15}NH_3$ (ν_2)	J.chem.Phys. 53(1970)1149	Harvard Univ. U.S.A.	F.Shimizu
$^{14}NH_2D$ (ν_2)	to be published in Japan.J.appl.Phys.	Univ. of Tokyo Japan	K.Shimoda et al.
H_2CO (ν_2) D_2CO (ν_2) *	J.mol.Spectrosc. 48(1973)354	N.R.C. Canada	J.W.C.Johns et al.
CH_3F	C.R.Acad.Sci. 276B(1973)733	Univ. of Lille France	F.Herlemont et al.
$^{12}CH_3F$ ($\nu_3,2\nu_3-\nu_3$) $^{13}CH_3F$ (ν_3)	J.mol.Spectrosc. 52(1974)38	N.R.C. Canada	S.M.Freund et al.
CD_3F (ν_3)	to be published in J.mol.Spectrosc.	N.R.C. Canada	G.Duxbury et al.
$CH_3{}^{35}Cl$ (ν_6)	J.Phys.Soc.Japan 38(1975)1106	Harvard Univ. U.S.A.	F.Shimizu
CH_2CF_2 (ν_4,ν_9)	IEEE Trans. MTT22(1974)1108	N.R.C. Canada	G.Duxbury
PH_3 (ν_2,ν_4)	J.Phys.Soc.Japan 38(1975)293	Harvard Univ. U.S.A.	F.Shimizu

* With CO laser

with a Stark-field sweep of up to several hundred e.s.u..

The situation in asymmetric top molecules is somewhat more complicated. The Stark shift of a slightly asymmetric top molecule comes mostly from electric dipole matrix element between a pair of K-type doubling. Therefore, lines of an asymmetric top molecule which has its permanent electric dipole moment in the direction of the axis of the medium moment of inertia (the b-axis) cannot be shifted by any significant amount even if the molecule is only slightly asymmetric. A fairly large Stark shift is expected when the accidental degeneracy happens to occur between a pair of levels with their J-values differing by ±1. But such chances are rare. There are several important molecules such as $D_2O(\kappa = -0.543, \mu_b = 1.87)$, $H_2Se(\kappa = 0.788, \mu_b = 0.24)$, $F_2O(\kappa = -0.930, \mu_b = 0.297)$, and $O_3(\kappa = -0.968, \mu_b = 0.58)$ which have their band frequencis near 1000 cm^{-1}. But they are excluded from the table.

The most-asymmetric and linear molecules cannot be a good candidate because they show very small second-order Stark shift. However, linear molecules in their excited states of degenerate vibration do show a fairly large second-order Stark shift because of their ℓ-type doubling.

Finally, the last condition is that the infrared-active bands of the candidate should be in the region from 900 cm^{-1} to 1100 cm^{-1}. Only those molecules which have their infrared-active *fundamentals* near 1000 cm^{-1} are included in Table I. If anharmonicity of vibrations of these molecules is not very large, however, the difference bands such as $2\nu_i - \nu_i$, $\nu_i + \nu_j - \nu_j$, $\nu_i + \nu_j - \nu_i$, should also be observable in this region even though their intensities are lower by a factor of 10^{-2}. Studies of these difference bands are particularly suitable to laser Stark spectroscopy because its sensitivity is very high. Rotational transitions in the vibrationally excited states of higher than about 2000 cm^{-1} above the ground state cannot be observed by using microwave spectrometers.

It can be seen from the above arguments that improvements in sensitivity and resolution of the laser Stark spectrometer should greatly relax the first condition imposed on the molecules to be studied profitably. A list of molecules which have been studied so far by Stark spectroscopy using CO_2 and N_2O lasers is given in Table II.

§2   Experimental Apparatus

The laser Stark spectrometer in our laboratory is basically similar to that described by Shimizu[2) and by Freund et al.[3), but it is different from theirs in some respects. A block diagram of the apparatus is shown in Fig. 1.

The laser is a 3-m flowing system and capable of oscillation on any lines from P(56) to R(52) in the 10.6 μm band, from P(56) to R(44) in the 9.4 μm band of CO_2, and from P(41) to P(2) and R(2) to R(40) of N_2O. In order to suppress pressure pulsations in the discharge tube ascribed to a rotary pump, a 40-ℓ reservoir is used between the discharge tube and the pump. The laser is feedback-stabilized at the maximum of the output by the use of a monochromator-detector-P.S.D.-P.Z.T. loop. The output is coupled out from zeroth-order diffraction of a metal grating at one end of the cavity. A subsidiary mirror is used to allow a constant deflection of the output beam while the grating is rotated.

The Stark electrode consists of a pair of metallized glass optical flats of 300 × 60 × 30 mm, and the gap between them is maintained exactly at 1.0000 mm by using a set of Johanson blocks calibrated by a gauge-block interferometer. The maximum deviation of the gap from the nominal 1 mm is estimated to be less than 1000 Å. A great care is taken in assembling the electrodes to get rid of dust and other contaminations.

The Stark voltage between the electrodes is measured by using a precision resistance divider of low temperature variation (5 ppm/K) and an A-D converter. The maximum rate of conversion of the latter is 20 samples/sec, and its accuracy is calibrated by a Weston standard cell during the course of measurements. Thus the accuracy of the Stark-field measurement is 10^{-4}.

Although only the precision and the repeatability of measurement of field strength are important in Stark spectroscopy, the above-mentioned accuracy allows us to determine an accurate value of the molecular dipole moment.

A maximum field strength of 350 e.s.u. can be applied to the sample gas without causing break-down.

The laser radiation transmitted through the Stark cell is detected by a liquid-helium-cooled Cu-Ge detector. The detector output is filtered by a quartz crystal filter, fed to a preamlifier, phase-sensitively detected and displayed on a strip-

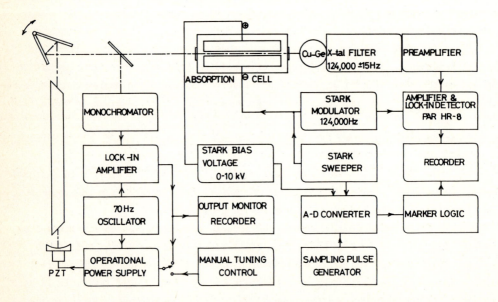

Figure 1. A block diagram of the CO_2 and N_2O laser Stark spectrometer.

-chart recorder together with a voltage marker from the A-D converter. Since the bandwidth of the crystal filter is 30 Hz and the bandwidth of the preamplifier is not narrower than 150 kHz, an increase of dynamic reserve by a factor of at least 70 than without the filter is attained. This allows us to operate the spectrometer under a condition of higher input power and consequently of quantum-noise limited regime.

A Stark-modulation voltage is generated by a crystal oscillator and its frequency is precisely tuned to the center frequency of the filter. A measurement on the noise spectrum of the output of the detector under incidence of the laser radiation show a 1/f variation between 0 to about 50 kHz. The detector noise decreases slowly with frequency beyond 50 kHz. Therefore, a rather high modulation frequency of 124 kHz is used in this experiment.

An integration time constant of about 30 ms and a scanning rate of the Stark field of about 1 s/ e.s.u. are used in observing allowed transitions. For observations of very weak transitions a time constant of 100 s and a scanning rate of as slow as 700 s/ e.s.u. are used. Under this condition the minimum detectable signal voltage at the amplifier input is 10 nV. Since the detector has a response of about 10 V at an input power of 100 mW and is linear in response up to this power level, we conclude that a minimum detectable absorption coefficient α_{min} of 3×10^{-11} cm^{-1} is attained in this spectrometer. The consumption rate of liquid helium in the cryostat of the detector is nearly doubled under a high input power of 500 mW.

§3 Stark Spectroscopy of NH_3

Stark spectroscopy of $^{14}NH_3$, $^{15}NH_3$, $^{14}NH_2D$, $^{14}NHD_2$, and PH_3 molecules have been performed. The study on $^{14}NH_3$ has been completed[4] and others are in progress.

Coincidences between the lines of the ν_2 band of $^{14}NH_3$ and the CO_2 laser line have been used in various nonlinear spectroscopic studies such as microwave-infrared double resonance[5], coherent transient[6], and laser pumped far-infrared lasers.[7] However, none of the accurate molecular constants other than that of the inversion doubling in the ground state are known. In 1969 about one hundred coincidences between the Stark components of $^{14}NH_3$ (ν_2) lines of $J \leq 7$ and the CO_2 and N_2O laser lines were reported by Shimizu[2], but no determination of the molecular constants was performed.

We have observed more than 1500 transitions of the ν_2 band in the region from 900 cm^{-1} to 1100 cm^{-1} with the Stark field of 0 - 350 e.s.u.. About 370 of them were identified as $M_J = J$ components and were used in a least squares analysis to yield a group of 32 molecular constants.

Absorption lines of the allowed transition in the ν_2 band are very strong and they can be detected under a pressure of as low as 10^{-5} Torr. Absorption lines of $^{15}NH_3$, $^{14}NH_2D$, and $^{14}NHD_2$ in their natural abundance and forbidden transitions in $^{14}NH_3$ can also be detected at a sample pressure of about 10^{-3} Torr using a large time constant and a slow scanning rate of the Stark field.

In the least squares analysis, the following expressions for Hamiltonian matrix of the ground and the ν_2 state are used:

$$\begin{pmatrix} E^a_{J+1} & a_{J+1} & 0 & b_{J+1} & & 0 \\ a_{J+1} & E^s_{J+1} & b_{J+1} & 0 & & \\ 0 & b_{J+1} & E^a_J & a_J & 0 & b_J \\ b_{J+1} & 0 & a_J & E^s_J & b_J & 0 \\ & & 0 & b_J & E^a_{J-1} & a_{J-1} \\ 0 & & b_J & 0 & a_{J-1} & E^s_{J-1} \end{pmatrix}$$

where

$$E_J^{a,s} = E_{rot}(J, K) \pm \frac{1}{2} E_{inv}(J, K),$$

$$a_J = -\mu \frac{\varepsilon KM}{J(J+1)},$$

$$\mu = \mu_0 + \mu_J J(J+1) + \mu_k K^2, \qquad (1)$$

$$b_J = \frac{\mu_0 \varepsilon}{J} \sqrt{\frac{(J^2 - K^2)(J^2 - M^2)}{(2J-1)(2J+1)}}.$$

$E_{rot}(J, K)$ is the rotational energy and is given by

$$E_{rot}(J, K) = BJ(J+1) - (B - C)K^2 - D_J J^2(J+1)^2 - D_{JK} J(J+1)K^2 - D_K K^4$$
$$+ H_J J^3(J+1)^3 + H_{JJK} J^2(J+1)^2 K^2 - H_{JKK} J(J+1)K^4 + H_K K^6.$$

The inversion doubling $E_{inv}(J, K)$ for the ν_2 state is given by

$$E_{inv}(J, K) = \Delta\nu' + \Delta B' J(J+1) + \Delta(B' - C')K^2 - \Delta D'_J J^2(J+1)^2 - \Delta D'_{JK} J(J+1)K^2$$
$$- \Delta D'_K K^4 + \Delta H'_J J^3(J+1)^3 + \Delta H'_{JJK} J^2(J+1)^2 K^2 + \Delta H'_{JKK} J(J+1)K^4 + \Delta H'_K K^6$$

$E_{inv}(J, K)$ for the ground state is given by Costain's exponential formula with 23 parameters including $\Delta K = \pm 3, \pm 6$ splitting terms[8]. Since these 23 parameters reproduce the observed microwave spectrum with sufficient accuracy, they are fixed in the least squares analysis. ε is the Stark field, μ is the dipole moment, and J, K, and M are the rotational quantum numbers. A single and a double primed parameters are of the ν_2 and of the ground state, respectively.

The result of the best fit is summarized in Table III. The number of places retained in the constant are determined by the requirement that the constants reproduce the spectrum to within the standard deviation of the fit. The numbers in the parenthesis correspond to one standard deviation of each constant.

Without assumption of the J and K dependence of the dipole moment operator as expressed in (1), a standard deviation which is much larger than the value quoted in Table III is found. This fact, together with the small estimated errors in μ_J', μ_K', μ_J'', and μ_K'', establishes the J and K dependence. A similar dependence of the quadrupole coupling constant was already noted by Gunther-Mohr et al.[9] from the analysis of hyperfine structure in the microwave inversion spectrum of $^{14}NH_3$. Effects of vibration and rotation on the internuclear distances of $^{14}NH_3$ were studied by Toyama, Oka and Morino[10]. A comparison of their numerical example and the observed dependence

Table III. Molecular Constants of $^{14}NH_3$.

Number of data Standard deviation Band origin	$N = 367$ $\sigma = 0.000885$ cm^{-1} $\nu_0 = 949.88108(19)$ cm^{-1}
Constants of the ν_2 state (in cm^{-1})	Constants of the ground state (in cm^{-1})
$B' = 9.980332(77)$	$B'' = 9.944331(74)$
$C' - 9D_K' = 6.11432(73)$	$C'' - 9D_K'' = 6.22075(73)$
$C' - B' - (C'' - B'') = -0.140447(26)$	
$D_J' = 9.2257(300) \times 10^{-4}$	$D_J'' = 8.4926(29) \times 10^{-4}$
$D_{JK}' = -1.84079(720) \times 10^{-3}$	$D_{JK}'' = -1.56601(700) \times 10^{-3}$
$D_K' - D_K'' = 2.2000(99) \times 10^{-4}$	
$H_J' = 3.560(400) \times 10^{-7}$	$H_J'' = 3.306(390) \times 10^{-7}$
$H_{JJK}' = -1.2809(1280) \times 10^{-6}$	$H_{JJK}'' = -1.0517(1250) \times 10^{-6}$
$H_{JKK}' = 1.5867(1460) \times 10^{-6}$	$H_{JKK}'' = 1.1918(1440) \times 10^{-6}$
$H_K' - H_K'' = -1.867(120) \times 10^{-7}$	
$\Delta\nu' = 35.68849(39)$	Inversion doubling in the ground state is
$\Delta B' = -0.180140(45)$	calculated by employing the constants and
$\Delta(C' - B') = 0.251626(51)$	the expressions given in reference (8).
$\Delta D_J' = -4.3253(160) \times 10^{-4}$	
$\Delta D_{JK}' = 1.18590(350) \times 10^{-3}$	
$\Delta D_K' = -8.0567(200) \times 10^{-4}$	
$\Delta H_J' = -5.735(161) \times 10^{-7}$	
$\Delta H_{JJK}' = 2.2838(550) \times 10^{-6}$	
$\Delta H_{JKK}' = -2.9899(625) \times 10^{-6}$	
$\Delta H_K' = 1.2850(240) \times 10^{-6}$	
$\mu' = 1.2477(28) + 6.18(1.40) \times 10^{-4} J(J+1)$ $- 9.31(1.60) \times 10^{-4} K^2$ (in debye).	$\mu'' = 1.47116(14) + 1.791(50) \times 10^{-4} J(J+1)$ $-3.434(60) \times 10^{-4} K^2$ (in debye).

of the dipole moment show that the magnitudes and signs of μ_J and μ_K are quite reasonable[4].

A correlation analysis between the observed Stark fields and the errors (calculated frequency-observed ones) shows no correlation nor anti-correlation between them; no systematic experimental error is found.

A possible effect of the anisotropic polarizability of the molecule is taken into account in a least squares fit, but no improvement in the standard deviation is found.

The final value of the standard deviation, $\sigma = 26.5$ MHz, is somewhat larger than the experimentally expected value of about 20 MHz. The discrepancy is mostly caused by the appearance of abnormally large errors of a few lines as shown in Figure 2.

There are 16 lines which have errors larger than $3\sigma = 80$ MHz, the statistical probability being less than 2×10^{-5}.

In addition to the allowed transitions, several weak $\Delta K = \pm 3$ transitions are observed and listed in Table IV.

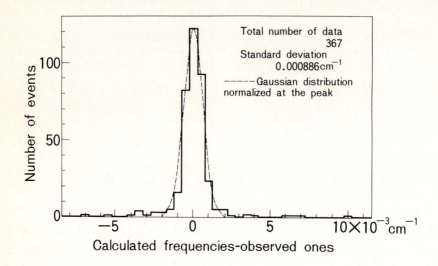

Figure 2. Error distribution of the least squares analysis.

Table IV. Observed $\Delta K = -3$ transitions.

Transition (J", K", M")	Laser	Stark field (e.s.u.)	
ss^NQ (3, 3, 3)	$CO_2R(8)$	136.47	
ss^NQ (3, 3, 2)	$CO_2R(8)$	203.26	
ss^NQ (3, 3, 3)	$N_2OR(36)$	38.93	
ss^NQ (3, 3, 2)	$N_2OR(36)$	58.09	In the first column, numbers
ss^NQ (3, 3, 1)	$N_2OR(36)$	112.37	in parenthesis are the rotat-
as^NQ (3, 3, 3)	$N_2OR(34)$	157.72	ional quantum numbers J", K",
as^NQ (3, 3, 2)	$N_2OR(34)$	240.19	and M" in this order. The
ss^NQ (5, 3, 5)	$CO_2R(10)$	83.98	superscript N denotes $\Delta K = -3$.
ss^NQ (5, 3, 4)	$CO_2R(10)$	104.95	In the second column, CO_2
ss^NQ (5, 3, 3)	$CO_2R(10)$	139.46	lines belong to the 10.6 μm
as^NQ (5, 3, 4)	$N_2OR(37)$	188.12	band.
as^NQ (5, 3, 5)	$N_2OR(37)$	150.29	

By using the data given in Table IV, the rotational constants about the c-axis are calculated as given in Table III.

There are also possibility of observing other $\Delta k = \pm 3, \pm 6$ transitions such as $k = \pm 2 \to \mp 1$, $k = \pm 4 \to \mp 2$. An intensive search has been made in the regions where these transitions should appear. However, many weak lines due to NH_2D, NHD_2 and $^{15}NH_3$ have prevented the identification of the above lines, because the spectrometer must then be operated under a condition of high sensitivity.

In order to discriminate lines due to the isotopic species, experiments with the enriched isotopic molecules have been carried out. An example of the spectrum of NH_2D is shown in Figure 3. A numerical analysis of the observed spectrum is under way.

§4 Discussion

Ovbiously, one of the real advantages of the Stark spectroscopy by CO_2 and N_2O lasers in a systematic spectroscopic study is its accuracy in frequency determination of absorption lines. The absolute frequencies of these laser lines are known to an accuracy of 25 kHz in CO_2[11] and 40 kHz in N_2O[12] lasers. These accurate frequencies are used as "bench-marks" distributed along the frequency coordinate; a distance between a bench-mark and an absorption line is determined by measuring the strength of the Stark field required to shift the absorption line into resonance of the laser line. The required precision of the measurement of Stark field depends both on the average distance $\overline{\delta\nu_b}$ between the bench-marks and on the average resolution $\delta\nu$ of observation; it is approximately given by $\delta\nu/\overline{\delta\nu_b}$.

Although the frequencies of the CO_2 and N_2O lasers are at the center of so called "finger-print region", where a large number of the characteristic absorption of various molecules prevail, the molecules which can be studied by the CO_2 and N_2O laser spectrometer are rather limited as can be seen from Table I. Observation and analysis of all the fundamental bands of a molecule are very desirable in a systematic spectroscopic study. At present, no good bench-marks other than the frequencies of the CO_2 and N_2O lasers are available in the infrared. Therefore, we must consider transferring these bench-marks into higher and lower frequency regionsof the infrared. This can be done most easily toward the higher frequency region by directly mixing up two or three output from Q-switched CO_2 and N_2O lasers by using nonlinear optical crystals such as T_e, $CdGeAs_2$, $CdSe$, $AgGaSe_2$, $ZnGeP_2$ and $AgGaS_2$. Since CO_2 and N_2O lasers oscillate at about 200 frequencies in the region from 900 cm^{-1} to 1100 cm^{-1}, a two-wave mixing will give about 20000 bench-marks in the region from 1800 cm^{-1} to 2200 cm^{-1} and a three-wave mixing will give 10^6 bench-marks in the region from 2700 cm^{-1} to 3300 cm^{-1}. The average distance $\overline{\delta\nu_b}$ between the bench-marks will be reduced and Stark spectroscopy can be carried out with greater ease.

Considerable progress has been made in the development of tunable coherent infrared sources such as spin-flip Raman laser, parametric oscillators, diode lasers

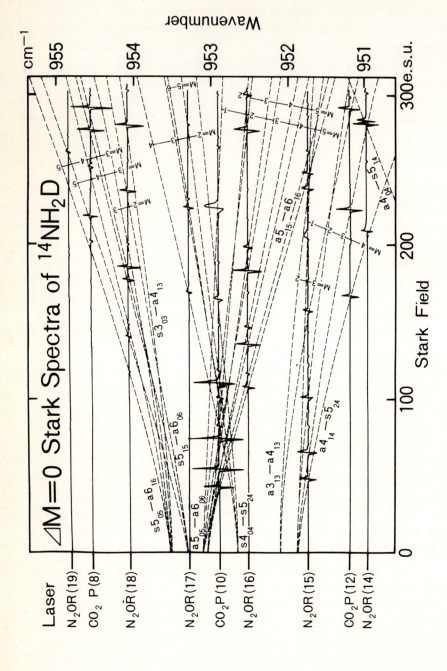

Figure 3. ΔM = 0 Stark spectra of $^{14}NH_2D$ displayed on an X-Y recorder. The Y-coordinate was adjusted to scale the wavenumber of each laser line.

and Zeeman-tuned lasers. However, a tunability always means an instability in frequency. Development of methods to lock a tunable oscillator on the bench-mark of accurately known frequency is imperative for practical applicability of the tunable sources to presicion spectroscopy.

Acknowledgments

One of the authors (Y. U.) would like to thank Fujio Shimizu for numerous and stimulating discussions. He is also grateful to Jun-ichiro Iwahori for his assistance throughout this work.

References

1) T. Shimanouchi : Tables of Molecular Vibrational Frequencies, Consolidated Volume I (U. S. Government Printing Office, Washington D. C., 1972).
2) F. Shimizu : J. Chem. Phys. 52 (1970) 3572.
3) S. M. Freund, G. Duxbury, M. Römheld, J. T. Tiedje, and T. Oka : J. Mol. Spectrosc. 52 (1974) 38.
4) Y. Ueda, J. Iwahori, and K. Shimoda : to be published in Japan. J. appl. Phys.
5) T. Shimizu and T. Oka : Phys. Rev. A2 (1970) 1177.
6) J. M. Levy, J. H. S. Wang, S. G. Kukolich and J. I. Steinfeld : Phys. Rev. Letters 29 (1972) 983.
7) T. Y. Chang : IEEE Trans. MTT22 (1974) 983.
8) E. Schnabel, T. Törring, and W. Wilke : Z. Phys. 188 (1965) 167.
9) G. R. Gunther-Mohr, R. L. White, A. L. Shawlow, W. E. Good, and D. K. Coles: Phys. Rev. 94 (1954) 1184.
10) M. Toyama. T. Oka, and Y. Morino : J. Mol. Spectrosc. 13 (1964) 193.
11) K. M. Evenson, J. S. Wells, F. R. Petersen, B. L. Danielson, and G. W. Day : Appl. Phys. Letters 22 (1973) 192.
12) B. G. Whitford, K. J. Siemsen, H. D. Riccius and G. R. Hanes : Opt. Comm. 14 (1975) 70.

RECENT ADVANCES IN TUNABLE INFRARED LASERS*

A. Mooradian

Lincoln Laboratory, Massachusetts Institute of Technology
Lexington, Massachusetts 02173 U.S.A.

This paper describes some recent results in understanding and developing tunable infrared lasers for use in high resolution spectroscopy and photochemistry and covers work done at M.I.T. Lincoln Laboratory on spin-flip Raman lasers, optically pumped semiconductor lasers, nonlinear frequency mixing in crystals, and optically pumped gas lasers.

Spin-Flip Raman Lasers

A number of recent experiments[1] have provided a better understanding of the spin-flip laser which is important for applications of this device in spectroscopy and photochemistry. Of particular importance is the pulsed operation in which some details of the temporal intensity behavior are explained for the InSb and InAs spin-flip lasers. Previous[2] results using pulsed CO_2 lasers to excite InSb spin-flip lasers reported external conversion efficiencies limited to about 1% because of spin saturation, but no detailed studies of the temporal characteristics of the spin-flip laser were made.

When a spin-flip laser in the form of a polished rectangular parallelepiped is excited with a pulsed laser that has a smooth (unmodelocked) intensity profile, the output usually has a complicated temporal behavior that is dependent in part upon the intensity of the pump laser. Figure 1 shows a series of oscilloscope traces of the output of an InAs spin-flip Raman laser operating at 2 K. The pump was a HF TEA laser with a smooth intensity profile and the crystals were 3 x 3 x 5 polished rectangular parallelepipeds. The traces were taken for increasing values of magnetic field from 72 to 85 kG. For this InAs crystal with 1.6×10^{16} electrons per cm^3 the quantum limit is reached for magnetic fields above 70 kG thereby reducing the threshold for stimulated spin-flip Raman scattering. Starting with the lowest magnetic fields, the spin-flip output is an initial transient pulse which decays rapidly to near zero intensity and is followed by a broader peak that grows and shifts to longer times for increasing magnetic fields.[3] Similar temporal behavior was observed for InSb spin-flip laser crystals when pumped by a CO_2 TEA laser with the exception that the secondary peak increased in amplitude and delay for increasing pump intensity. The initial transient is interpreted as normal laser relaxation oscillations that occur when a laser is switched on. While normally one would expect steady state laser oscillation to occur after the relaxation oscillations have died out, in both InAs and InSb complete quenching of the spin-flip output could occur before buildup of the second peak. This quenching is attributed to the substantial amount of free carrier absorption present at the pump wavelength (~ 1 cm^{-1} in the case of InSb) which results in scattering the free electrons from the bottom of the conduction band to high momentum states where they are effectively removed from participating in the spin-flip process. This quench-

---

*This work was sponsored by the Department of the Air Force, NSF/RANN, and the US ERDA under a subcontract from the Los Alamos Scientific Laboratory.

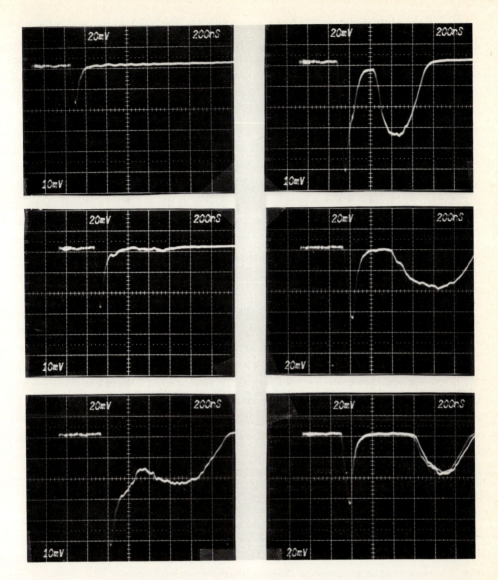

Fig. 1. Oscilloscope traces of the output of an InAs spin-flip Raman laser at 2 K. Input pump intensity is held constant. Traces from upper left downward are for increasing magnetic fields from 72 to 85 kG.

ing occurs along the focus of the TEM$_{00}$ mode pump beam within the crystal where the intensity is the greatest. In the wings of the Gaussian pump beam where the intensity is insufficient for collinear oscillation, buildup of parasitic bounce modes within the crystal occurs with a slower cavity buildup time. Special care in alignment can minimize but not eliminate these bounce modes. Scanning the intensity of the spin-flip output across the exit face of the crystal clearly indicated the difference between the collinear and bounce mode operation. The data of Fig. 1 was taken with a detector having insufficient response speed to resolve clearly the details of the initial relaxation oscillations as well as transverse mode beating during the bounce mode operation. This was clearly observed for the case of InSb and InAs when a high speed detector was used.

Bounce mode operation was effectively eliminated by the use of an external cavity on an antireflection coated InSb crystal.[1] In that case, only the initial relaxation oscillations were observed with no subsequent buildup of a secondary peak. It is clear that external cavity operation is necessary to effectively control the output of either a pulsed or a cw spin-flip Raman laser, especially when one wishes the mode volume to fill the entire crystal for maximum energy output.

One of the difficulties associated with the usefullness of a pulsed spin-flip laser using a cw discharge, rotating mirror Q-switched CO_2 laser as a pump source has been the frequency and amplitude stability as well as the spatial motion of the beam due to the rotating mirror of the Q-switch. These problems have been effectively overcome by the use of a cw discharge stable CO_2 laser which uses an internal cavity CdTe electro optic Q-switch to provide a stable, single-mode output pulse to pump a spin-flip laser. The CdTe modulator together with a CdS quarter-waveplate was capable of providing well controlled pulses at rates up to one kilohertz. The amplitude stability of an InSb spin-flip laser pumped with this laser approached that of the pump laser which was about 1-2%. The utility of the pulsed spin-flip laser for spectroscopic measurements can be greatly improved using such a pump source.

Optically Pumped Semiconductor Lasers

Despite the simplicity and convenience of semiconductor diode lasers for use in high resolution spectroscopy, there are some interesting uses which can be made of optically pumped semiconductor lasers. Recently,[4,5] some optically pumped infrared semiconductor lasers have been used to measure the Doppler limited absorption spectra of a number of gases. These optically pumped devices made use of relatively inexpensive, commercially available, GaAs diode lasers to excite cleaved platelets of several infrared semiconductor crystals. The GaAs diode lasers and the semiconductor platelets could both operate continuously near liquid helium temperature with tuning of the laser modes in the optically pumped samples occurring due to heating from the pump light. Since the GaAs laser photons are substantially higher in energy than the bandgaps of the semiconductors being excited, the electrons and holes that are excited high into the bands relax via phonon emission to provide the dominant heating mechanism for mode tuning. This inefficiency is one of the major disadvantages of optically pumped devices as compared to diode lasers. However, sufficient power for high resolution spectroscopy can be generated to resolve spectra even in real time such as that

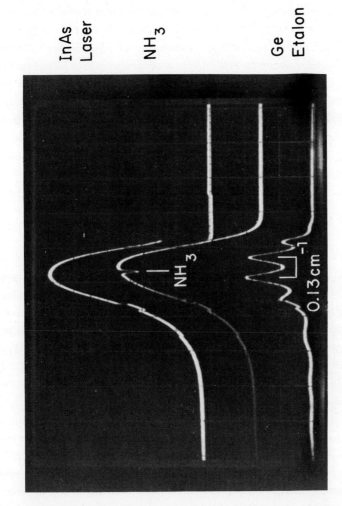

Fig. 2. Oscilloscope traces of (top) output of a thermally tuned single mode from a cw optically pumped (using a GaAs diode laser) InAs laser near liquid helium temperature; (middle) absorption spectrum of NH_3 at 2 Torr pressure; (bottom) transmission of laser mode through a low finesse, fixed Ge etalon. From reference 6.

shown[6] in Fig. 2 where an InAs laser at ~3 μm is excited by a cw GaAs diode laser near liquid helium temperature. The bottom trace in Fig. 2 shows the transmission of the InAs laser through a low finesse germanium etalon used as a relative frequency marker while the middle trace shows an absorption line of NH_3 at 2 Torr pressure.

An InSb laser has been optically excited using resonant pumping with a carbon monoxide laser. Because of the good energy match between the pump photons and the gap, power conversion efficiencies of 20% were achieved using colinear pumping geometries[5] with single longitudinal mode output powers of up to 20 mW. Using a wedged sample of InSb, continuous tuning across the gain bandwidth was obtained by moving the pump beam along the sample. The useful linewidth for such tuning was limited by the variation of cavity length over the mode diameter. Transverse pumping of InSb with a CO laser produced substantially lower output powers than colinear pumping due to the mode matching geometry. Perhaps one of the more important potential features of optical pumping is that of scalability where large area or volume excitation can be achieved without the associated problems of such high gain systems. Such large area pumping of a GaAs matrix in an external cavity has been reported using e-beam excitation.[7]

Nonlinear Mixing

Coherent radiation over a broad band in the infrared has been obtained by generating the sum and difference between a fixed frequency and a tunable laser in a nonlinear crystal. A practical system[8] has been demonstrated using a stabilized cw dye laser and a cw single-frequency argon-ion laser mixing in $LiNbO_3$ to generate a cw tunable output in the 2.2-4.2 μm region with a usable linewidth of 15 MHz. Infrared output powers in excess of one μW have been achieved for 10 mW and 100 mW of input dye and argon laser powers, respectively. Figure 3 shows a recorder trace of the spectrum[7] of N_2O taken using such a system. The marker lines at the bottom were taken by recording the transmission of a fixed Fabry-Petor etalon to the dye laser and are 300 MHz apart.

Operation further in the infrared for such a system is limited by the useful transmission of $LiNbO_3$ which cuts off at about 4.5 μm. This problem can be overcome by the use of infrared transmitting nonlinear crystals such as chalcopyrites which can have high nonlinear figures of merit than $LiNbO_3$ and useful transmission in the infrared. Of these, the most promising are $CdGeAs_2$ and $AgGaSe_2$. Conversion efficiencies (internal) of 25% in generating second harmonic radiation using a CO_2 TEA laser have been[9] achieved using $CdGeAs_2$. Output energies of 26 mJ were limited by the available energy from the pump laser. Present crystal growth capabilites should allow outputs of doubled CO_2 laser energy of a few hundred millijoules per pulse from a single crystal with average powers of a few watts. At present, $CdGeAs_2$ crystal technology is limited by absorption at shorter wavelengths which is due to impurities and to low yield of crack-free crystals.

Sum and difference frequency generation in these chalcopyrites has been demonstrated in the range of 2.5 - 18 μm using line selectable CO and CO_2 lasers. For many photochemical applications where it is sufficient only to have energy at a desired wavelength, the mixing of

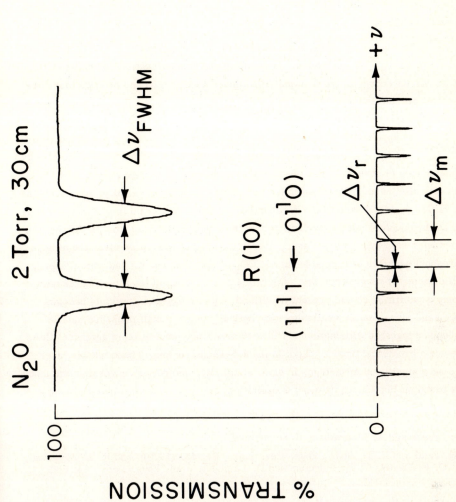

Fig. 3. Absorption spectrum of N_2O at Torr pressure near 3 μm taken using argon-dye laser difference frequency spectrometer. Relative frequency calibration markers (lower trace) were taken by measuring the transmission of the dye laser with a fixed etalon having a free spectral range of 300 MHz. The IR frequency resolution is 15 MHz. From reference 8.

two fixed frequency lasers is sufficient. For example, difference frequency generation between a CO and a CO_2 lasers using the four most abundant isotopes of carbon and oxygen can produce a line about every 200 MHz, which is less than the tuning range of a one atmosphere CO_2 laser. For continuous coverage using the mixing in chalcopyrites, a tunable infrared laser is required. One of the many possible schemes using high-pressure gases and chalcopyrite crystals is described in the next section.

Optically Pumped Gas Lasers

Recently,[11] a HBr laser was used to excite the 4.3 μm transition ($00°0 \rightarrow 00°1$) in CO_2 to produce laser action in the 9-11 μm region. In addition, the same vibrationally excited CO_2 was used to transfer its energy to the identical vibrational state in N_2O and produce laser action in N_2O at total pressures up to 33 atmospheres. At such high pressures there is broadening and complete overlap of adjacent rotational lines and continuous tunability in the 9-11 μm band. A more general application of this technique has been made[12] by using a frequency doubled CO_2 laser to excite the V = 1 mode of CO which could then transfer its energy into closely lying vibrational modes of several molecules. Figure 4 shows a simplified energy level diagram for such optically pumped CO transfer lasers. Laser action was produced in molecules such as CO_2, N_2O, CS_2, C_2H_2 and OCS. A crystal of $CdGeAs_2$ described in the previous section was used as the doubler. Of particular interest for a tunable laser system is the CO-CO_2 and CO-N_2O transfer schemes. These lasers operated pulsed at pressures up to one atmosphere with threshold energies of second harmonic CO_2 of less than one mJ and optical energy conversion efficiencies of several percent. Operation at pressures of ten atmospheres with present pumping capabilities should present no difficulties. As $CdGeAs_2$ crystals become better developed and more readily available, such optically pumped high-pressure tunable lasers could be used together with sum and difference frequency generation in another $CdGeAs_2$ crystal to provide a broadband tunable output in the 2.5-18 μm region. These tunable laser systems could have significant advantages in size, simplicity, and power conversion efficiency over schemes based on HBr lasers or e-beam excitation.

References

1. S. R. J. Brueck and A. Mooradian, to be published.
2. C. K. N. Patel and E. D. Shaw, Phys. Rev. Lett., **24**, 451 (1970); and R. L. Aggarwal, B. Lax, C. E. Chase, C. R. Pidgeon, and D. Lambert, Appl. Phys. Lett., **18**, 383 (1971).
3. Data taken in collaboration with R. S. Eng and H. Fetterman.
4. A. S. Pine and N. Menyuk, Appl. Phys. Lett., **26**, 231 (1975).
5. N. Menyuk, A. S. Pine, and A. Mooradian, IEEE J. Quantum Electronics, **11**, 477 (1975).
6. In Final Report "Developments of Tunable Semiconductor Lasers for Application to Air Pollution Measurements", NSF/RANN/IT/GI-34990/RR/75/4, A. Mooradian, A. S. Pine and N. Menyuk, May, 1975.

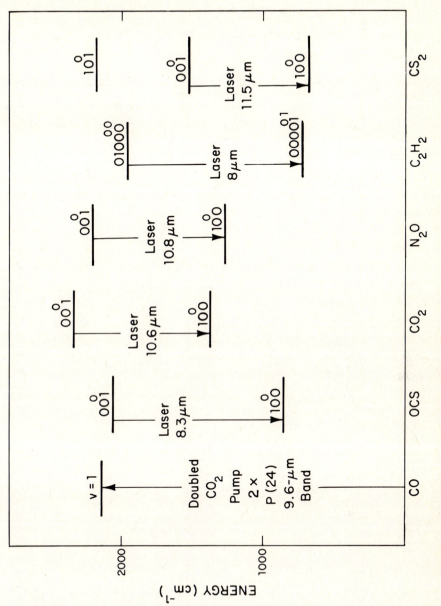

Fig. 4. Simplified energy level diagram for optically pumped CO vibrational energy transfer lasers. From reference 12.

7. O. V. Bogdankevich, S. A. Darznek, A. N. Pechenov, B. I. Vasiliev, and M. M. Zverev, IEEE J. Quantum Electronics, $\underline{9}$, 342 (1973).

8. A. S. Pine, J. Opt. Soc. of Am., $\underline{64}$, 1683 (1974).

9. H. Kildal and G. W. Iseler, to be published.

10. H. Kildal and J. C. Mikkelsen, Optic Commun., $\underline{10}$, 306 (1974).

11. T. Chang and O. Wood, Appl. Phys. Lett., $\underline{23}$, 370 (1973).

12. H. Kildal and T. F. Deutsch, reported in Conference on Laser Energy and Applications, Washington, D. C., May, 1975; and to be published.

A BROADLY TUNABLE IR SOURCE

Robert L. Byer, Richard L. Herbst, Robert N. Fleming
Applied Physics Dept., Stanford University
Stanford, California 94305.

INTRODUCTION

Improvements in the size and quality of nonlinear crystals has led to their increasing use in the generation of tunable ultraviolet, visible and infrared coherent radiation. This paper describes a widely tunable coherent source based on a Nd:YAG pumped $LiNbO_3$ parametric oscillator. When completed, the $LiNbO_3$ oscillator's basic tuning range of 1.4 μm to 4.2 μm will be extended to 2600 Å by frequency doubling and to 25 μm by frequency mixing.

SYSTEM DESCRIPTION

Two years ago[1] we noted that a 1.06 μm pumped angle tuned $LiNbO_3$ parametric oscillator's 1.4 to 4.2 μm tuning range could be extended by mixing in $AgGaSe_2$ to extend over the 3 μm to 18 μm infrared range. Since then, considerable progress has been made toward the realization of such a widely tunable infrared source.

A Nd:YAG laser pumped $LiNbO_3$ parametric oscillator was first reported by Ammann et al.[2,3] However, the lack of long, high optical quality $LiNbO_3$ crystals limited the oscillator's gain, tuning range, and efficiency. Two years ago at the Stanford Center for Materials Research crystal growth facilities, we began the investigation of new growth directions in $LiNbO_3$ that could possibly lead to longer, higher optical quality $LiNbO_3$ crystals. The investigation led to the growth of large high optical quality $LiNbO_3$ crystals[4] suitable for use in the 1.06 μm pumped angle tuned parametric oscillator and to the demonstration of a high gain, efficient parametric oscillator.[5] The tuning range and gain bandwidth of the parametric oscillator is shown in Fig. 1. From this figure it is immediately apparent that the $LiNbO_3$ parametric oscillator tunes over a very wide range (greater than 3 to 1 compared to typical dye laser bandwidths of 10%) and that bandwidth control is required to reduce the linewidth to values useful in infrared spectroscopy. It is also evident that rapid tuning is possible since only a $\pm 4°$ external crystal rotation angle is required for tuning over the entire 1.4 to 4.4 μm available spectral range.

With these considerations in mind we began a program to design and build a coherent spectrometer. The system will be pumped by a Nd:YAG Q-switched oscillator followed by a series of amplifiers to generate up to 450 mJ of TEM_{oo} mode, single axial mode 1.06 μm output at 10 pps. The 1.06 μm beam directly pumps the 5 cm long

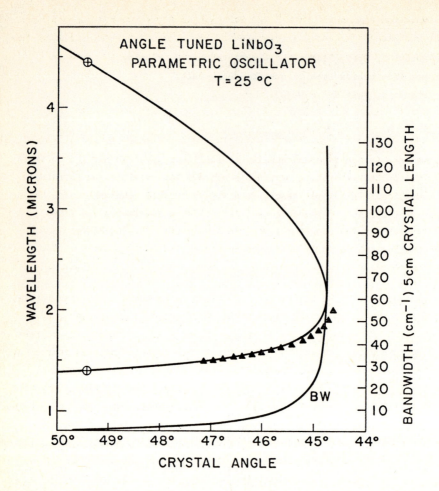

FIG. 1--Angle tuned LiNbO$_3$ parametric oscillator tuning curve and measured points and gain bandwidth for a 5 cm long crystal.

by 15 mm diameter LiNbO$_3$ crystal which is operated as a singly resonant parametric oscillator with the signal wave resonant. The crystal angle and oscillator line narrowing elements will be under computer control. The oscillator output is then used directly or frequency doubled in LiNbO$_3$, LiIO$_3$, and ADP to extend over the near infrared, visible and ultraviolet spectral range, or frequency mixed in AgGaSe$_2$, CdSe or N$_2$ vapor to extend over the infrared range out to 25 μm. To date we have operated the parametric oscillator at up to 40 mJ 1.06 μm input energy generating

16 mJ of output energy in the signal and idler waves for an energy conversion of 40%. We have demonstrated a method for frequency stabilizing the Nd:YAG source and have investigated three alternative linewidth narrowing methods for the parametric oscillator. These aspects of the source are discussed in more detail below.

ND:YAG SOURCE

The $LiNbO_3$ parametric oscillator places temporal, spatial, spectral and pulse energy requirements on the pump laser source if high conversion efficiency and narrow linewidth operation are to be achieved. The measured threshold energy density for the $LiNbO_3$ parametric oscillator ranges from .45 J/cm^2 for a pump spot size of w_p = .750 mm to .31 J/cm^2 for a pump spot size of w_p = 2.7 mm. This is well below the measured damage threshold of between 4-11 J/cm^2 for uncoated $LiNbO_3$ at 1.06 μm and between 20-30 J/cm^2 for SiO_2 anti-reflection coated $LiNbO_3$. These measurements were made for pulse lengths between 15 and 60 nsec and spot sizes between 25 μm and 280 μm. Therefore, a 15 mm diameter $LiNbO_3$ parametric oscillator crystal can conservatively handle between .45 J to 1.2 J of pump energy in a 2.7 mm spot size and still remain between 2 to 5 times below the anti-reflection coated energy density limit. We thus designed a Nd:YAG oscillator-amplifier chain shown schematically in Fig. 2 to provide approximately 0.5 J of diffraction limited pulse energy in a 20 to 30 nsec pulse at up to 10 pps.

To date we have operated the amplifier chain up through the second 1/4" Nd:YAG amplifier head. The measured 150 mJ output pulse energy agrees very well with the output energy predicted by an amplifier model calculation which utilizes a two dimensional spatial integration and a temporal integration to predict the spatial and temporal output of the system. The Faraday rotator isolator which uses FR-5 Hoya rotator glass, operates at 10 pps with 30 J required pulse input energy. Its measured open to close ratio is 2200 to 1. With the isolator in place, the amplifier chain is stable against feedback from even a high reflector 1.06 μm mirror. The 3/8" amplifier head and second Faraday rotator isolator are expected to be completed soon.

The Nd:YAG oscillator amplifier chain is operated without spatial filtering with mode filling to approximately the 1% intensity points on the transmitted Gaussian beam. The amplifiers operate with a small signal gain of near 28 and with the exception of the first 1/4" Nd:YAG amplifier stage, are operated well into

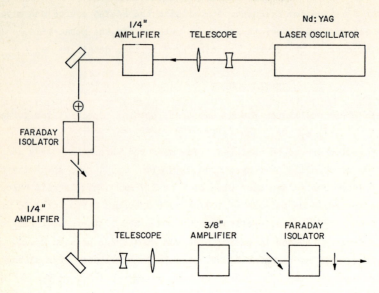

FIG. 2--Schematic of a 0.5 J per pulse Nd:YAG oscillator-amplifier system.

saturation. The high gain and saturation leads to leading edge steepening of the transmitted pulse and to a flat topped spatial intensity profile with a smoothly varying intensity falloff at the beam edges. The measured pulse to pulse stability is better than 2%. The high degree of pulse stability is due to operation with simmered lamps and to amplifier saturation. Simmering the lamps also leads to greater than 10^7 flash lamp pulses before replacement is necessary. These pump laser characteristics lead to stable, long lived parametric oscillator operation - an essential requirement for spectroscopic studies.

The parametric oscillator conserves both energy and momentum such that

$$\omega_p = \omega_s + \omega_i$$
$$\underline{k}_p = \underline{k}_s + \underline{k}_i \qquad (1)$$

The frequency conservation condition implies that if we desire a narrow linewidth output at ω_s and ω_i, then ω_p must be also narrow. Furthermore, ω_p must be frequency stabilized so that ω_s and ω_i are accurately known. The momentum conservation condition implies that the signal and idler wave momenta must always sum to equal that of the pump wave. Thus unlike optically pumped dye lasers which show isotropic gain and thus exhibit superfluorescence suppression and transverse mode control problems, the parametric oscillator has directional gain. This fact is of considerable practical importance since it allows high gain operation of the parametric oscillator with good spatial mode control. However, it also places mode control requirements on the pump laser and limits the geometry to collinear pumping.

We recognized early in the program the need for a single axial mode, frequency stabilized Nd:YAG oscillator source. A number of frequency stabilization methods were investigated before a particularly elegant fluorescence balance method using isotopic I_2 vapor fluorescence was discovered.

The frequency stabilization method uses a descriminant signal generated by 5320 Å induced fluorescence in separate I_2^{129} and I_2^{127} cells. Near the frequency doubled 1.06 μm line center, both iodine isotopes have absorption lines 1GHz wide that nearly overlap. Figure 3 shows the fluorescence signal generated in I_2^{129} and I_2^{127} cells for a 5320 Å pump wavelength as a tilted etalon is swept across the Nd:YAG gain line. These spectra are linear in frequency since the usual θ^2 etalon tuning is compensated by driving a General Scanner galvanometer which tilts the etalon with a square root voltage drive.

Figure 4 shows the difference spectrum of the iodine isotopic fluorescence signals. A very sharp derivative signal is evident where the I_2^{129} and I_2^{127} absorption peaks overlap. Figure 5 is an enlarged view of the discriminant at 607 MHz/cm scale. The axial mode interval of the Nd:YAG laser cavity is also indicated. This spectrum was taken single ended at 20 Hz repetition rate of the Nd:YAG oscillator source. The scan time was approximately two minutes with a 0.3

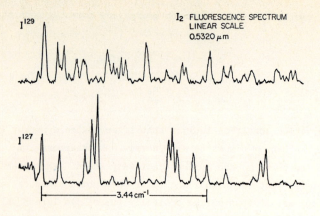

FIG. 3--I_2^{127} and I_2^{129} fluorescence spectra at 5320 Å for a linear frequency scale at a 3.44 cm^{-1} free spectral range.

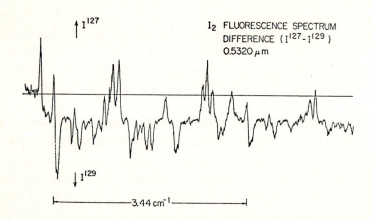

FIG. 4--I_2^{127} and I_2^{129} difference spectrum at 5320 Å.

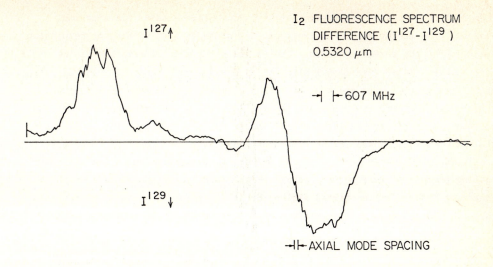

FIG. 5--Frequency stabilization discriminant on an expanded scale. The laser axial mode spacing is also resolved.

second averaging time. The frequency discriminant is 1.8 GHz wide thus offering the possibility of easily frequency stabilizing the 5320 Å wavelength to ± 18 MHz and the 1.06 μm wavelength to ± 9 MHz.

The stabilization system is simply a pair of I_2 isotopic cells whose induced fluorescence signal is separately monitored by a pair of silicon diode detectors. The output voltage from the silicon detectors goes to a difference amplifier, an integrator, a dc amplifier and then to a piezoelectric stack to control the Nd:YAG laser cavity length. The effective feedback bandwidth is the inverse pulse repetition rate of the Nd:YAG laser. We are presently completing spectroscopic measurements of the I_2^{127} absorption spectra at 5320 Å to identify the vibration-rotation transitions being pumped.

There are reasons to expect a non-stable axial mode output from the Q-switched, pulsed Nd:YAG oscillator source. They include frequency chirping due to thermal index changes in the Nd:YAG rod and rapid Q-switch phase changes. We therefore investigated the output spectrum of our Nd:YAG source with a 10 GHz free spectral range confocal scanning interferometer at 5320 Å, Fig. 6 shows the spectrum obtained. The measured 5320 Å axial mode linewidth is just within the resolution capability of the confocal etalon. The 50 MHz axial mode linewidth at 1.064 μm is close to the Fourier Transform limited linewidth for our 20 nsec Q-switched pulse length. This axial mode spectrum was taken over a 2 minute scan period with a .3 sec averaging time at 20 Hz laser repetition rate. It again demonstrates the amplitude stability of the laser source and in addition its spectral stability. With the 2 mm etalon used in these measurements, two 1.064 μm axial modes were oscillating (the center peak in Fig. 6 is due to sum generation in the nonlinear doubling process). We expect to obtain single mode operation with a 1 cm thick finesse of 7 fused silica tilted etalon for laser operation at the 2 mJ output level. The Nd:YAG oscillator-amplifier system is an ideal pump source for the $LiNbO_3$ parametric oscillator.

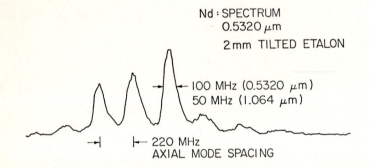

FIG. 6--Axial mode spectra of a frequency doubled, Q-switched Nd:YAG oscillator operating with a F=5 2 mm thick fused silica etalon.

$LiNbO_3$ PARAMETRIC OSCILLATOR

Initial results for the 1.06 μm pumped $LiNbO_3$ parametric oscillator have been reported previously.[4,5] We want to present here recent results including operation at higher energies, conversion efficiency measurements, tuning and line narrowing studies.

Figure 7 shows schematically the optical configuration of the parametric oscillator. The cavity design employing a beamsplitter was chosen so that line narrowing elements could be placed out of the high intensity 1.06 μm pump beampath. The output mirror is sapphire to allow transmission of the idler wave. All other components use fused silica substrates except the grating which is a 20 mm × 20 mm 600 groove/mm plastic replica grating from Bausch and Lomb.

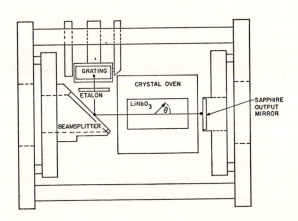

FIG. 7--Schematic of the 1.06 μm pumped, angle tuned $LiNbO_3$ singly resonant parametric oscillator.

The parametric oscillator cavity is resonant at the signal wavelength for two reasons. The first is that fused silica optics and standard coating materials can be used for the 1.4 to 2.1 μm wavelength range and that this fractional wavelength range is just within that available to a broad band optical coating. Thus a single set of optics is all that is required to tune over the entire parametric oscillator range of 1.4 μm to 4.2 μm. Second, the parametric oscillator gain is higher for a resonated signal wave than idler wave due to frequency and area factors in the parametric gain expression.

The input beamsplitter coating transmits 95% of the "p" polarized 1.06 μm pump wave and reflects 99% of the "s" polarized signal wave between 1.4 μm and 2.1 μm. The ThF_4-ZnS coating on a standard fused silica brewster window substrate has a measured damage intensity of 300 MW/cm^2 for a 20 nsec 1.06 μm beam. Hard coatings were also tried but were found to damage at a much lower intensity near 50 MW/cm^2. This beamsplitter is an important element in the present cavity design. However, it could be replaced by an air spaced calcite polarizer at a higher cost and increased loss at the signal wavelength.

The LiNbO$_3$ crystal is anti-reflection coated with SiO$_2$ centered at 1.7 μm. The SiO$_2$ index match to LiNbO$_3$ gives less than 0.3% reflectance at the coating center wavelength. The SiO$_2$ also increases the damage threshold as described earlier. The sapphire output mirror is coated with ThF$_4$-ZnS at 50% reflectivity at 1.7 to 1.9 μm with decreasing reflectivity between 1.9 μm to 2.1 μm to 40% and between 1.4 to 1.7 μm to 20%. A slightly higher reflectance seems desirable to improve operation efficiency. A higher reflectance does, however, increase the circulating signal power within the cavity which may lead to damage problems. The idler wavelengths between 2.1 and 4.2 μm are coupled out of the cavity with minimum loss through the sapphire mirror.

The cavity is shown in Fig. 7 with a tilted etalon and grating as line narrowing elements. We also have operated the oscillator with a thin (100 μm) fused silica spaced, air gap sapphire plate etalon plus a 1 mm fused silica etalon and with a two element LiNbO$_3$ birefringent filter plus a 1 mm thick fulsed silica etalon. The cavity is invar stabilized and is designed for cavity length control with a piezoelectric driven mirror position.

The LiNbO$_3$ crystal is temperature stabilized slightly above room temperature and held on a micrometer driven angular mount. The angular position is resettable to within one tenth of the gain linewidth or to a corresponding 8 seconds of arc. A stepper motor drives the micrometer to set the crystal angle.

The grating is also stepper motor driven through a sine drive arrangement. The grating is resettable to within ± .5 Å at 2.1 μm. The birefringent filter and tilted etalons are controlled by General Scanner galvanometers. The galvanometer drives allow 8 arc seconds absolute angular positioning with a response time of less than 0.1 sec. They allow a ± 4° scan range for a ± 1 amp input current. All control elements are accessible by computer through a CAMAC interface.

The design of the oscillator cavity provides a wide wavelength coverage without an optics change. It is a stable, compact cavity that allows full computer control of the tuning elements including the automatic insertion of the line narrowing tilted etalon when required for fine tuning. These features, including the use of the grating without a beam expanding telescope are unique to this device. Together with the wide tuning range, high conversion efficiency and virtually infinite lifetime of the $LiNbO_3$ crystal, the 1.06 μm pumped parametric oscillator forms the heart of a widely tunable spectrometer system.

The important operating characteristics of the $LiNbO_3$ parametric oscillator are tuning range, threshold energy density, conversion efficiency and linewidth. Also of concern is the required pump spot size and the maximum energy throughput of the cavity and $LiNbO_3$ nonlinear crystal.

Because of the large double refraction in $LiNbO_3$ at the $47°$ phasematching angle, a large pump beam spot size is needed to reduce the effect of pump beam walkoff on parametric gain. An estimate of the pump beam spot size can be obtained using[7,8]

$$w_p = (2\pi)^{-1/2} \rho \ell \qquad (2)$$

where ρ is the double refraction angle. For a 5 cm $LiNbO_3$ crystal at $47°$ with $\rho = 0.04$ rad, $w_p = 800$ μm. Using higher power lasers and maintaining constant intensity, the spot size increases such that the parametric gain approaches the plane wave limit given by[8] $G = \sinh^2 \Gamma\ell$, where

$$\Gamma^2 = \frac{2\omega_1\omega_2|d|^2}{n_s n_i n_p \epsilon_o c^3} \frac{P_p}{A_p} \qquad (3)$$

ℓ is the crystal length and $A_p = 1/2\, \pi w_p^2$. At degeneracy we have

$$\Gamma^2 = 2.3 \times 10^{-2} \frac{P_p}{A_p} \left(\frac{MW}{cm^2}\right) cm^{-2} \quad .$$

We have operated the $LiNbO_3$ parametric oscillator at 1.8 J/cm² for a 10 month period without crystal damage. The corresponding intensity for a 20 nsec pulse is 90 MW/cm² which gives a gain coefficient of $\Gamma\ell = 7.2$ for a 5 cm crystal or a total gain of $G = 1/4 \exp 2\Gamma\ell = 4.4 \times 10^5$. Under operation at such high gains, the parametric oscillator peak-to-peak stability reflects the 2% peak-to-peak stability of the pump laser.

In recent experiments we operated the oscillator at a pump spot size of w_p = 1.57 mm with a corresponding pump energy threshold of 7.3 to 7.5 mJ for a 7.5 cm and 12.1 cm cavity length. Figure 8 shows the relative parametric oscillator threshold for wavelength tuning vs that at degeneracy. The threshold is expected to increase as $(1 - \delta^2)^2$ where δ is the detuning factor given by $\delta = 2\omega_s/\omega_p - 1$. The measured relative threshold vs detuning increases more rapidly than expected due to increasing cavity losses for off degenerate operation.

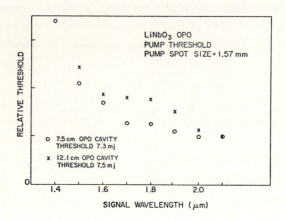

FIG. 8--Relative threshold vs wavelength tuning of the LiNbO$_3$ parametric oscillator.

Nonetheless, stable operation to 1.45 μm at a corresponding idler wavelength of 4.0 μm has been reached. With improvements in cavity coating design, operation to 1.40 and 4.4 μm should be possible. Beyond 4.4 μm LiNbO$_3$ itself becomes absorbing.

Figure 9 shows the measured parametric oscillator energy conversion efficiency vs the number of times above threshold for two cavity lengths. As expected from buildup time considerations, the longer cavity length requires a higher threshold energy. The relative conversion efficiency for both cavity lengths are identical. At 5.5 times above threshold the parametric oscillator operates at a total signal plus idler wave conversion efficiency of 40%. At this point, we have operated

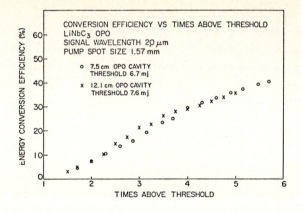

FIG. 9--Conversion efficiency vs times above threshold for the $LiNbO_3$ parametric oscillator operating at $\lambda_s = 2.0$ μm.

with 40 mJ of 1.06 μm energy at 10 pps and have generated 16 mJ of output energy at 1.9 μm and 2.41 μm. We were prevented from applying higher pump energies due to the damage limit of an inexpensive 45 degree reflector in the Nd:YAG amplifier chain. The 40% energy conversion efficiency and corresponding nearly 65% peak power conversion efficiency agrees very well with earlier predictions of conversion efficiency for a Gaussian beam pumped singly resonant parametric oscillator by Bjorkholm.[9] Bjorkholm showed that a peak power conversion efficiency of 80% should be reached for a singly resonant parametric oscillator at 10 times above threshold. The corresponding conversion efficiency at 5.5 times above threshold is 70% in very good agreement with our observations. The oscillator is operating at 1.07 J/cm^2 at 5.5 times above threshold. This is well below the expected damage level and illustrates the significance of the 5 cm $LiNbO_3$ crystals for convenient, highly efficient parametric oscillator operation.

Operation of the $LiNbO_3$ parametric oscillator at relatively large pump spot sizes required by the use of an angle tuned crystal, leads to linewidth control advantages. These advantages are enhanced by the oscillator's reproduction of the Nd:YAG pump beam spatial mode characteristics and the use of a plane parallel optical cavity.

We have investigated three primary linewidth control methods: a thin tilted etalon, a two element $LiNbO_3$ birefringent filter and a grating. All three methods lead to line narrowing of the parametric oscillator without experimental difficulty. In each case, a second 1 mm thick tilted etalon was used to further reduce the linewidth of the parametric oscillator to less than 0.1 cm^{-1} at the resonated signal wavelength. For a tilted etalon the insertion loss is given by[10]

$$\ell = \frac{R}{(1-R)^2} \left(\frac{4\alpha t}{nw}\right)^2 \tag{4}$$

where α is the tilt angle, t the etalon thickness, R the etalon reflectivity and w the beam radius and n the refractive index. The requirement for large spot sizes at the resonated wave is apparent. Figure 10 shows the spectral output of the oscillator with a 1 mm thick finesse of 5 etalon inside the oscillator cavity. The linewidth of the individual etalon modes at the signal wave is less than 0.2 cm^{-1} and is not resolved by the 1 meter spectrometer. The amplitude envelope of the etalon modes is determined by a single element birefringent filter narrowed oscillator linewidth. The resonated signal wave linewidth is less than 0.2 cm^{-1} while the nonreonant idler wave reproduces the 0.6 cm^{-1} linewidth of the Nd:YAG laser source.[11]

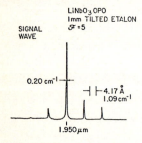
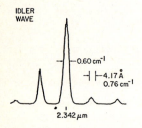

FIG. 10--Signal and Idler wave etalon narrowed output at 1.95 and 2.34 μm with a F=5, 1 mm thick fused silica tilted etalon. The wider idler wave spectrum reflects the linewidth of the 1.06 μm Nd:YAG pump source.

Prior to the operation of the parametric oscillator with a grating, we were concerned about possible grating insertion loss and damage. We therefore designed a two element $LiNbO_3$ birefringent filter as an alternative narrowing element.[12,13] The phase delay through a Brewster angle birefringent plate of thickness T at wavelength λ is given by

$$\delta = \frac{2\pi T}{\lambda \sin \theta_B} [n_o - n_e(\gamma)] \qquad (5)$$

where

$$\gamma = \tan^{-1} [\frac{n}{\cos A} - E] \qquad (6)$$

is the propagation angle relative to the crystal optic axis. Here E is the tilt angle of the optic axis out of the plane of the plate and A is the tuning rotation angle from the plane in the plate containing the incident beam to the optic axis, and n_o, $n_e(\gamma)$ and n are the ordinary, extraordinary and average indices of refraction at wavelength λ.

Consideration of the oscillator's operating bandwidth at degeneracy and the need for a relatively narrow bandwidth filter led to the design of a two element $LiNbO_3$ birefringent filter with element thicknesses of 1.5 mm and 3.0 mm. The c axis was chosen to lie in the plane of the plate to minimize temperature dependence and to provide a convenient tuning angle of 5 to 6° per free spectral range. The calculated free spectral range varies between 47 cm^{-1} at 1.4 µm to 48.2 cm^{-1} at 2.12 µm. The tuning rate is remarkably linear at near 8 cm^{-1}/degree rotation.

We fabricated the two element $LiNbO_3$ birefringent filter and mounted it on a General Scanner galvanometer for angle tuning control. Figure 11 shows the 3.3 cm^{-1} resultant bandwidth obtained with a two element filter. Also shown is the filter operating with a tilted etalon for additional linewidth control to less than 0.1 cm^{-1}. The strong modulation on the birefringent filter intensity profile is due to resonant feedback from the plane parallel sapphire output mirror acting as a partial etalon. The birefringent filter operated as expected and is a convenient, low loss primary line narrowing element.

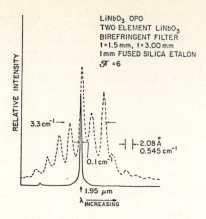

FIG. 11--LiNbO$_3$ parametric oscillator operating with a two element LiNbO$_3$ birefringent filter (dashed) and with an additional 1 mm thick fused silica tilted etalon (solid).

We were not optimistic about the use of a grating within the parametric oscillator cavity due to the high circulating signal wave intensity. However, preliminary measurements showed that at its blaze wavelength of 1.8 μm, an inexpensive 20 mm × 20 mm 600 ℓ/mm Bausch and Lomb plastic replica grating damaged at 1 GW/cm^2 for a 15 nsec pulse length. The corresponding burn density off the grating blaze at 1.06 μm was only 50 MW/cm^2 for a 50 nsec pulse length.

The resolution of a grating within an optical cavity with Gaussian beam spot size w is

$$\frac{\lambda}{\Delta \lambda} = \frac{m}{2d \cos \varphi} \pi w \qquad (7)$$

where m is the grating order, d the grating spacing and φ the angle of incidence for Littrow operation. Equation (7) can be re-written for the grating resolution in cm^{-1} as

$$\Delta \nu = \frac{1}{\pi w \tan \varphi} \qquad (8)$$

where the Gaussian beam radius w is in centimeters. For the present grating with $\omega \sim 20°$ and $w = .157$ cm we find $\Delta\nu = 5.57$ cm^{-1}. However, dynamic linewidth narrowing[14,15] of between 3 to 5 times tends to reduce the effective linewidth of the grating to near 1 cm^{-1}.

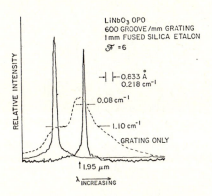

FIG. 12--LiNbO$_3$ parametric oscillator operating with a 600 groove/mm grating (dashed) and with an additional 1 mm thick fused silica etalon (solid) set at two tilt angles.

Figure 12 shows the output linewidth of the parametric oscillator with the grating alone (dashed curve) and with the addition of a 1 mm thick fused silica tilted etalon at two tilt angles. The grating linewidth is in close agreement with the expected value. The tilted etalon linewidth is less than 0.08 cm^{-1}, the limiting resolution of our 1 meter spectrometer. It should be mentioned that the intensity scales are arbitrary since the oscillator output energy is decreased only slightly by the insertion of the tilted etalon.

The use of a grating for primary line narrowing brings with it two distinct advantages; absolute wavelength control and wide tunability without overlapping orders. To demonstrate the second advantage we investigated the operation of the parametric oscillator near the degeneracy wavelength region where the oscillator bandwidth approaches 130 cm^{-1}. Figure 13 illustrates controlled signal and idler tuning to within 3 cm^{-1} of exact degeneracy. With the addition of a tilted etalon we were able to tune to within 0.2 cm^{-1} of degeneracy with a linewidth of less than 0.1 cm^{-1}. Furthermore, under these conditions either the signal or the idler wave could be resonated but not both. The oscillator operated singly resonant and remained stable in both amplitude and frequency. This is the first verification of a previous theoretical prediction of singly resonant operation under strong pumping conditions as opposed to doubly resonant operation under low gain conditions.[8]

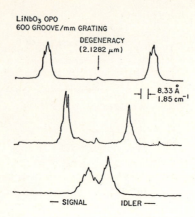

FIG. 13--LiNbO$_3$ parametric oscillator operation near degeneracy with a grating tuning element.

These line narrowing experiments demonstrate that highly stable, narrow linewidth operation is possible over the full tuning range of the parametric oscillator. In future experiments we plan to implement automatic tuning control at two levels. The first level is a continuous scan tuning where the LiNbO$_3$ crystal angle and the grating are synchronously tuned over the full 1.4 μm to 4.2 μm range at 1 cm^{-1} resolution. This tuning mode should be useful for optical pumping studies and survey spectroscopy. The second tuning mode is a high resolution scan where the grating and LiNbO$_3$ crystal are fixed at a center wavelength and the tilted etalon and cavity length are controlled to scan a single axial mode over a 1 cm^{-1} interval. Fourier transform limited linewidths of near 100 MHz are expected in the high resolution tuning mode.

CONCLUSION

The $LiNbO_3$ parametric oscillator described here forms the central element in a widely tunable coherent spectrometer system. Efficient extended frequency tuning by harmonic generation to the visible and ultraviolet is possible in angle phase-matched crystals of $LiNbO_3$ (1.4 μm to 0.7 μm), $LiIO_3$ (0.7 μm to 0.35 μm) and ADP (0.35 μm to 0.26 μm), and by mixing to the extended infrared in $AgGaSe_2$ (3 μm to 12 μm), CdSe (10 μm to 25 μm), GaP (20 μm to 200 μm) and $LiNbO_3$ (170 μm to 1 cm). We have recently carried out the doubling steps in $LiNbO_3$ and $LiIO_3$ and generated output wavelengths to 0.490 μm. We plan to conduct the mixing experiments in $AgGaSe_2$ and CdSe in the near future.

Two years ago, at the first tunable laser conference, one of us speculated on the possibility of a broadly tunable infrared source. Progress has been more rapid than expected in the intervening period. It is now possible to foresee a widely tunable, narrow linewidth coherent spectrometer under full automatic control operating prior to the next tunable laser conference.

ACKNOWLEDGEMENT

We wish to acknowledge support in this work by NSF-RANN, NASA, ARO and ERDA through LASL. We also want to acknowledge contributions to the work by Steve Brosnan, Hiroshi Komine and Michael Choy.

REFERENCES

1. R.L. Byer, "Parametric Oscillators", from Laser Spectroscopy, Ed. by R.G. Brewer and A. Mooradian, Plenum Pub. Co. New York, N.Y. (1974).

2. E.O. Ammann, J.D. Foster, M.K. Oshman and J.M. Yarborough, "Repetitively Pumped Parametric Oscillator at 2.13 μm", Appl. Phys. Letts. 15, p.131, (1969).

3. E.O. Ammann, J.M Yarborough and J. Falk, "Simultaneous Optical Parametric Oscillation and Second Harmonic Generation", J. Appl. Phys. 42, p.5618, (1971).

4. R.L. Byer, R.L Herbst, R.S. Feigelson and W.L. Kway, "Growth and Application of [01·4] $LiNbO_3$", Optic Commun. 12, p.427, (1974).

5. R.L Herbst, R.N Fleming and R.L. Byer, "A 1.4 to 4 μm High Energy Angle Tuned $LiNbO_3$ Parametric Oscillator", Appl. Phys. Letts. 25, p.520, (1974).

6. We wish to acknowledge helpful discussions with R. Deslattes, NBS, Gaithersburg, Md.

7. G.D. Boyd and D.A. Kleinman, "Parametric Interaction of Focussed Gaussian Light Beams", J. Appl. Phys. 39, p.3597, (1968).

8. R.L. Byer, "Optical Parametric Oscillators", in Quantum Electronics, Ed. by H. Rabin and C.L Tang, (Academic Press, New York, to be published); see also R.L Byer, "Nonlinear Optical Phenomena and Materials", Ann. Rev. Mat. Sci. vol. 4, p.147, (1974).

9. J.E. Bjorkholm, "Some Effects of Spatially Nonuniform Pumping in Pulsed Optical Parametric Oscillators", IEEE J. Quant. Elect. QE-7, p.109, (1971).

10. M. Hercher, "Tunable Single Mode Operation of Gas Lasers Using Internal Tilted Etalons", Appl. Opt. 8, p.1103, (1969).

11. S.E. Harris, "Tunable Optical Parametric Oscillators", Proc. IEEE, 57, p.2096, (1969).

12. A.L. Bloom, "Modes of a Laser Resonator Containing Tilted Birefringent Plates", Journ. Opt. Soc. Am. 64, p.447, (1974).

13. G. Holton, O. Teschke, "Design of a Birefringent Filter for High Power Dye Lasers", IEEE Journ. Quant. Elect. vol. QE-10, p.577, (1974).

14. D.C. Hanna, B. Luther-Davies, R.C. Smith, "Single Longitudinal Mode Selection of High Power Actively Q-switched Lasers", Opto-electronics, 4, p.249, (1972).

15. J.B. Atkinson and F. Pace, "The Spectral Linewidth of a Flashlamp-Pumped Dye Laser", IEEE Journ. Quant. Elect. vol. QE-9, p.569, (1973).

BROADLY TUNABLE LASERS USING COLOR CENTERS

L. F. Mollenauer
Bell Telephone Laboratories
Holmdel, New Jersey   07733

I. Introduction

Certain color centers in the alkali halides have made possible broadly tunable, optically pumped, cw, "dye-like" lasers for the near infrared. Like their dye counterparts, the color centers have homogeneously broadened emission bands that permit a tuning range of several tens of percent with any given material. However, there are no bleaching or aging effects during normal operation, and the required pump power is on the order of 30 times smaller than that required for the most efficient dye lasers.

The luminescence bands of a few of the centers suitable for laser action are shown below:

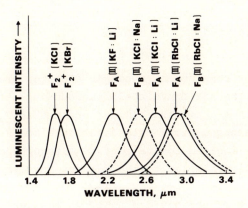

Figure 1   Luminescence bands.

Only two examples of the F_2^+ centers have been shown in Fig. 1. In fact, these can be made in any alkali halide. When all hosts are included, the F_2^+ luminescence bands cover the range 0.9 μm ≲ λ ≲ 2 μm continuously. The total tuning range for color center lasers (0.9 μm ≲ λ ≲ 3.3 μm) is of fundamental importance to molecular spectroscopy, pollution detection, fiber optic communications, and the physics of semiconductors. It is in terms of this special tuning range that the color center devices have their greatest advantage, since the region is one completely inaccessible to organic dyes.

The centers to be discussed[1] are based on a simple anion (halide ion) vacancy in an alkali halide crystal having the simple rocksalt structure:

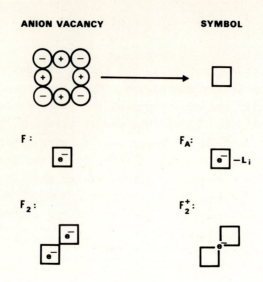

Figure 2

The ordinary F-center consists of a single electron trapped at such a vacancy. If one of the six immediately surrounding metal ions is foreign, say a Li^+ in a potassium halide, one has an F_A center.[2] Two F-centers adjacent along a [110] axis constitute the F_2, and, of course, the F_2^+ is its singly ionized counterpart.[3] Although not shown in Fig. 2, the F_B centers are very similar to the F_A, but involve two foreign ions instead of just one.[4]

The F_A centers are divided into two classes: those of type I behave like the ordinary F-center, whereas those of type II have a radically different relaxation behavior. The ordinary F-center and the $F_A(I)$ centers are not suitable for laser action, primarily on account of very low emission cross-sections. However, such is not the case for the $F_A(II)$, the F_2^+, or the F_2. Since the complexities introduced by two electrons make the ultimate suitability of the F_2 hard to predict, in this paper we will concentrate on the two simpler, single-electron systems, the $F_A(II)$ and the F_2^+.

II. Optical Pumping Cycles and Gain

Following optical excitation, the $F_A(II)$ center relaxes to a double well configuration, as shown below:

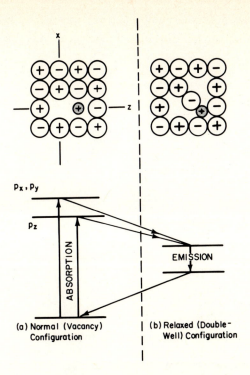

Figure 3 $F_A(II)$ Center

The relaxed system is somewhat analogous to the H_2^+ molecular ion, in this case with an additional negative charge between the two attractive centers. The oscillator strength for the luminescence transition is quite large ($f \sim 0.2$). The relaxation times are quite short ($\tau \sim 10^{-13}$ sec), such that the cycle shown constitutes a nearly ideal four level system.

The radical change in configuration is accompanied by an equally radical Stokes shift: the absorption bands are in the visible, whereas the emission energies are typically less than 0.5 eV. Note also that in the normal configuration, the foreign ion causes P_z orbitals to be distinguished from P_x and P_y. The extra absorption band thus created greatly increases the probability of overlap with a convenient pump source.

The quantum efficiency, η, of $F_A(II)$ center luminescence in KCl:Li is about 40 percent for T ≲ 77°K, and decreases slowly with increasing temperature until it approaches zero at 300°K.[5] Nevertheless, laser action has been obtained for T as high as 200°K.[6,7] Behavior of $F_A(II)$ centers in other hosts ought to be similar, although η has not yet been measured for these.

The F_2^+ center is even more closely analogous to the H_2^+

molecular ion. The energy levels of the F_2^+ can be predicted very closely from a model of an H_2^+ ion embedded in a dielectric continuum.[3] Below is an energy level diagram of the F_2^+:

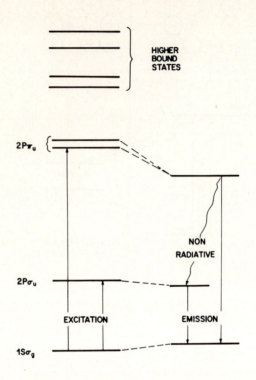

Figure 4 F_2^+ Energy Levels

There are two strong transitions: the $1S\sigma_g \to 2P\sigma_u$ in the infrared, and the $1S\sigma_g \to 2P\pi_u$ in the visible. The emission of the visible transition is seriously quenched at all but very low temperatures (T≲50°K) by competition from the lower energy emission. For this and other reasons the pump cycle of the visible transition is not suitable for laser action.

However, the infrared pumping cycle exhibits an emission quantum efficiency that is temperature independent, and probably 100 percent, although the absolute efficiency has not yet been measured. The absorption and emission bands for the infrared transition of the F_2^+ in KCl are shown below:

KCl F_2^+ CENTER: INFRARED TRANSITION

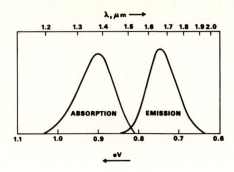

Figure 5

Since the Stokes shifts are small in this case, the oscillator strength of the luminescence is thought to be the same as that of the absorption, $f \sim 0.2$. The infrared cycle ought to be highly suitable for laser action.

It is instructive to compare the optical gains possible with the above-mentioned centers. For a Gaussian band of full width at the half-power points $\delta\nu$, the gain coefficient at the band peak, α_o, can be calculated from the well known formula:

$$\alpha_o = \frac{N^* \lambda_o^2 \eta}{8\pi n^2 \tau_\ell} \frac{1}{1.07 \, \delta\nu} \tag{1}$$

where N^* is the density of centers in the relaxed-excited state, λ_o is the wavelength at the band center, n is the host index, η is the quantum efficiency of luminescence, and τ_ℓ is the measured decay time. Values of α_o thus calculated for the various centers in the common host KCl are presented in the table below. A value of $N^* = 10^{16}/cm^3$ has been assumed in all three cases. (This represents the largest N^* that can be used for the ordinary F center, without incurring serious interaction among the excited centers. However, no such problem exists for the other two types.)

ASSUME $N^* = 10^{16}/cm$ IN ALL CASES
HOST: KCl; n = 1.5

QUANTITY	F	F_2^+	$F_A(II)$	UNITS
λ_0	1.0	1.68	2.7	μm
τ_β/η	600	200	200	nsec
$\delta\nu$	6.3	1.69	1.45	10^{13} Hz
α_0	0.04	3.5	4.2	cm^{-1}

Table 1

α_0 is ominously small for the ordinary F center. In fact, self-absorption by the excited F center reduces the true gain to a net loss. On the other hand, the values of α_0 calculated for the other two types are more than ample, and furthermore, no self-absorption exists in either case.

III. <u>Construction and Performance of a Color-Center Laser</u>

To date we have constructed and operated two tunable cw lasers using $F_A(II)$ centers.[7,8] The basic cavity configuration of the later and more sophisticated version is shown below. The parameters were f = 25 mm, d_2 = 600 mm, t = 1.72 mm, and 2θ = 20°; ϕ = Brewster's angle.

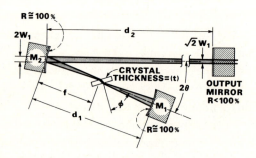

Figure 6 Basic Cavity Configuration

The focused, folded cavity configuration of Fig. 6 will be recognized as that used in many dye lasers, with a crystal slab substituted for the dye cell or jet stream. A more complete schematic of the laser is shown below:

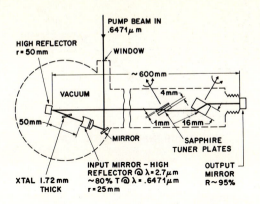

Figure 7   Laser Schematic

The entire cavity was surrounded by a vacuum enclosure, as indicated by the dashed lines. The vacuum was required for two reasons: first, to provide thermal insulation for the crystal, and second, to prevent atmospheric absorption (especially from H_2O) from interfering with laser action. The cylindrical can surrounding the crystal and spherical mirror section was open at the top, and the vacuum seal was completed by a removable liquid nitrogen storage can. The crystal was mounted on a cold finger, and referenced to the other components via a thin spider of thermally insulating material. All other optical components were at room temperature.

To measure the gain capabilities of the KCl:Li laser, an output mirror having R = 50 percent was used. The required single pass gain was compared with that calculated for the input power at threshold (130 mW) and the beam waist cross-section area (0.83×10^{-5} cm^2). The calculated gain was 1.47, in excellent agreement with the required value of $\sqrt{2}$.

The maximum energetic efficiency of the KCl:Li laser should be about 10 percent, since the ratio of pump to luminescence photon energies is 5, and only about one-half the centers will be oriented such that they can radiate into a linearly polarized laser mode. Thus, with an output mirror transmission of 1.6 percent and a measured intracavity loss of 5 percent, the net efficiency should be about 2.2 percent. The behavior shown in Fig. 8 below is consistent with that estimate.

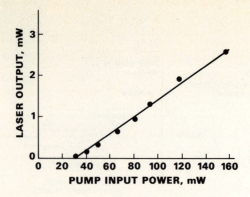

Figure 8 P_{out} versus P_{in}, KCl:Li F_A(II) Laser

In Fig. 9 below, the reciprocal pump power required at threshold is compared with the luminescence band shape. Note that the laser tunes as far as the 25 percent power points of the band.

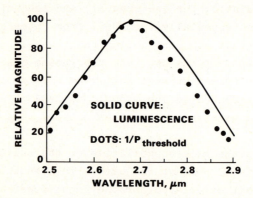

Figure 9 Tuning Characteristic, KCl:Li F_A(II) Laser

The spectral purity obtained with the birefringence tuner alone and for the laser operating far above threshold is shown below.

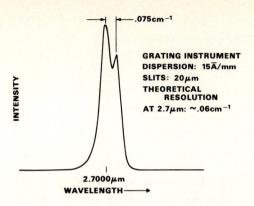

Figure 10 Spectral Purity, $F_A(II)$ Center Laser

Figure 10 most probably indicates simultaneous operation on two axial mode frequencies. True single frequency operation could always be obtained by addition of an intracavity etalon.

To sum up, the performance outlined above satisfied our most optimistic expectations. It should be mentioned that we have obtained quite satisfactory performance with the $F_A(II)$ center in RbCl:Li as well. Experiments to test the F_2^+ as a laser material will be performed soon. The host will be KCl and the pump will be a Nd:YAG laser operating at 1.34 μm. (See Fig. 5.) The cavity configuration will be essentially that described above.

IV. Preparation of Color Centers

For the most part, preparation of the color centers discussed above is not difficult. To aid the newcomer, we briefly outline the required steps and list some important references.

In all cases, ordinary F-centers are created first, usually through the process of additive coloration.[9,10] The formation of F_A or F_2 centers then results from a simple aggregation process,[11,2] as follows: First, thermal ionization of optically excited F-centers results in the formation of pairs of F' centers (the F' is a two-electron center) and empty vacancies. At sufficiently high temperatures (T≳-50°C) the empty vacancies wander through the lattice until they meet either an F-center or a foreign metal ion. Recapture of an electron from the F' by the vacancy then leads to formation of F_2 centers in the first instance, or to formation of F_A centers in the second.

If the foreign metal ion concentration is several orders of

magnitude greater than that of the F-centers, an essentially complete conversion can be carried out, with F_A centers as the exclusive end product. However, the creation of F_2 centers cannot be carried to completion without an accompanying creation of higher aggregates, such as F_3, F_4, etc. Thus, the optimum conversion to F_2 will necessarily involve a finite residue of F-centers.

F_2 centers are converted to F_2^+ by subjecting the F_2 centers to ionizing radiation.[12] To make the process efficient, traps must be provided for the excess electron. The most successful scheme[13] for creating traps involves the U center, which is an H^- ion trapped at an anion vacancy. U centers are formed by baking a crystal containing F-centers in an atmosphere of H_2. The U centers absorb only in the hard UV. Pumping of the U band at crystal temperatures below $\sim 200°K$ results in formation of empty vacancies (the desired traps) and interstitial H^- ions in pairs.

U centers can be converted into ordinary F-centers by gentle x-raying or by pumping with UV at room temperature. The process of temporarily "storing" F-centers as U centers is quite helpful in the manufacture of laser quality crystals. The slow cool-down possible after $F \rightarrow U$ conversion allows for the complete annealing out of strain and the elimination of strain-induced birefringence. Furthermore, the clarity of the crystal containing U centers greatly aids in the inspection for defects during crystal cutting and polishing.

V. A Distributed Feedback Laser

For certain applications, such as polluation monitoring, it would be desirable to have a very inexpensive laser that would be fixed-tuned to a predetermined frequency, such as a prominent absorption line of a given molecular species. Figure 11 suggests one way that such a device might be made from color centers. It would make use of the principle of distributed feedback. That is, if either the index n or the gain coefficient α is modulated spatially with period d, there will be strong feedback at those wavelengths that satisfy the Bragg condition:

$$n\lambda = 2d \qquad (2)$$

without the need for external mirrors.

The device shown in Fig. 11 would use a modulation of α itself, obtained by means of a periodic variation in the F_A center concentration.

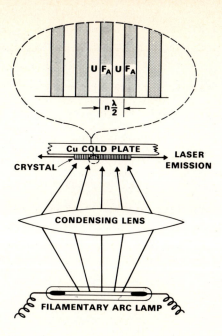

Figure 11 Distributed Feedback Laser

The required grating could be written into the KCl:Li or other similar crystal by taking advantage of the photochromic conversion process U → F. The beam of an ultraviolet laser can be split and made to interfere with itself to form an interference pattern of the desired period. For a grating of a few cm length, the required pump intensities at threshold should be on the order of 30-100 W/cm^2. Such intensities should be attainable from an arc lamp.

REFERENCES

1. For a comprehensive review of color centers in alkali halides, see W. B. Fowler in <u>Physics of Color Centers</u>, edited by W. B. Fowler (Academic Press, New York, 1968), Chapter 2.

2. For an extensive treatment of F_A centers, see F. Lüty, in <u>Physics of Color Centers</u>, edited by W. B. Fowler, (Academic Press, New York, 1968), Chapter 3.

3. M. A. Aegerter and F. Lüty, Phys. Stat. Sol. $\underline{43}$, 244 (1971).

4. N. Nishimaki, Y. Matsusaka and Y. Doi, J. Phys. Soc. Japan $\underline{33}$, 424 (1972).

5. G. Gramm, Phys. Lett. $\underline{8}$, 157 (1964).

6. B. Fritz and E. Menke, Solid State Comm. $\underline{3}$, 61 (1965).

7. L. F. Mollenauer and D. H. Olson, Appl. Phys. Lett. $\underline{24}$, 386 (1974).

8. L. F. Mollenauer and D. H. Olson, J. Appl. Phys. $\underline{46}$, 3109 (1975).

9. H. Rögener, Annalen der Physik, $\underline{29}$, 386 (1937).

10. C. Z. van Doorn, Rev. Sci. Instr. $\underline{32}$, 755 (1961).

11. H. Härtel and F. Lüty, Z. Physik $\underline{177}$, 369 (1964).

12. M. A. Aegerter and F. Lüty, Phys. Stat. Sol. $\underline{43}$, 227 (1971).

13. M. A. Aegerter and F. Lüty, 1971 International Conference on Color Centers in Ionic Crystals, Abstract 47.

THE OXYGEN AURORAL TRANSITION LASER SYSTEM
EXCITED BY COLLISIONAL AND PHOTOLYTIC ENERGY TRANSFER[*†]

J. R. Murray, H. T. Powell, and C. K. Rhodes[‡]

Lawrence Livermore Laboratory
Livermore, California 94550

ABSTRACT

The properties of laser media involving the auroral transition of atomic oxygen and analogous systems are examined. A discussion of the atomic properties, collisional mechanisms, excitation processes, and collisionally induced radiative phenonema is given. We find that crossing phenomena play a particularly important role in governing the dynamics of the medium.

I   Atomic Properties and Interactions

It is now clearly established that certain metastable atomic species which are open shell systems exhibiting the characteristics of a two electron spectrum have several properties that are unusually well suited for high power lasers.[1,2] The states that appear most appropriate for energy storage are the ^1S_o terms which correspond to the highest levels of np^2 and np^4 configurations (two electron or two hole systems, respectively). Aside from the order of the fine structure levels of the lowest ^3P manifold, the energy level structures of these two configurations are identical. As an example, figure (1a) shows the terms of the oxygen atom relevant to our discussion. The three states (^3P, ^1D, and ^1S) arising from recoupling of the electrons in the $2p^4$ configuration are shown in addition to other levels originating from excited configurations[3] (i.e. ^5S, $^3S^o$, ^5P, and ^3P). The corresponding data are illustrated in Figure (1b) and Figure (1c) for sulfur and selenium, respectively. At this point we will concentrate our discussion on the first three members of column VI (O, S, and Se) as these appear to be the most favorable candidate systems.

---

[*]Work partially performed under the auspices of the United States Energy Research and Development Administration.

[†]Work partially performed at Stanford Research Institute.

[‡]Present address: Molecular Physics Center, Stanford Research Institute, Menlo Park, California 94025.

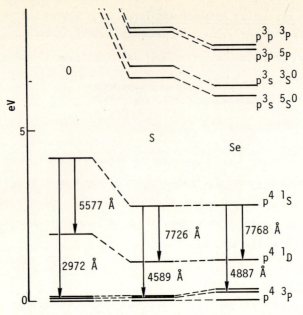

Figure 1. Partial energy level diagram characteristic of column VI atoms illustrating the term values pertaining to the np^4 configuration as well as selected excited states arising from excited configurations. The particular term values of O, S, and Se are shown in (a), (b), and (c) respectively.

The radiative and collisional characteristics of these metastable systems are remarkable. We begin with the free system radiative properties which are compiled in Table I with the assumption that the optical cross sections listed correspond to Doppler broadened transitions at a temperature of $300°K$. Note that for selenium, hyperfine effects and the normal isotopic abundance of selenium isotopes, provides

Table I

ATOMIC PROPERTIES

Atom	Transition	Wavelength (Å)	Lifetime (sec)	Cross Section (cm^2)
O	$^1S_0 \rightarrow {}^1D_2$	5577	~ 0.7	9×10^{-20}
O	$^1S_0 \rightarrow {}^3P_1$	2958	~ 15	5.2×10^{-22}
S	$^1S_0 \rightarrow {}^1D_2$	7726	~ 0.56	4.5×10^{-19}
S	$^1S_0 \rightarrow {}^3P_1$	4589	~ 2.9	3×10^{-20}
Se	$^1S_0 \rightarrow {}^1D_2$	7768	~ 0.43	4.8×10^{-19} (9.7×10^{-19})*
Se	$^1S_0 \rightarrow {}^3P_1$	4887	~ 0.13	4.0×10^{-19} (8×10^{-19})*

*Selected single isotope rather than normal isotopic abundance.

an additional mechanism for the variation of the optical cross section. The $^1S_o \to$ $^1D_2$ transition arises from the electric quadrapole (E2) transition moment while the $^1S_o \to {}^3P_1$ is a magnetic dipole transition (M1) which violates the spin rule^4. We observe that for selenium the magnetic transition is stronger than the electric quadrapole transition, a fact which serves to contrast with the lighter two species.

The collisional behavior of the ^1S_o term of these atoms is exceptional. Generally, collisions can transfer energy into kinetic energy of motion, internal excitation, or radiation. In terms of the ^1S_o state, we desire low cross sections for the former two inelastic processes as these represent nonradiative losses of the stored energy. We will see later that collisions can have a dramatic and desirable influence on the radiative properties of the medium and allow a convenient mechanism for adjusting the coupling of stored energy to the electromagnetic field. Table II and Table III illustrate, respectively, the quenching rate data for oxygen and sulfur

Table II

DEACTIVATION RATES OF ATOMIC OXYGEN METASTABLES*

Collision Partner	$O(^1S_o)$	$O(^1D_2)$
N_2	$< 5 \times 10^{-17}$	5.5×10^{-11}
N_2O	1.5×10^{-11}	2.2×10^{-10}
NO	4×10^{-10}	2.1×10^{-10}
$O(^3P)$	7.5×10^{-12}	--
CO_2	4×10^{-13}	2.2×10^{-10}
CO	9.4×10^{-14}	7.5×10^{-11}
Ar	5.2×10^{-18}	10^{-12}
Xe	6.7×10^{-15}	10^{-10}

*Rates given in cm^3-molecule^{-1} sec^{-1} for 300°K. For further details and references see the material in Ref. (2). The majority of these rates are reviewed in R. J. Donovan and D. Husain, Chem. Rev. **70**, 489 (1970).

^1S_o and $^1D_2$ states in collision with a variety of partners. Data for the corresponding states of selenium do not currently exist, although we anticipate roughly similar behavior. Two basic facts are apparent: the ^1S_o states relax slowly in collisions with several systems while the corresponding $^1D_2$ states quench orders of magnitude more rapidly. We will see that this difference in behavior is produced by the presence of an available surface crossing in the $^1D_2$ reaction channel and the absence of such a crossing mechanism in the ^1S_o case.

Table III

DEACTIVATION RATES FOR ATOMIC SULFUR METASTABLES*

Collision Partner	$S(^1S_o)$	$S(^1D_2)$
OCS	4×10^{-13}	8×10^{-11}
CO	$< 3.5 \times 10^{-16}$	2.2×10^{-11}
CO_2	$< 6 \times 10^{-17}$	2.6×10^{-11}
CS_2	8×10^{-10}	--
Ar	$< 3.5 \times 10^{-17}$	2×10^{-12}

---

*Rates given in cm^3-molecule^{-1}-sec^{-1} for 300°K. Data concerning the sulfur rates are contained in R. J. Donovan, Trans. Faraday Soc. 65, 1419 (1969); R. J. Donovan, L. J. Kirsch, and D. Husain, Trans. Faraday Soc. 66, 774 (1970); O. J. Dunn, S. V. Filseth, and R. A. Young, J. Chem. Phys. 59, 2892 (1973).

At high excited particle densities ($\sim 10^{17}$ cm^{-3}) inelastic collisions between two excited species will impose a fundamental limitation on the maximum achievable stored energy density. For example, a mechanism of this nature is known to be a limiting process in the rare gas excimer systems[5]. In this case, we require a low probability for inelastic scattering in an encounter between two excited ^1S_o systems. Specifically, we would like the dominant channel to be elastic such as that given by the following reaction involving two oxygen atoms.

$$O(^1S_o) + O(^1S_o) \rightarrow O(^1S_o) + O(^1S_o) \quad (1)$$

Such collisions redistribute the kinetic energy of relative motion, but leave internal motions unaffected.

Dominance of the elastic channel for the scattering of two electronically excited atoms establishes requirements on the nature of the interatomic potential and on the relationship of that potential to other curves of the diatomic system. We now examine these characteristics for the case of O_2. S_2 and Se_2 will be similar to O_2 as will the heteronuclear systems such as SSe except for the absence of gerade and ungerade symmetry. We expect the elastic process given by reaction (1) to dominate the scattering under circumstances for which no mechanism exists to (a) convert internal energy to kinetic energy of relative motion and (b) to resonantly excite another mode of internal excitation. An example of the former is the reaction

$$O(^1S_o) + O(^1S_o) \rightarrow O(^1S_o) + O(^1D_2) + \Delta KE \quad (2)$$

while the process

$$O(^1S_o) + O(^1S_o) \rightarrow O_2^+(X^2\Pi_g) + e^- \quad (3)$$

is characteristic of the latter. For an analysis of these reactions we refer to the potential energy curves available for O_2 shown in Figure (2) as derived from the

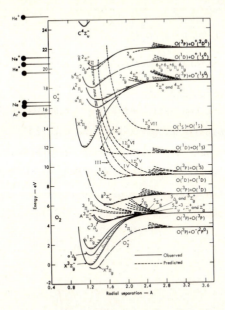

Figure 2. Energy levels of the oxygen molecule as compiled from the literature.[6,7]

appropriate sources.[6,7] We note immediately that the $^1\Sigma_g^+$ curve arising from the $O(^1S) + O(^1S)$ interaction is of strongly repulsive character, and therefore, does not cross or even significantly approach any other curve of the diatomic system for kinetic energies below 2 electron-volts. Since normal thermal velocities correspond to approximately 0.025 electron-volts, the repulsive nature of this curve prevents the atoms from entering a region where transitions to other states of the O_2 system could occur. For example, no crossing with the ionic $O_2^+(X^2\Pi_g)$ curve is available at thermal energies, a fact which prevents reaction (3) from having an appreciable amplitude. In this $^1\Sigma_g^+$ state the system is isolated from other channels and elastic scattering is expected to dominate.

Quite the opposite behavior is characteristic of the $^1D_2$ states, a fact which is strongly suggested by its relatively rapid quenching. The reaction

$$O(^1D_2) + N_2(X) \rightarrow O(^3P) + N_2(X) \tag{4}$$

has been considered extensively in the literature.[8,9] In this case, the presence of a strong attraction enables crossing with other potentials rising from below.

This crossing now provides an efficient mechanism for the conversion of electronic energy into the energy of nuclear motions. For N_2 this energy can appear as both vibration and translation. Experimental data verifying this conclusion have been obtained for the reaction[10]

$$O(^1D_2) + CO(X) \rightarrow O(^3P) + CO^\dagger(X) \tag{5}$$

where $CO^\dagger(X)$ denotes a vibrationally excited system.

II Collisionally Stimulated Radiation

Under ambient conditions consisting of a high pressure of rare gas (e.g. 25 atmospheres of krypton) the radiative properties of the excited system are grossly changed from those of the free atom. These changes are manifestations of the collisional influence on the radiative properties of the free atoms. Both the radiative rate and the bandwidth of the fluorescence are substantially modified. The upper spectrum in Figure (3) illustrates the spectrum of the fluorescence observed in an

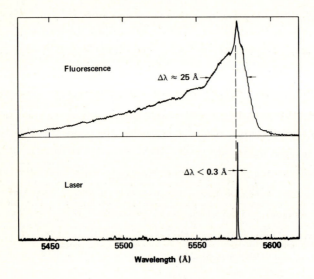

Figure 3. The spectra (densitometer traces) of fluorescence and stimulated emission from KrO observed in a mixture of 5 torr O_2 and 25 atmospheres of Kr. The upper spectrum illustrates the observed fluorescence while the lower shows the stimulated spectrum.

electron beam excited mixture consisting of 5 torr O_2 and 25 atmospheres Kr. A structured peak slightly displaced from the position of the atomic line and a long blue wing are observed. For comparison, the lower spectrum in Figure (3) shows the spectrum of the stimulated emission observed from this same mixture.

The emitting species is the quasi-molecule KrO^* and can be regarded as a manifestation of the process

$$O(^1S) + Kr(^1S) \rightarrow O(^1D) + Kr(^1S) + \gamma. \tag{6}$$

Such collisionally induced emission is generally characteristic of these oxygen metastables for collisions with closed shell systems of both atomic and molecular varieties[11] (e.g. Ar and N_2). The heavier atoms sulfur and selenium will exhibit analagous behavior. It is important to note that for the quenching of $O(^1S)$ by the rare gases induced radiation is the dominant channel and accounts for the <u>total</u> quenching rate.[12] The collision dynamics discussed earlier prevent a contribution from other inelastic channels, and thus, allow the radiative mechanisms to dominate the inelastic scattering. Under the circumstances the net rate of emission via the molecular process is then directly proportional to the collision rate, or equivalently the density. The linewidth is determined by the relationship between the induced emission probability and the associated level shifts and is, hence, independent of density. The stimulated emission cross section, which is proportional to the ratio of these two quantities, then scales as the density of the medium. We state finally that considerable variation in this parameter is achievable, since the collision induced rates for the rare gases cover several orders of magnitude.[12]

III <u>Kinetic Processes</u>

The desired ^1S_o can be excited by both radiative, collisional, and presumably, combined radiative/collision processes. Radiative mechanisms such as

$$\gamma + N_2O(\widetilde{X}) \rightarrow O(^1S_o) + N_2(X) \tag{7}$$

and

$$\gamma + OCS(\widetilde{X}) \rightarrow S(^1S_o) + CO(X) \tag{8}$$

are known to exhibit high quantum yields (> (80%) and provide an attractive means for excitation, since the quanta required may be efficiently generated from the rare gas continua.[13] The Ar_2 continuum is an excellent match for the N_2O spectrum while the Kr_2 continuum has a similarly superb overlap with the corresponding OCS absorption.[14] The radiative processes for the oxygen system are thoroughly examined in Ref. (2). A most fortuitous match also exists for the Xe_2 continuum at $\sim$ 1715 Å with the strong band[15] in OCSe which is currently believed to lead to photolytic production of $Se(^1S_o)$.

Several collisional mechanisms exist which are capable of producting the desired excited atoms[2]. We list below a number of mechanisms which can play a role for oxygen, a case for which there is considerable data stemming from atmospheric studies.[16]

$$O(^3P) + e^- \rightarrow O(^1S) + e^- \tag{9}$$

$$O_2^+ + e^- \rightarrow O(^1S) + O(^3P, ^1D) \tag{10}$$

$$O_2 + e^- \rightarrow O(^1S) + O(^3P, ^1D) + e^- \tag{11}$$

$$O + O + O \rightarrow O(^1S) + O_2 \tag{12}$$

$$O_2 + N^+ \rightarrow O(^1S) + NO^+ \tag{13}$$

$$O(^3P) + N_2(A^3\Sigma_u^+) \rightarrow O(^1S) + N_2(X^1\Sigma_g^+) \tag{14}$$

The cross section given theoretically for process (9) is $\sim 3 \times 10^{-18}$ cm^2 at 9 eV while the competing reaction generating $O(^1D)$ has a value $\sim 3 \times 10^{-17}$ cm^2 at the same energy.[17] For oxygen bearing impurities in high pressure rare gases excited by relativistic electron beams another mechanism has been proposed.[18] It is under these circumstances that stimulated emission[19] has been observed from ArO*, KrO*, and XeO*. An example of the process suggested by Huestis is

$$Ar_2^* + O(^3P) \rightarrow Ar + Ar + O(^1S) \tag{15}$$

which is similar mechanistically to that proposed for an atomic metastable (Ar*) reaction by Golde and Thrush.[20] The entrance channel switches to the attractive ionic curve which, on account of its deep binding, crosses the exit channel and thus provides a pathway for process (15). Crossings again play a dominant role.

IV Concluding Remarks

We conclude by observing that the column VI materials O, S, and Se are attractive candidates for high energy laser systems, a finding which is based on both the free system and collisional properties of these atoms. In this discussion the behavior of surface crossings emerged in a central position for both excitation and quenching mechanisms. Indeed, collisions were seen in two complementary roles, governing the channeling of energy flow and in influencing the radiative properties of the medium.

References

1. J. R. Murray and P. W. Hoff in High Energy Lasers and Their Applications, edited by Stephen F. Jacobs, Murray Sargent, III, and Marlan O. Scully (Addison-Wesley Publishing Co., Inc., Reading, Massachusetts, 1974).

2. J. R. Murray and C. K. Rhodes, The Possibility of a High-Energy-Storage Visible Laser on the Auroral Line of Oxygen, UCRL-51455, 1973; available from NTIS, Springfield, Va.

3. Charlotte, E. Moore, Atomic Energy Levels as Derived from the Analyses of Optical Spectra, NSRDS-NBS 35 (USGPO, Washington, D.C., 1971).

4. Roy H. Garstang in Atomic and Molecular Processes, edited by D. R. Bates (Academic Press, New York, 1962) p. 1.

5. Charles K. Rhodes, IEEE J. Quantum Electron. QE-$\underline{10}$, 153 (1974); C. W. Werner, E. V. George, P. W. Hoff, and C. K. Rhodes, Appl. Phys. Lett. $\underline{23}$, 139 (1973).

6. F. R. Gilmore, J. Quant. Spectrosc. Radiat. Transfer $\underline{5}$, 369 (1965).

7. H. F. Schaefer, III and F. E. Harris, J. Chem. Phys. $\underline{48}$, 4946 (1968).

8. J. C. Tully, J. Chem. Phys. $\underline{61}$, 61 1974).

9. George E. Zahr, Richard K. Preston, and William H. Miller, J. Chem. Phys. $\underline{62}$, 1127 (1975).

10. M. C. Lin and R. G. Shortridge, Chem. Phys. Lett. $\underline{29}$, 42 (1974).

11. S. V. Filseth, P. Stuhl, and K. H. Welge, J. Chem. Phys. $\underline{52}$, 239 (1970).

12. M. Atkinson and K. H. Welge, Collisionally Induced Emission of $O(^1S)$, to be published.

13. J. J. Jortner, L. Meyer, S. A. Rice, and E. G. Wilson, J. Chem. Phys. $\underline{42}$, 4250 (1965); C. W. Werner, E. V. George, P. W. Hoff, and C. K. Rhodes, Appl. Phys. Lett. $\underline{25}$, 235 (1974); E. E. Huber, Jr., D. A. Emmons, and R. M. Lerner (private communication).

14. Graham Black (private communication).

15. J. Finn and G. W. King, J. Mol. Spectrosc. $\underline{56}$, 39 (1975); ibid., 52 (1975).

16. J. W. Chamberlain, Physics of the Aurora and Airglow (Academic Press, New York, 1961); A. Omholt, The Optical Aurora (Springer-Verlag, Berlin, 1971).

17. R. J. W. Henry, P. G. Burke, and A. L. Sinfailam, Phys. Rev. $\underline{178}$, 218 (1969).

18. D. Huestis (private communication).

19. H. T. Powell, J. R. Murray, and C. K. Rhodes, Appl. Phys. Lett. $\underline{25}$, 730 (1974).

20. M. F. Golde and B. A. Thrush, Chem. Phys. Lett. $\underline{29}$, 486 (1974).

SYNCHRONOUS MODE-LOCKED DYE LASERS FOR PICOSECOND
SPECTROSCOPY AND NONLINEAR MIXING

L.S. Goldberg and C.A. Moore*
Naval Research Laboratory
Washington, D.C. 20375

I. INTRODUCTION

Mode locking of organic dye lasers was first accomplished by pumping with the pulse train from high power mode-locked Nd:glass and ruby lasers.[1-4] By setting the cavity lengths of the two lasers equal or in integer relationships, the gain in the dye medium was impulsively driven in synchronism with the circulation period of the dye cavity radiation, resulting in the emission of a train of short duration pulses. Interest in this method diminished with the advent of passive mode locking of flashlamp-pumped and cw dye lasers using saturable absorbing dyes.[5-7] However, by its very nature synchronous mode locking affords an important experimental advantage: the tunable short pulses generated from the dye laser can possess a high degree of time synchronism with the intense short pulses from the pump laser source. This result was recently shown in our laboratory in streak camera studies with a Nd:glass pumped system.[8] The pulse synchronism feature is particularly useful for applications in picosecond spectroscopy and nonlinear frequency mixing since it enables one to illuminate a material system with intense picosecond pulses at two independent frequencies, where one (or both, if two dye lasers were synchronously pumped) can be continuously tuned and given an adjustable time delay.[8]

Synchronous pumping has recently been shown effective for mode locking of cw dye lasers.[9-12] Continuous modulation of the dye gain was provided by mode-locked He-Ne or argon-ion pump lasers with pulse widths of ∿ 200 - 300 psec. Tunable dye output pulses as short as a few psec duration have been reported.[10] Since no saturable absorber is necessary with the synchronous pumping technique, the mode-locked output can be continuously tuned over an extensive lasing range

---

* NRC-NRL Postdoctoral Research Associate

without pulse degeneration. Because the dye laser is driven well above threshold, it can operate with improved stability as compared to passive mode locking.

In this paper we shall describe studies and applications of a synchronous mode-locked dye laser which is pumped by intense 25 psec pulses from a passively mode-locked, frequency doubled Nd:YAG laser. The performance of the dye laser replicates the main features of the Nd:YAG pump, namely high repetition rate, fundamental-mode beam quality, high peak power, and uniform pulse trains. Through use of Fabry-Perot tuning elements we have obtained efficient generation of short pulses of transform-limited bandwidth which can be tuned over a broad range in several laser dyes.[13] The synchronism of the pump and dye pulses has been applied to the operation of a high gain short-pulse dye amplifier, and to the generation of narrow-bandwidth tunable picosecond pulses in the UV and IR spectral regions by nonlinear mixing.[14]

II. SYNCHRONOUS DYE LASER

Figure 1 shows a schematic of the experimental arrangement. A repetitively pulsed, passively mode-locked Nd:YAG laser provided 1.064 μm pulse trains of 3 mJ energy in a TEM_{oo} mode beam. After frequency doubling in a KD*P crystal the pulse train energy at 532 nm was approximately 1 mJ. The individual pulses in the pump train were of approximately 25 psec duration and separated by 5.6 nsec. The pulses have a transform-limited bandwidth of ~ 1 cm^{-1}, which is of particular advantage for the nonlinear mixing experiments. The dye laser cavity was formed by a dichroic 1 m radius input mirror and a $\sim$25% transmitting flat output mirror mounted on a translation stage to allow matching to the length of the pump cavity. The longitudinal pumping geometry and gentle focusing enabled excitation of the fundamental-mode volume in the dye, contained in a 2 mm-path Brewster-angle flow cell. The resulting energy density of the pumping pulses, which reached $\sim$50 mJ/cm^2 at the peak of the train, was sufficient to drive the dye medium highly into saturation. A Nd:YAG amplifier could be used to provide a factor of 5-10 greater incident pump energy, in which case a 3 m radius input mirror was substituted to allow pumping of a larger mode volume in the dye. Lasing was studied in dyes of rhodamine B, cresyl violet perchlorate, and carbazine 122 in alcohol or water-Ammonyx LO solutions. Tilted, narrow-gap (3-6 μm) Fabry-Perot etalons provided primary spectral narrowing and tuning of the

dye emission. By addition of a second etalon of larger gap spacing
(100 µm), the spectrum was further narrowed to attain transform-
limited short-pulse operation.

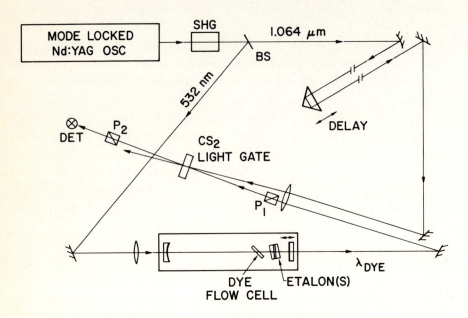

FIGURE 1. Schematic of the dye laser and experimental arrangement.

In Figure 1, a CS_2 Kerr-effect light gate is shown which is
operated by the intense 1.064 µm pulse trains remaining after frequency
doubling. This represents one example of the utility of synchronous
pulses from the pump laser. The light gate in this instance was used
to determine the relative timing and duration of the dye and pump
pulses exiting the dye laser. A picosecond streak camera was used for
temporal measurements at higher resolution on individual pulses in
the train.

Figure 2 shows an oscillogram of the pulse-train emission from
the dye laser interleaved with the input pump pulse train. The initial
delay before dye lasing occurs indicates the relatively few cavity
transits required for pulse build up. Despite the pulse-burst mode of
operation, efficiencies as high as 35% have been obtained for the
conversion of pump to dye pulse train energy using rhodamine 6G. The
output of the dye laser is in a TEM_{oo} mode beam, linearly polarized
parallel to the pump.

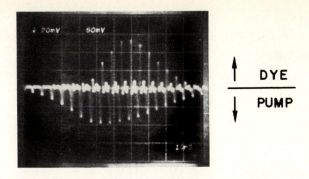

FIGURE 2. Dye laser and 532 nm pump pulse trains.

The lasing efficiency and temporal character of the dye pulse emission show a strong dependence on the relative cavity lengths of dye and pump lasers. Studies of this dependence revealed a characteristic peaking in the output energy over a length change of several mm. At the cavity setting corresponding to maximum output energy ($\pm$ 0.5 mm), measurements with the light gate and streak camera showed that the dye pulses were generated in time coincidence with the pumping pulses over the entire train and that their duration was approximately 30 psec, somewhat longer than the pump. Figure 3 presents light gate data obtained at three dye cavity lengths relative to matching. There is a distinct asymmetry in the appearance and timing of the dye pulse emission for cavities which are, respectively, shorter and longer than matching. In particular, temporal overlap between dye and pump pulses is lost for a slight lengthening of the cavity from match. These characteristics are a result of an asymmetry in the conditions for generating the pulses, since the circulating dye cavity radiation will arrive at the dye cell either progressively earlier or later than the appearance of gain.[8]

The formation of shortened pulses in an amplifying medium following rapid creation of a large gain has been discussed in studies of super-radiant dye emission by picosecond excitation.[15-16] The large gain and exponential nature of the amplification process lead to a sharpening of the amplified spontaneous emission. At sufficiently high dye intensities the high stimulated-emission rate will rapidly deplete the upper-level population resulting in a further shortening of the pulse. These considerations apply to pulse generation in the

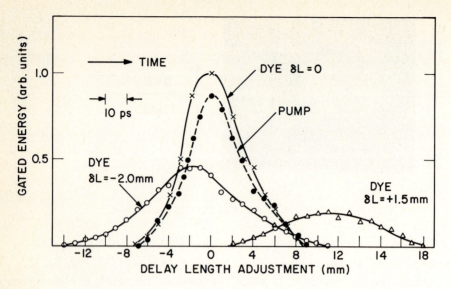

FIGURE 3. Light gate measurement of pulse synchronism.

synchronous dye laser through repeated amplification in successive cavity transits.

We have found for the matched-cavity case that the duration of the dye pulse does not shorten below that of the pump because the population of the upper lasing level is being continuously replenished by additional absorption from the intense co-propagating pump pulses. Evidence for the population recycling-effect comes from oscilloscope observations of the residual pump radiation transmitted through the dye. They show an induced attenuation in the trailing portion of the pump train that occurs only during lasing. The induced attenuation exhibits a cavity-length dependence that correlates with observations on lasing efficiency and pulse temporal overlap (Fig. 3), demonstrating that the recycling occurs on the time scale of individual pulses in the train. A careful examination was therefore made with the streak camera of the dye pulse width for a cavity lengthened slightly from match, where the dye pulse becomes fully delayed relative to the pump in its passage through the gain medium. The gain depletion could then operate free of recycling effects to shorten the pulse. Figure 4 presents streak camera densitometry which shows a shortening of the dye pulse emission to 12 psec. This was obtained for a cavity lengthened 0.8 mm from match, which represents only a small loss in synchronism over the full train. The three pairs of pulses seen in the streak photo inset correspond to independently captured events.

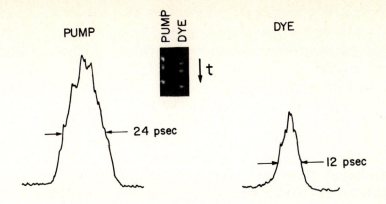

FIGURE 4. Streak camera display of shortened dye output pulses.

Synchronous mode-locked output was obtained over a broad spectral region from 549-727 nm. The temporal character of the dye pulse emission was not observed to change significantly as the dye laser was tuned in wavelength. Figure 5 shows the individual tuning ranges and relative output energies determined for the various lasing dyes when pumped at 532 nm. The solution concentrations were optimized for maximum output energy at $3\text{-}4 \times 10^{-4}$ M for R6G and RB, 6×10^{-4} M for CVP, and 1×10^{-3} M for C122 (5 mm-path cell). Each dye is seen to lase predominantly in the high gain regions near the fluorescent emission maximum. The very short wavelength extent of tuning results from the reduced ground-state absorption in the highly inverted dye medium.

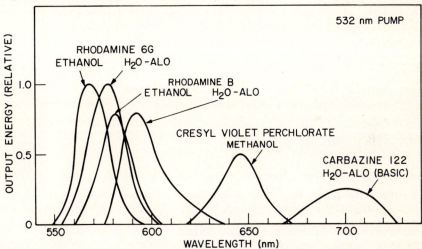

FIGURE 5. Spectral tuning ranges of dyes synchronously pumped at 532 nm.

The spectral width of the dye pulse emission was efficiently narrowed with use of the Fabry-Perot etalon tuning elements. Tilting of the narrow-gap etalon for tuning necessitated occasional readjustments of cavity length. Use of a Lyot filter in place of the narrow-gap etalon for tuning should eliminate this requirement. Figure 6 shows densitometry of spectra taken of the full train at the peak of the R6G emission using (a) an etalon of 6 μm gap (finesse 7), and (b) etalons of 6 μm and 100 μm gap (finesse 7). The peak output pulse-train energies obtained in the two cases were, respectively, 70% and 35% of the output energy from the untuned cavity. The structure seen in the single-etalon spectrum varied randomly from shot to shot. Such structure was also observed for an untuned cavity and may be attributable to hole-burning effects. The use of both etalons compressed the lasing bandwidth to a stable, structureless line of 2 cm^{-1} width, without lengthening the pulse. The spectrum showed no apparent dependence on cavity length. For the 12 psec pulses the 2 cm^{-1} bandwidth corresponds closely to the transform limit.

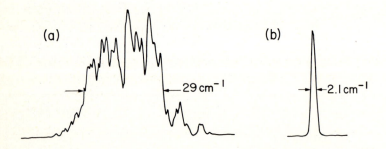

FIGURE 6. Spectral bandwidth of dye laser output with (a) single etalon, and (b) two etalons.

III. SHORT-PULSE DYE AMPLIFICATION

The suitability of synchronous pumping to short-pulse dye amplification was studied. The experimental arrangement was similar to that shown in Fig. 1. The light-gate components were replaced by a 5 mm-path dye amplifier cell containing R6G at a concentration of 4×10^{-4} M. A Nd:YAG amplifier and additional doubling crystal were placed in the delay path to provide amplifier-pump pulse trains at 532 nm. The delay path was adjusted to pass the pulses from the etalon-narrowed dye oscillator through the amplifier cell just following the pumping pulses. This arrangement allows prompt extraction of the deposited energy. By strong attenuation of the input dye beam a peak small-signal gain of 100 was measured. With the unattenuated dye beam lightly focussed into the amplifier cell, a saturated gain of about 4 was obtained. Streak camera studies of the dye pulse temporal profile before and after amplification showed no apparent changes under the saturated gain conditions. Thus, by pumping successive dye amplifier stages with energetic pulses from the amplified Nd:YAG laser the short-pulse output from the dye laser could be scaled upwards considerably in energy.

IV. NONLINEAR MIXING

Nonlinear sum and difference mixing experiments were conducted using the synchronous outputs from the dye laser and the Nd:YAG laser at 1.064 µm and 532 nm. These experiments demonstrated the efficient generation of narrow-bandwidth tunable picosecond pulses in the UV from 270 to 432 nm and in the IR from 1.13 to 5.6 µm.[14]

In comparison with an alternative method using parametric generation,[17-19] the mixing process does not require build-up of a pulse from parametric noise, thus enabling good conversion efficiencies to be obtained without requiring extremely high power densities. Additionally, the bandwidth obtained by mixing reflects that of the two input beams and can be considerably narrower than the phase-matching bandwidth that characterizes a parametric generator's output. Both methods can be susceptible to pulse lengthening by effects of group velocity walkoff.

The experimental arrangement was similar to that shown in Fig. 1 except for replacement of the light-gate components with the nonlinear crystal, and an optional second doubling crystal in the 1.064 µm delay path. A Nd:YAG amplifier was used to enable efficient mixing without focusing in the crystal. Angle-tuned crystals of KDP and ADP were used to sum mix the dye with 1.064 µm and 532 nm, respectively; $LiIO_3$ was used for difference mixing with both 1.064 µm and 532 nm; and ADP was used for frequency doubling the dye.

Figure 7 shows oscilloscope traces of the pulse trains involved in sum mixing of the dye and 1.064 µm to produce UV. For this process the energy conversion from the dye pulse train to the UV measured 20%. For difference mixing of the dye and 1.064 μm pulse trains an energy conversion of 2% was obtained. These conversion efficiencies were fairly uniform over the full dye tuning range, since they depend primarily on the power density of the input 1.064 μm beam. The output energy, however, has a wavelength dependence that reflects the spectral lasing efficiencies (Fig. 5) of the various dyes used.

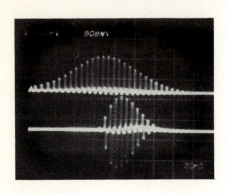

FIGURE 7. Pulse trains in uv sum-frequency generation.

Figure 8 summarizes the spectral coverage for picosecond pulse generation demonstrated in these experiments. The spectral coverage obtainable with the Nd-pumped system could be increased even further by use of lasing dyes pumped by the third harmonic, by 3-wave mixing in other crystals for middle and far IR generation, and by 4-wave mixing[20,21,19] in atomic vapors. In other experiments in our laboratory, two dye lasers have been synchronously mode locked to demonstrate tunable picosecond pulse generation in the vacuum uv by 4- wave mixing in Sr vapor.[22]

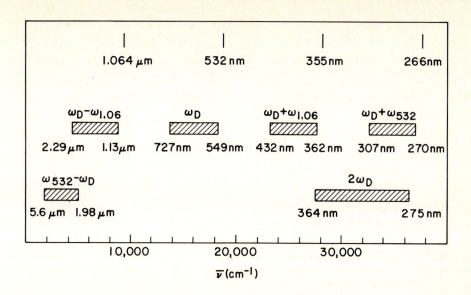

FIGURE 8. Spectral coverage of the synchronous mode-locked dye laser with nonlinear generation.

REFERENCES

1. W.H. Glenn, M.J. Brienza, and A.J. DeMaria, Appl. Phys. Lett. 12, 54 (1968).
2. D.J. Bradley and A.J.F. Durrant, Phys. Lett. 27A, 73 (1968); D.J. Bradley, A.J.F. Durrant, G.M. Gale, M. Moore, and P.T. Smith, IEEE J. Qu. Elect. QE-4, 707 (1968).
3. B.H. Soffer and J.W. Linn, J. Appl. Phys. 39 5859 (1968).
4. L.D. Derkachyova, A.I. Krymova, V.I. Malyshev, and A.S. Markin, JETP Lett. 7, 572 (1968).
5. W. Schmidt and F.P. Schäfer, Phys. Lett. 26A, 558 (1968).
6. E.P. Ippen, C.V. Shank, and A. Dienes, Appl. Phys. Lett. 21, 348 (1972).
7. For a review of dye-laser mode locking see: C.V. Shank and E.P. Ippen, in Dye Lasers, edited by F.P. Schäfer (Springer, New York, 1973).
8. T.R. Royt, W.L. Faust, L.S. Goldberg, and C.H. Lee, Appl. Phys. Lett. 25, 514 (1974).
9. P.K. Runge, Optics Comm. 5, 311 (1972).
10. C.K. Chan and S.O. Sari, Appl. Phys. Lett. 25, 403 (1974).
11. J.M. Harris, R.W. Chrisman, and F.E. Lytle, Appl. Phys. Lett. 26, 16 (1975).
12. H. Mahr and M.D. Hirsch, Optics Comm. 13, 96 (1975).
13. L.S. Goldberg and C.A. Moore, Appl. Phys. Lett. (to be published).
14. C.A. Moore and L.S. Goldberg, (to be published).
15. M.E. Mack, Appl. Phys. Lett. 15, 166 (1969).
16. C. Lin, T.K. Gustafson, and A. Dienes, Optics Comm. 8, 210 (1973).
17. T.A. Rabson, H.J. Ruiz, P.L. Shah and F.K. Tittel, Appl. Phys. Lett. 21, 129 (1972).
18. A. Laubereau, L. Greiter and W. Kaiser, Appl. Phys. Lett. 25, 87 (1974).

19. A.H. Kung, Appl. Phys. Lett. $\underline{25}$, 653 (1974).
20. A.H. Kung, J.F. Young, and S.E. Harris, Appl. Phys. Lett. $\underline{22}$, 301, 1972.
21. P.P. Sorokin, J.J. Wynne, and J.R. Lankard, Appl. Phys. Lett. $\underline{22}$, 342 (1973); R.T. Hodgson, P.P. Sorokin, and J.J. Wynne, Phys. Rev. Lett. $\underline{32}$, 343 (1974).
22. T.R. Royt, W.L. Faust, and C.H. Lee (unpublished).

PHOTOCHEMISTRY AND ISOTOPE SEPARATION IN FORMALDEHYDE

A.P. Baronavski, J.H. Clark, Y. Haas, P.L. Houston, and C.B. Moore
Chemistry Department
University of California
Berkeley, Ca 94720

The application of lasers in molecular spectroscopy, photochemistry, chemical kinetics, and isotope separation has opened many new problems and areas of investigation in chemistry. Laser experiments on formaldehyde illustrate several of these applications.

Properties of the first excited singlet state of formaldehyde have been studied with ultraviolet lasers[1] and via two photon absorption[2] with a visible laser. Carbon isotope shifts have been resolved in many vibronic bands.[3] Measurements of single vibronic level lifetimes have been made for H_2CO, D_2CO, and HDCO.[1,4] The energy states and rates of formation of the dissociation product CO have been studied following excitation of a single vibronic level.[5] These results have called into question the model of spontaneous predissociation of formaldehyde[6]

$$H_2CO(S_1, v_a) \longrightarrow H_2 + CO \qquad (1)$$

$$\longrightarrow H + HCO \qquad (2)$$

through internal conversion to dissociative vibrational levels of the ground singlet state. The original experiments on isotope separation of D from H[7] have been followed by separation of ^{13}C from ^{12}C.[3] Practical laser separation of carbon and oxygen isotopes is likely in the near future.

I. Excitation Spectra

Photochemical isotope separation requires an isotopic shift in the absorption spectrum. The high resolution spectra of $H_2^{12}CO$ and $H_2^{13}CO$ in the range 280-355 nm have been studied by fluorescence excitation spectroscopy. The laser beam (Molectron N_2-pumped dye laser, linewidth 2-3 GHz, average power $\leq 10^{-6}$ W, repetition rate 25 Hz) passed through two cells, each containing a pure isotopic species, and the fluorescence at wavelengths longer than 400 nm was monitored as a function of excitation wavelength. Using a gated electrometer detection circuit, spectra of levels with fluorescence quantum yields as low as 10^{-6} could be observed with good signal-to-noise at gas pressures of a few torr. A typical excitation spectrum is shown in Figure 1.

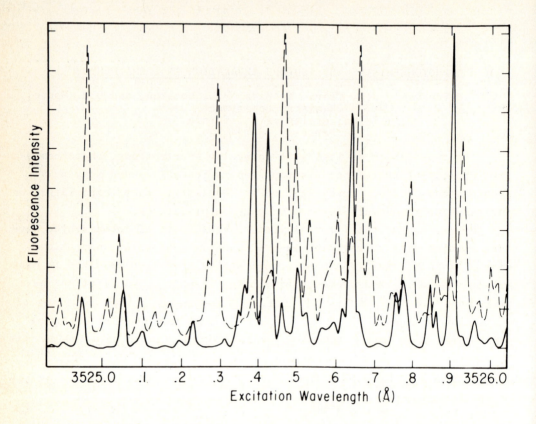

Fig. 1. Fluorescence excitation spectrum of $H_2^{12}CO$ (———) and $H_2^{13}CO$ (----).

It can be seen that the isotopic shift exceeds the Doppler linewidth as well as the laser bandwidth. The spectra correlate quite well with conventional absorption spectra taken with a comparable resolution at $\lambda \sim 352$ nm.

The fluorescence excitation spectrum is truly proportional to the absorption spectrum only if the fluorescence quantum yield is the same for all the levels of both isotopes. Direct absorption measurements using the same laser system will allow us to determine the relative $^{13}C:^{12}C$ fluorescence quantum yield and the dependence of quantum yield on rotational state for a given vibrational band. Estimates for absolute fluorescence quantum yields of some single rovibronic levels should also be possible.

II. Lifetimes of single vibronic levels

Lifetimes of single vibronic levels of the $^1A_2$ electronic state of H_2CO, D_2CO, and HDCO have been measured using a tunable ultraviolet laser in the region 3500 Å to 3080 Å.[1] The results in Figure 2.

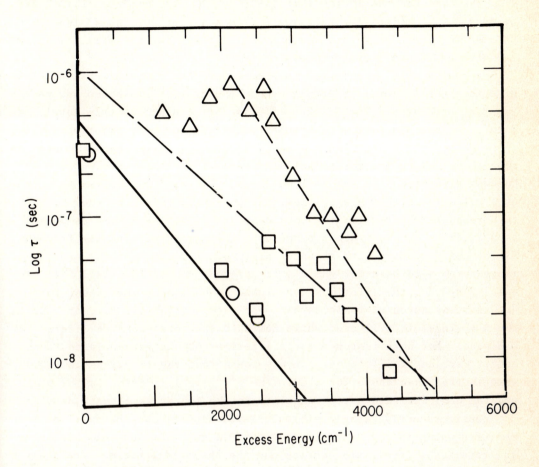

Fig. 2. Excited state lifetime vs excess vibrational energy for H_2CO (O), HDCO (☐), and D_2CO (Δ).

clearly show that the lifetimes become shorter as the frequency of excitation increases. A large deuterium isotope effect is observed with lifetimes for the three isotopic species in the order D_2CO > HDCO > H_2CO.

In the vibrationless level of the $^1A_2$ state, the lifetime of D_2CO is found to be $\sim$ 5 μsec, close to the expected radiative lifetime of that state, while H_2CO is found to have a lifetime of 366 nsec and HDCO $\sim$ 1 μsec. As the excitation energy is increased, lifetimes for H_2CO range from 282 nsec for the 4^1 level at 3532 Å to 20 nsec for the 2^24^1 level at 3262 Å. Lifetimes for D_2CO range from 4.6 μsec for the 4^1 level to 53 nsec for the 2^34^3 level, while HDCO lifetimes for the same levels are 290 nsec and 8 nsec, respectively.

Recently, Miller and Lee[8] have extended the lifetime measurements for H_2CO to the 2^25^1 level at 3000 Å using an ultrafast flashlamp and photon counting apparatus. They find a lifetime of $\sim$ 6 nsec for this level. Their data also indicate a strong dependence of lifetime on excitation energy.

Further experiments have been carried out at wavelengths shorter than 3000 Å where the rotational structure in the absorption spectrum becomes diffuse. High resolution (600,000) absorption spectra were taken on a 3.4 m spectrograph and linewidths were measured. In order to obtain reasonable error limits, band contours of one band were computed as a function of linewidth. Linewidths were measured directly from the experimental spectra with an uncertainty of ± 20%. Lifetimes of H_2CO ranged from $\sim$ 14 psec for the 2^64^1 level to $\sim$ 4 psec for the $1^12^54^1$ level. For D_2CO, the lifetimes were found to be longer than those for H_2CO for any given vibronic level. For example, the lifetime of D_2CO in the 2^74^1 level is $\sim$ 13 psec while for H_2CO the corresponding lifetime is $\sim$ 5 psec. For both isotopes these lifetimes may be extrapolated to the longer wavelength lifetimes. It was found that the lifetimes for HDCO were shorter than those for H_2CO. The 2^74^1 level lifetime is $\sim$ 2 psec and the HDCO data did not extrapolate to the longer wavelength data.

Until recently, single vibronic level fluorescence decay times were measured using fairly broad excitation (0.1 nm). Because of this, several rotational levels are excited and the observed lifetimes are a weighted average of the individual rovibronic level lifetimes. Preliminary results indicate that single rovibronic level lifetimes vary within one vibronic band, thus causing the fluorescence quantum yields within a band to vary. A determination of these lifetimes would help relate fluorescence excitation spectral intensities to absorption coefficients.

Fluorescence lifetimes of the first excited singlet state of formaldehyde give information about the initially excited state and its decay. The variation in the decay rate with vibrational and rotational energy state and with isotopic substitution provides a severe test for theories of radiationless transitions. Unfortunately, the decay rates give no direct information on the relative importance of the several available decay channels.

III. Photochemical analysis by product observation

Observation of S_1 lifetimes, as outlined in the preceding section, yields important information on the originally excited singlet state and offers strong suggestions as to the kinetics of photodissociation in formaldehyde. The complementary techniques outlined in this section use observation of the appearance rate, yield, and quantum state of the photochemical product to investigate this dissociation mechanism more fully.

The CO photochemical product of UV laser excitation of formaldehyde has been monitored by its infrared fluorescence and by its absorption of a cw CO laser. In the former case, fluorescence from CO($v = 1$) was observed using an Hg:Ge detector, while, in the latter case, the ultraviolet and CO lasers were made to overlap spatially in a 1-m cell of formaldehyde. Pulsed excitation of the formaldehyde then produced CO which absorbed the CO laser and caused a change in signal intensity at an Au:Ge detector. By tuning the CO laser to various vibrational transitions and observing the relative signal intensity at each line, a measure of the vibrational distribution of the nascent CO product could be obtained. For each detection scheme, a variety of ultraviolet laser sources was used for excitation of either H_2CO or D_2CO. These lasers included a doubled ruby laser (347.2 nm, Korad), a nitrogen laser (337.1 nm, Molectron), and a flashlamp pumped dye laser with internal doubling, tunable from 350-265 nm (Chromatix). The results of these studies fall into three categories outlined below.

1). The vibrational distribution of the CO product was found to vary with the ultraviolet excitation wavelength. At 337.1 and 347.2 nm the distribution was CO($v = 0$), 90% and CO($v = 1$), 10%. At shorter wavelengths near 300 nm, CO was observed in vibrational levels as high as $v = 5$. However, the ratio of the vibrational excitation of the CO product to the total energy available to the molecular products was always found to be rather low, on the order of 1-3%.

2). The yield of CO was found to be linear in ultraviolet pulse energy and formaldehyde pressure. The latter of these facts indicates that the quantum yield for CO production does not change in the pressure region studies, 0.1-10 torr. Product yields of CO were also measured as a function of addition of foreign gases. For nitrogen and argon, the CO yield decreases slowly with increasing pressure, indicating the possibility of some quenching mechanism. However, for NO and O_2 the yield increases rapidly and then levels off. This sensitivity to NO and O_2 may indicate that intersystem crossing is important in the dissociation mechanism. However, the present study cannot rule out a rapid reaction of NO or O_2 with the HCO radical.

3). The appearance rate for CO was found to be pressure dependent and equal to 0.96 μsec^{-1} $torr^{-1}$ for CO produced from D_2CO dissociation at 337.1 nm. The zero-pressure appearance rate was very slow, less than 0.1 μsec^{-1} and possibly zero, and less than the zero-pressure disappearance rate of formaldehyde S_1 at 337.1 nm (>0.8 μsec^{-1}). These facts indicate that formaldehyde does not dissociate directly from S_1 but, rather, proceeds through some intermediate state whose lifetime at zero-pressure is considerably longer than that of S_1. Possibilities for this intermediate include a) highly excited vibrational states of S_0, b) the triplet state, c) a collision complex such as H_2COH or $(H_2CO)_2^*$.

The results outlined above indicate that the dissociation mechanism of formaldehyde is more complex than has been previously assumed.[6] Dissociation does not proceed in a single step from S_1, but involves an intermediate state which has a number of final channels available to it. The use of CO product state detection has aided considerably in investigating this mechanism. CO is not the only photochemical product from formaldehyde. Further information concerning the dissociation could be gained by monitoring the H, HCO, and H_2 products. An understanding of formaldehyde photochemistry should not only increase our fundamental knowledge of such processes but should aid in making practical applications of this knowledge as well.

IV. Isotope separation

Our research into the photochemical processes occuring in formaldehyde has led us to exploit this photochemistry to achieve separation of hydrogen[7] and carbon[3] isotopes. Since a single photon both selectively excites and dissociates the molecules into stable products (H_2

and CO) which are easily separated from the starting materials, highly efficient isotope separation of the ^{13}C, ^{14}C, ^{17}O, and ^{18}O isotopes can be obtained.[9]

Experiments to demonstrate isotope separation have been carried using a commercially available tunable dye laser (Chromatix CMX-4).[3] The narrow-band (10GHz), frequency-doubled laser light is tuned to a wavelength near 304 nm, where ^{12}C absorption predominates over that of ^{13}C. The excitation selectivity is monitored by observing the total fluorescence from each of two isotopically pure samples as the laser is tuned. Once a frequency where the fluorescence signal from the pure $H_2^{12}CO$ sample is more than 20 times that of the pure $H_2^{13}CO$ sample is reached, photolysis cells are introduced into the path of the laser beam. After photolysis times of about 1 hour, the photolyzed samples are introduced into a mass spectrometer and the carbon isotope ratio of the photolysis products is measured. The results of two such experiments are described in Table 1.

Table I. Mass spectral results of ^{13}C:^{12}C separation.

Initial Ratio 12/13	Pressures		Excitation Ratio 12/13	Photolysis Time (min)	Final Ratio 12/13*	Enrichment Factor**
	Total H_2CO (torr)	NO (torr)				
1.0	4.3	0	>40	140	6.5	6.5
1.0	4.3	3.2	>40	140	10.5	10.5
0.1	2.2	0	>27	100	1.4	14
0.1	2.2	2.4	>27	100	8.1	81

*   ± 20%

** final isotope ratio/initial isotope ratio

The high enrichment factors obtained in these experiments are realized for a laser linewidth 5 times broader than the Doppler linewidth, and for unoptimized gas temperature and pressure conditions. Furthermore, the experiments were carried out in a pressure regime where collisions dominate the photochemistry.

Since the primary impact of laser isotope separation lies in the possibility of drastically reducing the cost of isotopically pure materials, it is interesting to assess the practical parameters of a ^{13}C separator based on the formaldehyde process. Table 2 lists values of some of the important system parameters for a ^{13}C separator.

Table II. System parameters for a ^{13}C isotope separator based on laser-induced predissociation of formaldehyde

System Parameter	Comments
^{13}C production rate: 5 kg/yr	Approximates current demand
Product purity: > 90% ^{13}C	Comparable to presently available materials
Laser: 5 watts ave. power λ = 305 nm $\Delta\nu$ = 2 GHz efficiency = 10^{-4}	Achievable given present dye laser state-of-the-art. Chance coincidence with a fixed frequency laser considerably enhances prospects of success.
Cell length: 100 m	Easily achievable by multiple-pass techniques. This length provides 90% photon utilization even after removal of 90% of the available $H_2^{13}CO$ by photolysis.
Energy flux: 6 J/cm^2	Necessary flux to remove ~ 95% of available $H_2^{13}CO$.
Laser power cost: 1 \$/gm	Assumes 10^{-4} efficiency
Capital cost: 12 \$/gm	Assumes \$300k paid in 5 years
Operating costs: 10 \$/gm	Assumes total operating costs \$50k per year.
Total production cost: 23 \$/gm	Current cost 70 \$/gm.

As the table suggests, even presently available laser technology is sufficient to allow commercial exploitation of the technique.

V. Acknowledgments

This research was supported by the National Science Foundation, the University of California Board of Patents, and the U.S. Army Research Office-Durham.

References

1. E.S. Yeung and C.B. Moore, J. Chem. Phys. 58, 3988 (1973).
2. E.S. Yeung and C.B. Moore in Fundamental and Applied Laser Physics (ed. M.S. Feld, A. Javan, and N.A. Kurnit; John Wiley and Sons, 1971), p. 223.
3. J.H. Clark, Y. Haas, P.L. Houston, and C.B. Moore, Chem. Phys. Lett., to be published.
4. A.P. Baronavski, A. Hartford, Jr., and C.B. Moore, J. Mol. Spectrosc., submitted.
5. P.L. Houston and C.B. Moore, in preparation.
6. E.S. Yeung and C.B. Moore, J. Chem. Phys. 60, 2139 (1974).
7. V.S. Letokhov, Chem. Phys. Lett. 15, 221 (1972).
   H.M. Bazhin, G.I. Skubnevskaya, N.I. Sorokin, and Y.N. Molin, JETP Lett. 20, 18 (1974).
   J.B. Marling, Chem. Phys. Lett., to be published.
   E.S. Yeung and C.B. Moore, Appl. Phys. Lett. 21, 109 (1972).
8. R.G. Miller and E.K. Lee, Chem. Phys. Lett., to be published.
9. V.S. Letokhov, Science 180, 451 (1973).
   C.B. Moore, Acc. Chem. Res. 6, 323 (1973).

SEPARATION OF URANIUM ISOTOPES BY SELECTIVE PHOTOIONIZATION*

Benjamin B. Snavely, Richard W. Solarz
and Sam A. Tuccio

Considerable progress has been made during the past year in the development of selective photoionization processes for the separation of uranium isotopes. Uranium enrichment by selective photoionization has been scaled from the microscopic level reported by Tuccio et al[1] in June 1974 to the milligram per hour rate[2]. This progress has been supported by developments in the understanding of the uranium spectrum resulting from the application of tunable dye lasers as spectroscopic tools. In this paper, recent results of experiments on the laser photoseparation of uranium isotopes are reported.

The high density of uranium energy levels in the 10,000 cm^{-1} to 50,000 cm^{-1} energy range provides the opportunity for selective photoionization of uranium by a number of different schemes. Some of these are represented in Figure 1. In Figure 1A is shown a two-step process in which the isotopically selective step, $h\nu_1$, is followed by the absorption of photons with energy $h\nu_2$ to produce ions of the desired isotope. In the diagram, $h\nu_2$ is shown terminating on an autoionization state or discrete state, above the ionization continuum of the atom to take advantage of the relatively large absorption cross section associated with these states.

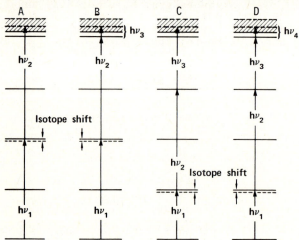

FIG. 1 Alternative Excitation Schemes for the Isotopically Selective Photoionization of Uranium

A potentially serious problem with the two-step photoionization process is that a small value of the photoionization cross section may preclude efficient utilization of the laser producing $h\nu_2$. This could seriously impair the economics of such a process. A variation of the two-step process is shown in Figure 1B. In this case, photons of energy $h\nu_2$ excite the atoms to an energy level slightly below the

---
*This work was performed under the auspices of the U. S. Energy Research and Development Administration

ionization continuum. Ionization is accomplished by the absorption of a third photon of energy $h\nu_3$. This scheme was suggested by Nebenzahl and Levin[3] as a means by which small photoionization cross sections could be utilized effectively in a separation system. By using an efficient infrared laser, such as a CO_2 laser, to provide the photons for the ionization step, process economics may not be affected adversely by a small photoionization cross section.

Another type of three-step process is diagrammed in Figure 1C. If the photons, or energy levels, can be properly chosen, it would be possible to ionize with photons approximately equal energy, that is:

$$h\nu_1 \cong h\nu_2 \cong h\nu_3 .$$

This places the required laser wavelengths in a spectral region which is readily accessible to presently available tunable lasers. A variation on this scheme, Figure 1D, utilizes the infrared photon $h\nu_4$ to overcome a small photoionization cross section.

In all of the diagrams, photons have been used to accomplish the ionization. If, however, the desired isotope is excited to a state which lies slightly below the ionization energy, there are alternatives for the ionization step. These include: field ionization by an electric field or by electron impact. These techniques have also been proposed as means of overcoming effectively small photoionization cross sections.

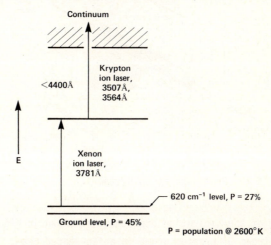

FIG. 2 Excitation Scheme Used in the Photoseparation of Small Macroscopic Quantities of Enriched Uranium

The excitation scheme used in an experiment carried on at Livermore to separate macroscopic quantities of enriched uranium is shown in Figure 2. Two lasers are used for the selective photoionization. A xenon ion laser operating 3781 Å excites uranium atoms from a metastable state at 620 cm^{-1} above the ground state, to an energy level at 27,068 cm^{-1}, approximately one-half of the ionization energy. Excited atoms are then ionized by the ultraviolet output of a krypton ion laser operating simultaneously on two lines at 3507 Å and 3564 Å. These lines were chosen for convenience and do not necessarily carry the atom to an autoionization state. The terminal state is considerably higher in energy than any of the autoionization states which have been found to date. Therefore, the cross section for this process[1] is expected to be of the order of 10^{-17} cm^2. The xenon laser was chosen for the isotopically selective excitation step because of the fortuitous coincidence between the xenon laser line at 3781 Å and a uranium-atomic transition.

In the experiment, the atomic vapor source was operated at a temperature of 2600°K. At this temperature, approximately 27% of the atoms are in the 620 cm^{-1} state. The ground state contains approximately 45% of the atoms. If a selective photoionization process is to utilize the greatest fraction of the uranium atoms in the vapor stream, then it is necessary to provide a second laser for excitation from the ground level. Since the purpose of this experiment was proof of principle, and not large scale operation, a laser accessing the ground state was not used.

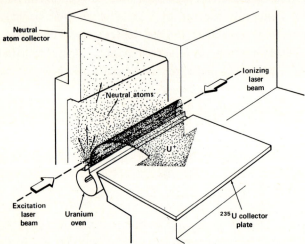

FIG. 3 Geometry of the Photoseparation System

The geometry of the separation experiment is shown in Figure 3. The uranium oven produces a stream of uranium atoms in a fan-shaped beam. The xenon laser beam, passing near the slit as shown, produces the isotopically selective excitation. Ionization of the excited atoms is then produced by the krypton laser beam traveling in the opposite direction. The ^{235}U ions thus produced, are collected on a beryllium plate by an electric field produced by maintaining the collector plate at a negative potential of 1500 volts with respect to the uranium oven. Neutral atoms are collected on a plate provided for the purpose as shown.

The xenon laser beam power in this experiment was approximately 70 mW. The krypton laser had a power of about 1-1/2 watts. The circulating power was increased by placing the uranium vapor within the cavity of the krypton laser. In order to accomplish this, the output mirror of the xenon laser served as one of the reflectors for the krypton laser. The circulating power in the ionizing laser beam was between 30 and 50 watts.

The deposition pattern of uranium ions on the collector plate is shown in Figure 4 as determined from densitometer tracings of an autoradiograph of the plate. Approximately 1 mg of material enriched to slightly over 1% in ^{235}U was deposited on this plate. In subsequent experiments, approximately 4 mg of uranium enriched to between 2.5 and 3% has been obtained during a run lasting approximately 2 hours.

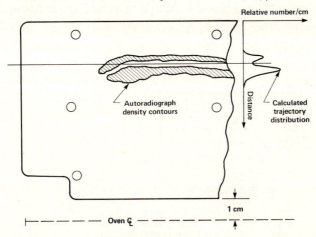

FIG. 4 Uranium Ion Distribution on the Beryllium Collector Plate

Calculations have been carried out to describe the ion trajectories in the collector system. The two-band pattern observed on the beryllium plate is predicted by these calculations and results from distortion of the electric field by the neutral atom collector structure. At the right hand side of Figure 4 is shown the calculated trajectory distribution as a function of distance across the beryllium collector plate.

The quantity and the enrichment of material collected in these experiments was determined primarily by the power of the krypton laser. The number of thermal ions produced by the hot uranium vapor source corresponds to about 0.1% of the total number of atoms in the vapor stream. With the laser powers used in this experiment, enrichment greater than approximately 3% cannot be expected. There was no provision in these experiments for suppression of thermally generated ions to improve the enrichment.

Knowledge of uranium atom spectroscopic parameters is essential in the assessment of the economic feasibility of enrichment processes based on uranium metal vapor. The value of the photoionization cross section is of particular importance in this regard and a system has been assembled to study the photoionization spectrum. In this apparatus, a uranium beam is produced by an atomic vapor source based on the alloy URe_2[1]. The uranium beam is excited to a selected energy level by the tunable output of a CW dye laser. The dye laser, with an output power of about 30 mW, produces sufficient intensity in the beam region to nearly saturate the chosen transition. In a scheme which is very similar to that used to demonstrate selective two-step photoionization, a second, ultraviolet, optical beam intercepts the excited uranium beam. The ultraviolet beam is produced by a 2500 watt mercury arc filtered by a monochromator. As the ultraviolet wavelength is scanned, the rate at which uranium ions are produced is measured. Uranium 238 ions are filtered out from spuriously produced ions of other types by a quadrupole mass filter. By measuring the ^{238}U ion current as a function of the ultraviolet photon wavelength, it is possible to determine the photoionization spectrum related to the selected state of uranium.

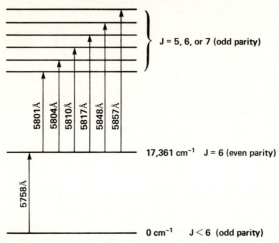

FIG. 5 High Lying Odd Parity Levels of Uranium

A photoionization spectrum obtained at a resolution of 8 Å is shown in Figure 5. This spectrum represents photoionization from a state at 16906 cm^{-1} of the $^7M_7$ level. This is an even-parity state reached from the odd parity $^5L°_6$ ground state by the tunable CW dye laser. The $^7M_7$ level is a level of the f^3dsp electron configuration. From this state, it is expected that autoionization states can be reached corresponding to the electron configuration f^3dp^2.

Several distinct peaks are seen in the photoionization spectrum. In the experiments reported in Reference 1, an average cross section for the region from 3200 Å to 2100 Å of about 10^{-17} cm^2 was deduced for photoionization transitions from the same state. From the spectrum of Figure 5, it appears that the peak cross section for transitions to autoionization states from the $^7M_7$ level is approximately

10 times larger than the background cross section at short wavelengths. The implied value of $\sim 10^{-16}$ cm^2 is large enough to be used in large scale uranium separation systems.

Subsequent measurements at higher resolution have been obtained and it has been found that the peaks shown in Figure 5 have a half width (FWHM) of approximately 3 Å. Photoionization spectra such as that shown in Figure 5 yield quite accurate values of ionization potential of the uranium atom. The value deduced from these experiments is 6.15 ± .02 eV.

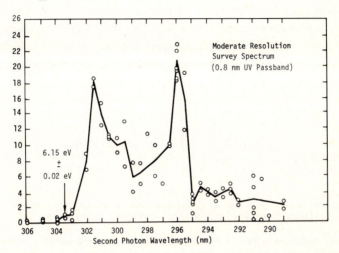

FIG. 6 Autoionization Spectrum from the $^7M_7$ Level at 16,904 cm^{-1} in Uranium

One possible excitation scheme for selective photoionization involves a two-quantum transition to an isotopically selected state. There are practical advantages for such a process in that the effective linewidth of the uranium transition could be narrowed to less than the doppler linewidth[4]. This, in turn, could result in more efficient utilization of the excitation laser. To take advantage of such an excitation scheme, odd-parity levels of the uranium atom lying between 30,000 cm^{-1} and 40,000 cm^{-1} would be required. Very few odd-parity terms in this energy region have been observed and experiments have been undertaken to identify such levels[5]. The experimental arrangement for these experiments was similar to that used in studying the photoionization spectrum. In this case, however, a tunable nitrogen-laser-pumped dye-laser was used for the excitation step. In one set of experiments an even-parity level at 17,360 cm^{-1} was excited using the CW dye laser. The pulsed dye laser operating at over a wavelength range of approximately 5800 to 5860 Å was used in the second and subsequent ionization step excitations. Transitions to odd parity states are readily observed since the same photons which populate the odd parity levels can also produce ionizing transitions from the odd parity states. Assuming relatively little structure in the photoionization spectrum in the energy region accessed by the photons used, the ion current produced as a function of pulsed laser wavelength should show peaks corresponding to the odd-parity terms. Six of the odd parity levels found by transitions from the 17,360 cm^{-1} level are shown in Figure 6. Utilizing this information, experiments are now underway to measure the two-photon cross sections for transitions from the ground state to these high lying odd-parity states.

Several experiments of importance in the assessment of the feasibility of metal vapor processes for laser photoseparation of uranium isotopes have been described. At this point, there are many additional experiments which need to be carried out before the problems of scaling and economic feasibility of metal vapor processes can be fully addressed.

References:

1. S. A. Tuccio, J. W. Dubrin, O. G. Peterson, and B. B. Snavely, "Two-Step Selective Photoionization of ^{235}U in Uranium Vapor", presented at VIII International Conference on Quantum Electronics, San Francisco, June 1974.

2. S. A. Tuccio, R. W. Foley, J. W. Dubrin, and O. Krikorian, "Macroscopic Isotope Separation of Uranium by Selective Photoionization", presented at the Conference on Laser Engineering and Applications, Washington, D.C., May, 1975.

3. I. Nebenzahl and M. Levin, "A Process for Isotope Separation", Patent#DT2312194, Federal Republic of Germany.

4. T. W. Hänsch, K. C. Harvey, G. Meisel, and A. L. Schawlow, Optics Comm. __11__, 50 (1974).

5. Richard Solarz, Lee Carlson and Charles May, "Autoionization Spectra and High Lying Odd Parity Levels in Uranium Vapor", presented at the Conference on Laser Engineering and Applications, Washington, D.C., May, 1975.

LASER ISOTOPE SEPARATION

C. Paul Robinson
Los Alamos Scientific Laboratory
University of California
Los Alamos, New Mexico 87545

The application of lasers for isotope separation is having a revolutionary effect on a number of scientific disciplines. Laser separation schemes include two principal areas: laser spectroscopy and laser-induced chemistry. Although the major portion of the Los Alamos work involves classified schemes for uranium enrichment, we are investigating prototype processes for the other elements. Results of a fundamental nature have been obtained in molecular spectroscopy, both experimentally and theoretically. Besides just applying the exceptional laser bandwidths to resolve single states in molecular absorptions; quantum number assignments, force-field determinations, and group theory analyses have contributed new and more complete knowledge of molecular structure than was previously possible. Characteristics of laser-induced chemical reactions are being studied, including the accompanying processes of energy transfer and level kinetics. In addition to offering rich data for quantum chemical analyses, the selectivity possible in the laser studies provides sufficient detail to elucidate many details of ordinary thermal chemical reactions. Lasers are certain to become one of the most important tools for chemical research in the next 5 to 10 years.

This selectivity is, of course, crucial to the many envisioned applications of laser induced chemistry, especially for laser isotope separation. Other applications include uniform initiation of chemical explosions for chemical lasers; purification of materials by selective excitation and scavenging of impurities. Application of laser-induced chemistry to partition nuclear radioactive wastes may prove quite useful. Probably the most important ultimate application is the synthesis of new compounds through chemical replacement on laser activated sites in the principal reactant. Several examples have recently been presented. The probability of this concept being applied on an industrial scale seems assured in both organic and inorganic chemistry.

The possibility of applying lasers for inducing chemical reactions began at Los Alamos in 1970 in the Chemical Laser Research and Development Group, which was part of the Laser Fusion effort. The experiments of Lyman and Jensen on the infrared photolysis of N_2F_4 and the stimulated chemistry of SF_6 and H_2 were among the first to show vibrational enhancement of chemical reaction rates over thermal rates. Figure 1 shows the experimental setup of those experiments.

When powerful CO_2 pulses were made incident on a cell of SF_6 and hydrogen, an HF probe laser at 90° showed the appearance of HF molecules following the laser initiated chemical reaction. Comparison of the rate of HF appearance with the rate predicted by a computer code which included V-T and V-V relaxation processes, showed that the reaction was proceeding 100 times faster (10,000 times faster at early times) than an equilibrium rate with the energy addition provided by the laser absorption. An initial surprise in those measurements was that, whereas the HF probe laser was tuned to $1 \rightarrow 0$ transitions in order to detect the appearance of ground state HF, the signals initially showed HF laser gain before going absorbing. This observation became the key to our prediction of isotope selection using infrared lasers to promote selective dissociation of molecules which I will discuss in more detail later in this talk.

From the evidence that laser pulses could produce rate enhancements of 100-fold and greater when the chemical reactants were promoted to excited vibrational states, we recognized that similar schemes using tunable lasers could result in direct isotope separation process by tuning the laser to selectively excite a single isotope. The reduced mass effect on the frequencies of the vibrational states would directly allow mass discrimination. The single stage isotope separation factor in such experiments would be equal to the rate enhancement. Thus, extremely high enrichment ratios, 100 or greater, might be achieved.

In late 1971 a multidisciplinary task force was assembled at Los Alamos to investigate possible applications of lasers for isotope separation, especially for uranium. Members of this group conceived a large number of possible techniques to achieve this goal and since January of 1972, a large and growing experimental effort has emerged. Although it is not appropriate at this time to discuss that work, a companion program in enriching lighter isotopes has met with some success which I would like to discuss with you. After a general discussion of options and problems in laser selective processes, I will discuss in particular the purely scientific fallout that has resulted from a new science being vigorously pursued simultaneously with the major new tools which are being developed for the studies.

Figures 2 and 3 show the basic processes to be considered for promoting chemical (or physical) change from vibrational or electronic states. Though conceived first for isotope separation, they apply in a straightforward and less complicated way to the general laser-induced chemistry problem.

Besides the usual linear absorption relationship for a two-level system which are adequately described by the Einstein relationships, a number of new phenomena arise when a laser is employed as the light source. Using pi pulses or suitably frequency-chirped pulses one can promote the entire manifold to the excited state. There are also processes in which the absorption cross-section depends on the incident laser power. Multiple sequential absorption or true multiphoton absorption can promote the system up the vibrational ladder to higher levels of excitation. Power broadening of the absorption linewidth by the laser fields enhance these processes.

We have measured power dependent cross-sections for both SF_6 and BCl_3 and our work as well as recent Soviet work indicates that as many as 30-40 photons/molecule can be promptly absorbed.

For the levels here one attempts to find isolated isotopic spectral features. Besides the reduced mass effect of vibrational and vibronic states, further isotope specific states are available due to: (1) nuclear spin and the associated magnetic moment; (2) nuclear shape and the associated electric quadrupole moment; (3) finite nuclear size. These effects range in size from about 0.001 to 10 cm^{-1} and in some cases can be uncovered or enhanced by external electric or magnetic fields. Once an isolated optical feature has been selected, it is used as a staging ground on which to base a separation process.

After selectively exciting a particular state, whether it be the first state above ground or well up in the level manifold, a number of options exist for effecting the isotope separation. Some of the alternatives which one can attempt to produce are shown here. Effects which will naturally compete with the desired action are shown also. Thus, in choosing a second step in a separation scheme, one must carefully consider the possible consequences of mechanisms which reduce the yield or compromise the enrichment factor. In general, after the initial optical selection or tagging of one isotope in a mixture, one should promote as <u>permanent</u> a change as is possible, as rapidly as possible.

The relative importance of the various processes shown will depend upon the mixture of other molecules present besides AB. The state can reradiate in fluorescence back to the ground state or another state. It can collisionally transfer to a vibrational state of another molecule or can vibrationally relax directly or stepwise with conversion into translational energy with any other body. If one has found a suitable reactant partner, the molecule AB can undergo a chemical reaction to a new form.

Especially if the selected state is an electronic state the possibility exists for predissociation into a new chemical form provided the initial state lies higher in energy than a dissociation channel of the molecule. Ionization by a collision partner either by electron capture or electron transfer is also a possible consequence for an electronic state as well as in some special cases for vibrational states. Penning ionization through collision with an excited molecule (intentionally produced) or its analogy for collisionally induced dissociation are further options.

Finally, the selected state can serve as the lower state for a second photon which would promote the molecule to a still higher state. Because of the energy discrimination provided by transition to the first state, the tuning requirements on the second photon are greatly relaxed, so that ultraviolet photons of relatively wide frequency bandwidth can be applied to achieve selective dissociation or ionization. The latter are the much discussed two-photon selective processes.

Following either chemical reaction or dissociation, it may be necessary to chemically scavenge the fragments to prevent back reaction or subsequent reactions. If the fragment is weakly bound, the possibility of chemical exchange reactions through collision with the original molecular form can scramble the enrichment. For processes which result in ionization, the probability of ion exchange looms large because of the long range of the electric field increasing the cross section for interaction of this type.

With all of the options for processes as well as the deleterious interactions that can occur at each step, one can appreciate the intricacies of quantum chemistry and energy transfer kinetics which must be examined to guarantee success in any given undertaking. At the outset, one must obtain optical spectral data of a higher quality than has been obtained previously. Some of the devices and techniques that have been developed in our program to date represent a revolution in molecular spectroscopy, both in obtaining data and theoretically describing the structures resolved. A sampling is given on the next figures.

On Fig. 4 is shown some hot-band absorptions of the CO_2 molecule taken with a conventional spectrometer of quite good resolution. Notice that while there are some structures, these cannot be identified as quantum states, nor can quantum numbers be assigned. The region shown here was scanned with a laser diode (developed with our support at MIT-Lincoln Laboratory) whose resolution is 5×10^{-5} cm^{-1}. The spectral data obtained in shown in Fig. 5. Assignments are made for these individual rotational lines of a Q branch; identified as the 010 to 0$\overset{\circ}{2}$0 transition. Direct fitting of the line spacing (shown in Fig. 6) yields a ΔB (change in rotational constant between the upper and lower state) of 7.76×10^{-4} with residuals of about 1/10 of a milliwavenumber. By simply varying the pressure of this cell, we can read directly the pressure broadening effect on each rotational line.

Next let me skip over to a case of much higher complication, indeed thought to be so complicated as to be beyond present understanding by the authors of the standard textbooks, as well as in earlier measurements of absorption spectra by both conventional and laser means. Figure 7 shows two absorption curves for SF_6, a seven-atom polyatomic; the lower curve is by Brunet and Perez and has been in the literature for some time without being fully analyzed. It is the ν_3 band whose vibration-rotation transitions are nearly coincident with the carbon dioxide laser lines near 10.6 μm. The upper curve is one of the first laser diode spectra to be measured and was performed by Hinkley, et al. The conclusion made was that these were individual states being seen, although of course they were too complicated for theoretical analysis.

Note however, the proposal we advance for the description of the previously unexplained structures of the original curve (see Fig. 8). It is suggested that SF_6 is in actuality a quite simple molecule, showing a classic P, Q, R structure. At 300°K many states are excited, each of these anharmonic states possessing a P, Q, R structure. The resultant summation of curves leads to the data here, where we identify the Q branch of the ground state, first excited state, second excited state, and third and fourth excited states. The integrated populations of the spikes agree almost perfectly with that predicted by a simple Boltzmann distribution.

With respect to Hinkley, et al's conclusion let me now present some laser diode data we have taken, where we have taken the SF_6 to low temperatures in a cold cell, (see Fig. 9). This data is of major importance for a sulfur isotope separation I will show you later. The SF_6 is measured in a cold cell at 135°K so that almost all of the molecules are initially in the ground state. Note the classic P, Q, R structure, similar to a triatomic molecule, where the individual rotational lines are resolved for the first time and we have made the quantum number assignments as shown. Yet there is even more richness in the spectrum as evidenced by the higher J lines. Although using a Hamiltonian keeping terms through fourth order to accurately determine 15 spectral constants of the SF_6 molecule, the elucidation of the Coriolis splitting of the rotational lines into individual symmetry components is another perhaps more fundamental first. Figure 10 presents higher resolution scans of P (17, 18, and 19) with the identified splittings. What is quite fundamental is the inconsistency which was solved by some of our theorists. Applying the usual group theory approaches with respect to molecular structure the relative intensities or statistical weights were inconsistent with the new data. A new group theoretical approach was proposed that systematically introduces sets of group operations that have internal and external geometric interpretations. Thus for the problem at hand correct statistical weights are predicted by this new theory which appears to set the group theoretical analysis on a solid basis. We have new data for methane, an intermediate molecule in complication and find excellent comparisons there also.

In continuing our work in laser spectroscopy we have established the Laser Applied Spectroscopy Laboratory, which includes diode lasers, optical parametric oscillators, and tunable dye lasers. These tools allow the study of absorption and emission spectra from particular excited states, as well as energy transfer reactions and vibrationally enhanced chemistry. Whereas most of our own chemistry studies have concentrated on molecules that can absorb CO_2 laser radiation near 10 μm, new nonlinear mixing process lasers (being developed under contract with R. L. Byer at Stanford University) will allow us to study states without regard to coincidence with existing lasers.

Lastly, I should note several examples of laser isotope separation. The enrichment of the boron isotopes was first demonstrated in our Laboratories. We originally used a two-photon dissociation of BCl_3 using a low power CO_2 laser and a filtered uv flashlamp (see Fig. 11). A scavenger (O_2) was necessary to prevent back reaction of the fragments. Yet isotopic enrichment was obtained, which scaled identically with the previously measured V-V transfer rate as a function of BCl_3 pressure.

Now I would like to discuss our recent separation experiments. We proposed, during the summer of 1973, a new method of isotope separation involving dissociation or chemical reaction using only a single infrared laser. The idea was to utilize a strong laser field to promote a polyatomic molecule up the vibrational manifold. In Fig. 12 is shows the vibrational manifold with the usual anharmonic state distribution. Tuning of the laser frequency to match the initial step for one isotope will greatly increase the transition probability for that isotope to walk up the vibrational ladder. For polyatomic molecules the level structure is especially suited to walking up the ladder (see Fig. 13). This absorption mechanism is a complicated one combining level density effects, resonantly enhanced multiphoton transitions, A. C. Stark shifts and the breakdown of selection rules at high molecular excitations (distortions). We have recently enriched the isotopes of boron, chlorine, silicon, carbon, and sulfur using only CO_2 laser pulses. The boron and chlorine were simultaneously enriched using the molecule BCl_3. Silicon used SiF_4 and CCl_2F_2 (Freon 12) was used for carbon and chlorine. The sulfur used SF_6. Both were laser dissociation reactions with H_2 used as an F atom scavenger. Enrichments ranging from 500 to 3300 percent were demonstrated. Figures 14 and 15 show sample data of the sulfur and silicon work.

I would be remiss were I not to report to you that following our B and Cl experiments, but during the course of our SF_6 work, we learned that Ambartsumyian, et al. (our Soviet counterparts) had achieved sulfur enrichment by our proposed technique also, with even higher enrichments reported. We congratulate them in that effort and look forward to a continued competition in enriching many other isotopes. In looking to economic aspects of this work, ^{34}S costs approximately \$1000/g, whereas the energy requirement alone for our process is approximately 40¢/g.

In closing, may I say that Laser Isotope Separation is a dynamic and productive field, yielding equally important results in fundamental science as in useful applications. I predict even more important results in both areas for many years to come as a result of the activity in this field.

FIGURE CAPTIONS

1. Experimental Set-up of Original SF_6-H_2 Experiments.
2. Isotope Separation Processes
3. Isotope Separation Processes
4. CO_2 Absorption: Conventional Resolution
5. Diode Scan of CO_2 Absorption
6. Fit to Obtain ΔB (010-020)
7. SF_6 Absorption: (Brunet & Perez; Hinkley)
8. Identification of Peaks in SF_6 Room Temperature Data
9. SF_6 High Resolution Data (135°K)
10. Coriolis Splitting of SF_6 P lines
11. Set-up of Original BCl_3 (2 photon) Experiment
12. Absorption Mechanism: Multiple Photon Experiments
13. Level Densities in Polyatomic Molecules
14. SF_6 Separation Results
15. SiF_4 Separation Results

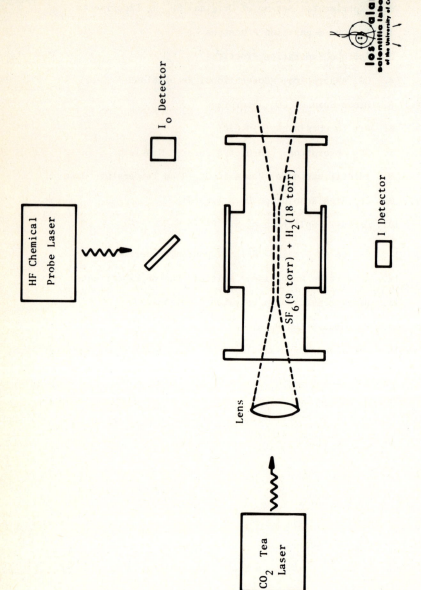

J. L. Lyman & R. J. Jensen. Chem. Phys. Lett., 13, 421 (1972) and J. Phys. Chem., 77, 883 (1973)

FIGURE 1

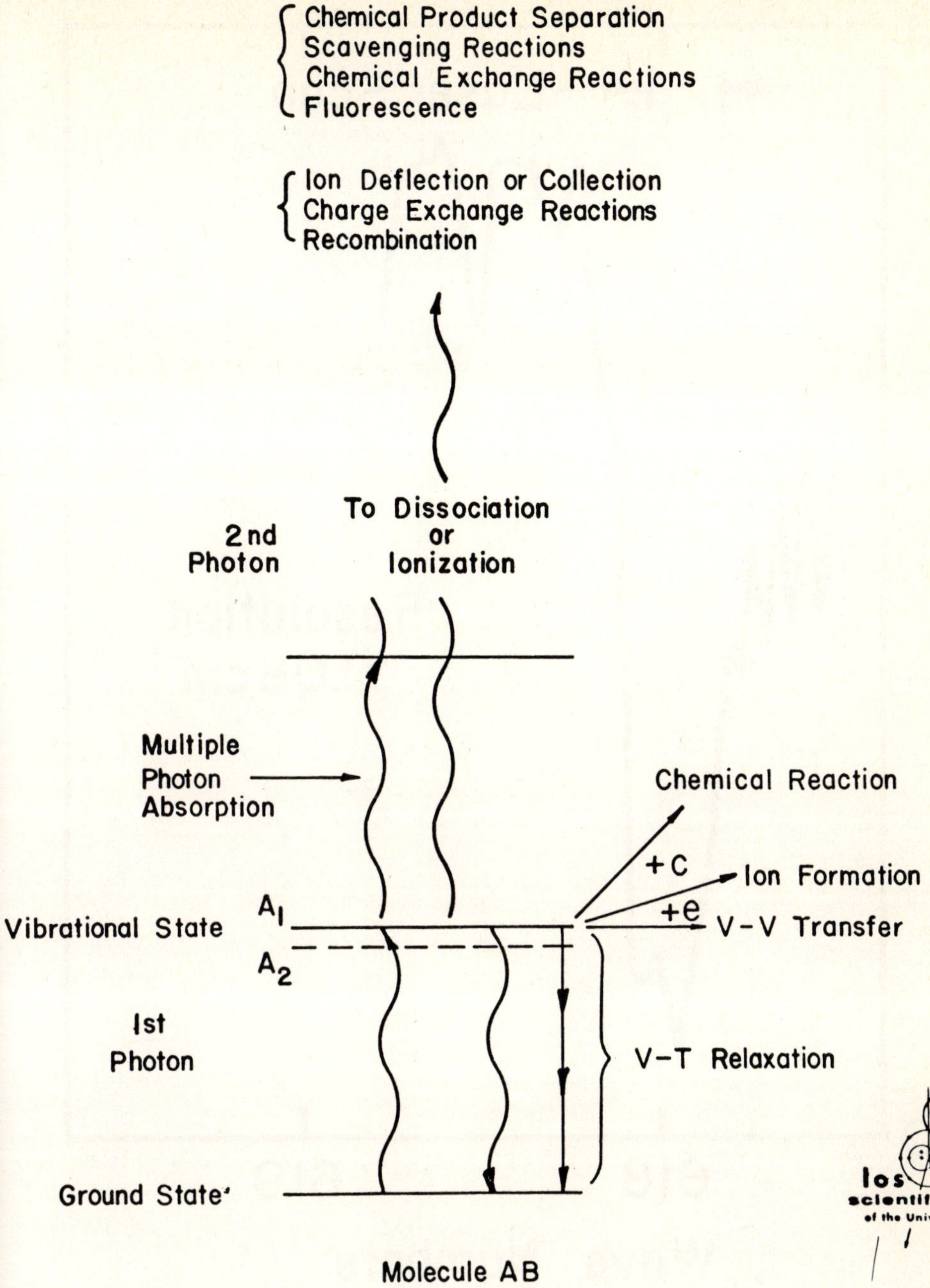

Basic Processes for Laser Isotope Separation

FIGURE 2 & 3

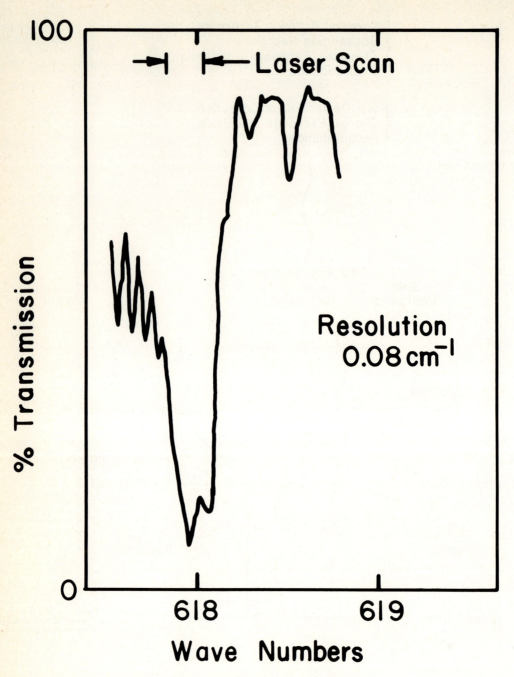

L. Clay, University of Michigan, Thesis

FIGURE 4

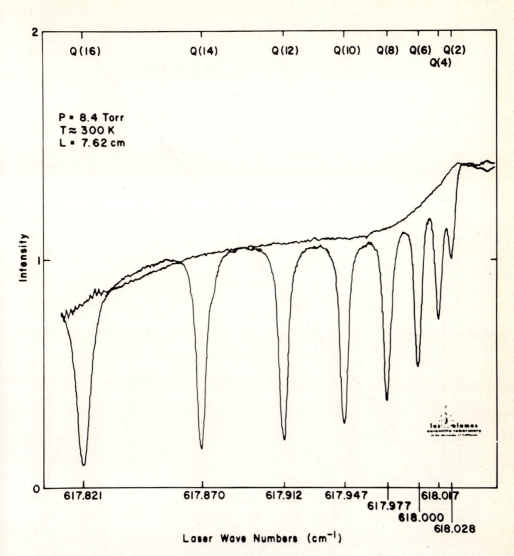

J.P. Aldrige, R.F. Holland, H. Flicker, K.W. Nill, T.C. Harmon, J. Molec. Spectros., to be published.

L6-VG-086A

FIGURE 5

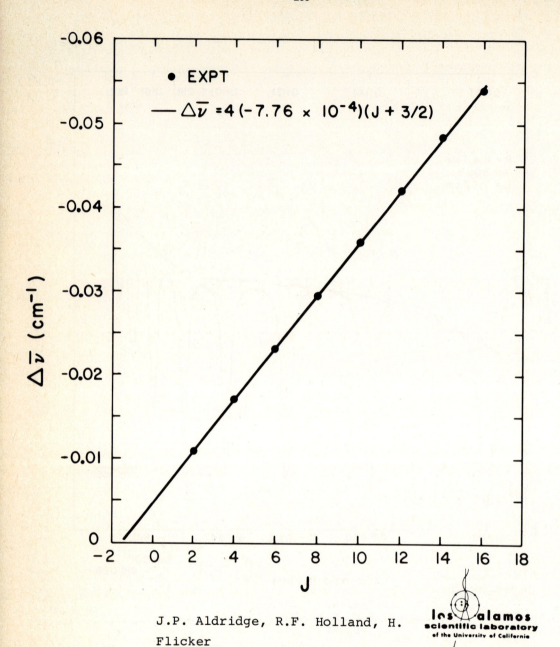

J.P. Aldridge, R.F. Holland, H. Flicker

J.P. Aldridge, R.F. Holland, H. Flicker, K.W. Nill, T.C. Harmon, J. Molec. Spectros., to be published

FIGURE 6

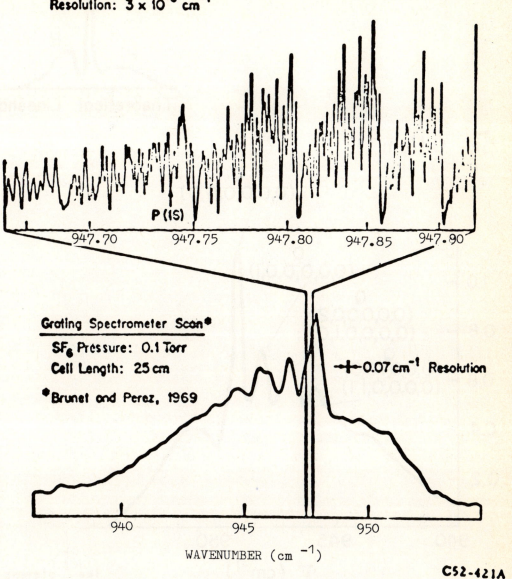

INFRARED SPECTRUM OF ν_3 BAND OF SF_6

FIGURE 7

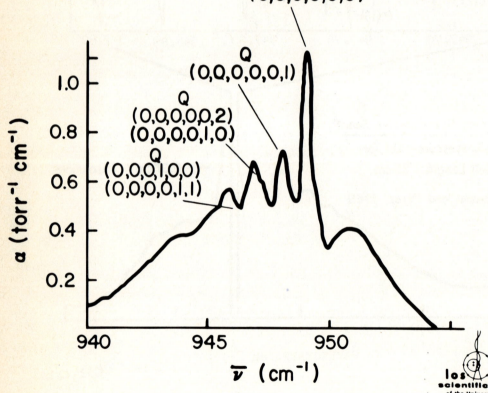

FIGURE 8

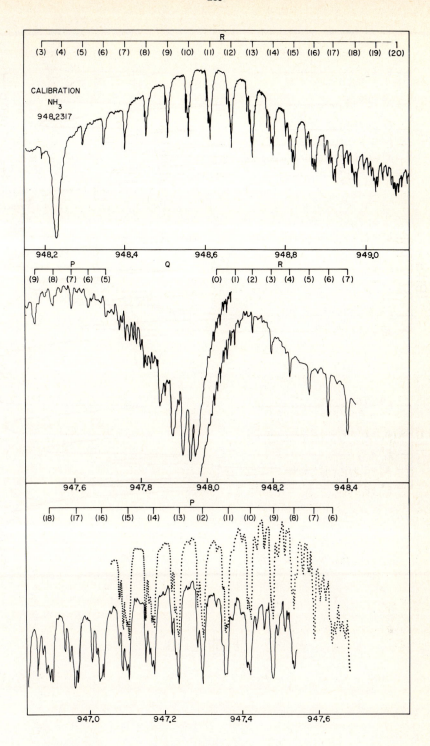

FIGURE 9

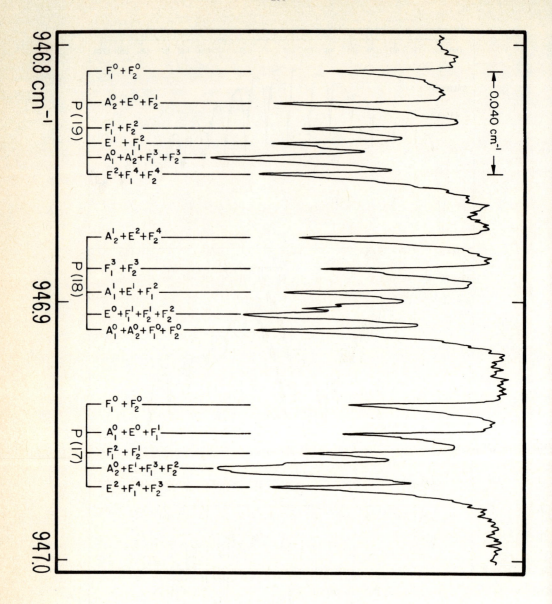

FIGURE 10

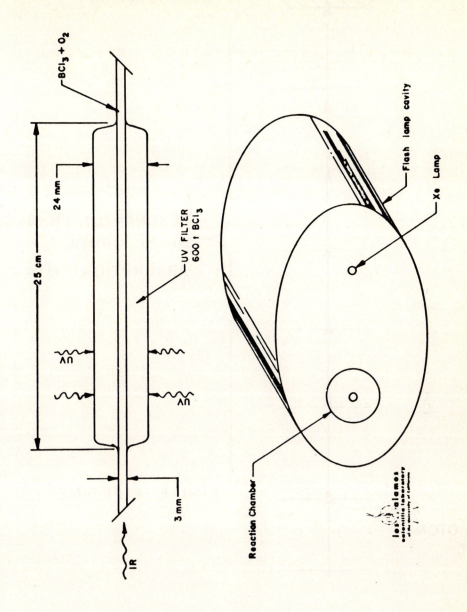

FIGURE 11

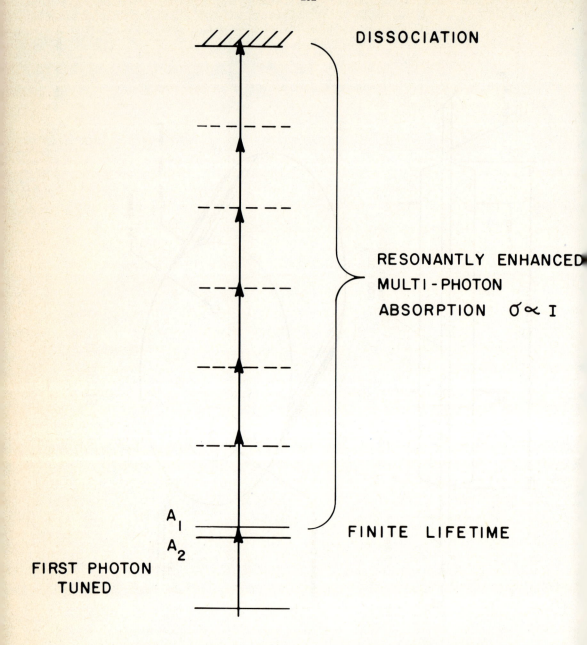

FIGURE 12

PROPOSED ABSORPTION MECHANISM

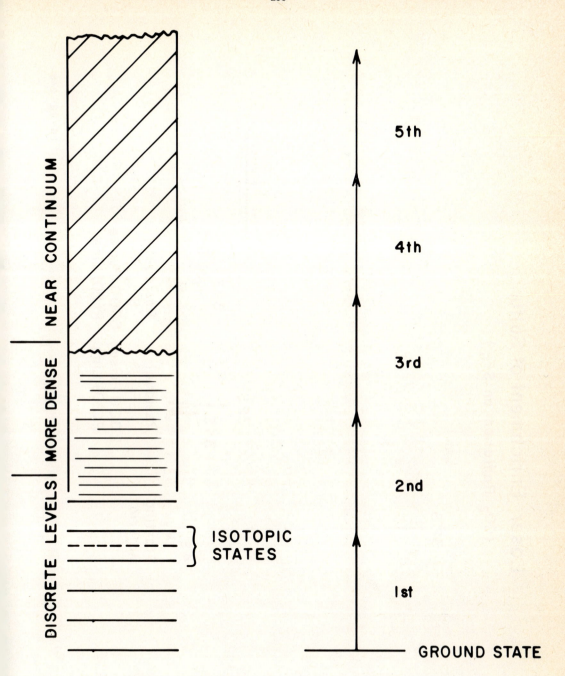

LEVEL SPACINGS IN POLYATOMICS

FIGURE 13

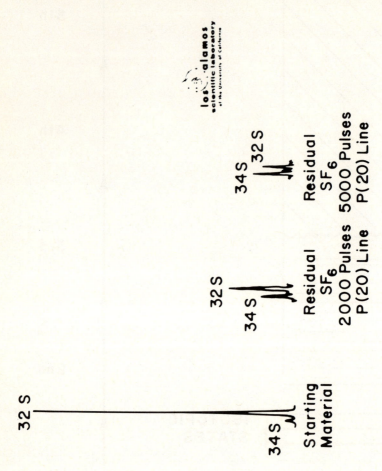

FIGURE 14

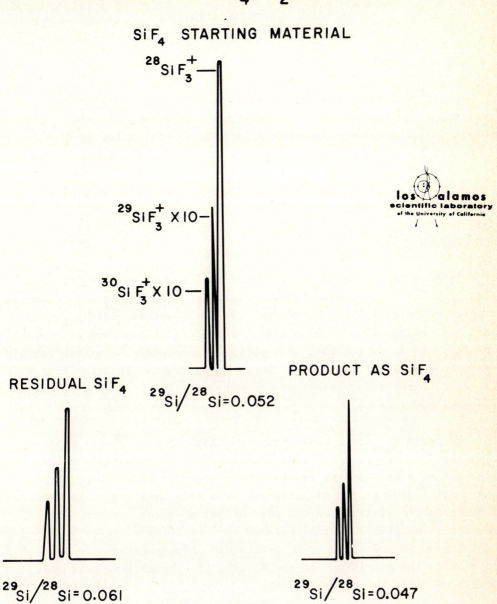

FIGURE 15

ISOTOPIC ENRICHMENT IN LASER PHOTOCHEMISTRY

Richard D. Deslattes, Michel Lamotte,[*] Harry J. Dewey, Richard A. Keller,
Samuel M. Freund, Joseph J. Ritter, Walter Braun and Michael J. Kurylo
National Bureau of Standards
Washington, D.C. 20234

ABSTRACT

Laser technology has permitted more efficient study of the chemistry of molecules in excited states. The extent to which isotopic specificity is preserved in going from initial excitation to final product is a valuable diagnostic for excited state chemistry. This report summarizes initial results from several areas of investigation, each of which suggests that laser stimulation may offer more than simple rate enhancement.

INTRODUCTION

Laser stimulation of unimolecular processes - two step photo-ionization and multi-photon molecular fragmentation - has been addressed in other reports at this conference. We have been especially concerned with laser stimulation of bimolecular reactions. These are attractive for study since, in contrast with photo-fragmentation, they may exhibit a gain in energy efficiency. Further, in contrast with unimolecular processes, bimolecular reactions can more readily be found which lead to stable products.

We begin by noting some early expectations and results on laser stimulated processes. Thereafter, we mention the relatively simple situation of single-photon excited ozone reaction rates where there remain still unclarified mechanisms. From this, we turn to the effect of electronic excitation in the case of thiophosgene reacting with substituted ethylenes. Finally, some work on infrared laser influenced chemistry on the reaction of $BC\ell_3$ with H_2S is summarized. In each of these cases the aim is one of emphasizing the progress we have made and the problems which still remain.

---

[*] Present Address: Department of Chemical Physics, Faculty of Sciences of Bordeaux, Talence, France. Work done while a guest worker at the National Bureau of Standards.

In approaching more complicated chemical systems, it is probably necessary at the present time to accept the fact that chemical engineering cannot be replaced by Schrödinger engineering. It is thus unrealistic to expect that chemistry of the excited states of these systems can readily be formulated with a great degree of formal rigor.

EARLY EXPECTATIONS AND OBSERVATIONS

In general, it was anticipated that excited reactants would speed up chemical reactions. It was further expected that this enhancement might be large for the case of electronic excitation and smaller but, in some cases, appreciable for vibrational excitation. It was also expected that the extent of rate enhancement by excitation would depend on the nature of the reaction. For instance, an already fast reaction might have its rate changed only to a small extent by excitation whereas, in a slow reaction, one would hope to enhance the rate of reaction by orders of magnitude.

These expectations can be formulated in various ways. Consider a reaction:

$$A + BC \xrightarrow{\kappa} AC + B \qquad (1)$$

with

$$\kappa = A e^{-E_a/kT} .$$

The effect of vibrational excitation on this reaction might have been thought of as reducing the activation energy to $(E_a - h\nu)$ or increasing temperature to some T_{eff}. This point of view was sufficiently widespread that several reports on rate enhancements quoted results as fractional utilizations of the available vibrational energy.

In contrast to this expectation, we note the results of a recent study on the temperature dependence of the rate constant for $O_3^\dagger$ + NO (1). Vibrationally excited O_3 indicated here by $O_3^\dagger$ is conveniently pumped by 9.6 μ radiation from a CO_2 laser. This excites the ν_3 assymmetric stretch which is coupled to the symmetric stretch and bending modes. The details of this coupling are only partially understood.

It turns out that when one makes an Arrhenius plot of the excited state rate constant and the thermal rate constant the results are shown in Fig. 1.

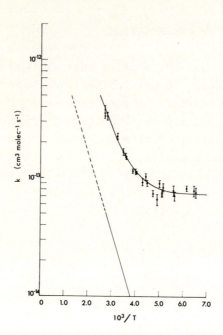

Fig. 1. The straight line is an Arrhenius representation of the thermal NO + O_3 data of Herron and Huie. Their measurements extend over the solid portion of the line. The curved line and data points are the data of Kurylo and Braun for the $O_3^\dagger$ + NO reaction. These are a composite of both a reactive and deactivative channel.

The fact that the excited state rate constant does not continue to decrease with increasing 1/T (above 4.0) is indicative of direct quenching of $O_3^\dagger$ in NO collisions. The principal results are, however, obvious. First there is a considerable rate enhancement for $O_3^\dagger$ compared to O_3 in the thermal process. Second, the high temperature slopes are characterized by the same value of E_a. According to recent thermal results by Huie and Herron (2), they are within 1%.

Thus the expectation that the role of vibrational excitation is to reduce E_a to $(E_a - h\nu)$ or $(E_a - \alpha h\nu)$, $\alpha \sim 1$ is not realized in this case. Formally, this means that the rate enhancement is found in the pre-exponential factor A. What this means is not altogether clear but it suggests that $O_3^\dagger$ behaves as though it were chemically distinct from O_3.

REACTION OF EXCITED $CSCl_2$ (3)

Thiophosgene ($CSCl_2$) is a particularly attractive molecule for electronic excitation studies. Its spectrum in the visible is well characterized and vibronic assignments are essentially complete. Isotope shifts between isotopic configurations of $^{12}C^{32}S^{35}Cl_2$, $^{12}C^{32}S^{37}Cl^{35}Cl$ and $^{12}C^{32}S^{37}Cl_2$ are quite easily resolved and a simply constructed dye laser, i.e., a nitrogen pumped dye laser with grating and beam expander, is quite sufficient for selective pumping. We have also made use of a coincidence between an absorption in $^{12}C^{32}S^{37}Cl_2$ and the 4658Å line of an argon ion laser.

Thiophosgene is known to react very slowly with olefins in the dark but more rapidly in the presence of light. A number of substituted ethylenes were tried as reaction partners but only two, diethoxyethylene (DEE) and tetramethyl ethylene (TME) were found to sufficiently fast in the presence of laser irradiation.

The following discussion is confined to DEE. Under laser illumination the reaction proceeds quite rapidly with product showing up first as a fog in the reaction cell and finally as liquid on the wall. Typically the reaction was studied at pressures on the order of a few Torr of each of the reactants. The reaction was essentially complete in a few hours with only a few milliwatts average power irradiation.

The route by which the stimulated reaction appears to go is the following:

$$\underset{Cl}{\overset{S^*}{\underset{\|}{C}}}\underset{Cl}{} \quad + \quad \underset{H}{\overset{OEt}{C}}=\underset{H}{\overset{OEt}{C}} \quad \rightarrow \quad \underset{Cl}{\overset{S}{\underset{\|}{C}}}-\underset{H}{\overset{OEt}{C}}-\underset{OEt}{\overset{H}{C}}-Cl$$

The product has a low vapor pressure and condenses from the gas phase starting materials. Quantum efficiencies can exceed 50%. Abundance changes produced by the two types of irradiation are indicated in Table I.

Table I

Isotope Enrichment of Thiophosgene

	$CS^{35}Cl_2$	$CS^{35}Cl^{37}Cl$	$CS^{37}Cl_2$
Natural abundance	.55	.38	.07
After irradiation of $CS^{35}Cl_2$	.35	.545	.105
After irradiation of $CS^{37}Cl_2$	.61	.34	.04

The one step enrichment of the 35-35 species is quite large. For convenience, abundance changes were measured in the unreacted gas phase residue. Isotope enrichment was studied as a function of extent of reaction for a range of starting reactant concentrations.

If the enhanced reaction takes place from the state initially pumped, then a relatively simple kinetic model can be constructed.

$CS^{35}Cl_2^*$ ($^1A_2$)

$+ CS^{35}Cl^{37}Cl \xrightarrow{k_1} CS^{35}Cl^{37}Cl^*$ ($^1A_2$) $+ CS^{35}Cl_2$ energy transfer

$+ DEE \xrightarrow{k_2}$ Product (35-35)   reaction

$+ DEE \xrightarrow{k_3} DEE + CS^{35}Cl_2$ quenching

$\xrightarrow{k_4} CS^{35}Cl_2 + h\nu$ fluorescence

$+ CSCl_2 \xrightarrow{k_5} CSCl_2^*$ ($^3A_2$) $+ CSCl_2$ ISC

$+ DEE \xrightarrow{k_6} CSCl_2^*$ ($^3A_2$) $+ DEE$ ISC

$+ CSCl_2 \xrightarrow{k_7} 2\, CS^{35}Cl_2$ ($^1A_1$)   quenching

From this model, an expression can be derived for the enrichment as a function of the extent of reaction. All of the rate constants with the exception of k_1 have been measured by various techniques. For a particular set of reactant concentrations, the enrichment as a function of the extent of reaction can be modeled on a computer and a value of k_1 extracted to fit the data. Unfortunately, the k_1 derived in this manner changes as the reactant concentrations are varied over the limited range permitted by the low vapor pressure of DEE. This means that something is the matter with our model and other possibilities must be considered.

Alternatively, we have explored the hypothesis that the stimulated reaction may not take place from the state initially pumped but rather from a state formed by a subsequential collisional process. The kinetic modelling begins to get more complicated for this more sophisticated model and higher order possibilities. There are, however, some simple observations which are informative.

Suppose, for instance, that the chemically active state is a long-lived triplet state reached by collisionally induced intersystem crossing (ISC). This hypothesis was eliminated by observing that the quantum yield was not increased by the addition of NO (a compound known to enhance ISC). Direct irradiation into the triplet manifold results in a process no longer first order in the reactants with a decrease of the quantum yield to less than 0.1.

These results have led us to consider the possibility that the chemical reaction might proceed from high vibrational levels of the ground electronic state of $CSCl_2$ formed by collisional quenching of the originally excited species. Although the vibrational hypothesis is attractive no direct evidence exists to support it.

THE BCl_3 + H_2S SYSTEM (4)

Work was begun on CO_2, TE laser-initiated BCl_3 reactions following the report of Karlov which indicated stimulated reaction with acetylene (5). The acetylenic borane product is interesting in that its synthesis is normally quite difficult. In the course of attempting to understand the poor yields obtained in repeating this work, it was found that the product was photolyzed by the pumping radiation. A search was carried out for other reaction partners where the expected products and hopefully intermediates as well would be transparent to the pumping light.

H_2S was a likely reaction partner for several reasons. Among these are that BCl_3 and H_2S react to yield a solid product B_2S_3 by a slow, many-stage thermal route. A principal gas phase intermediate, $HSBCl_2$ was reported to have absorption bands well removed from those of BCl_3.

The $^{10}B-Cl$ stretching vibration in BCl_3 is clearly resolved from that for $^{11}B-Cl$, and the absorptions coincide with several lines from a CO_2 laser.

Typical experimental conditions were: approximately 2 Torr BCl_3 and 2-10 Torr of H_2S (or D_2S). Mixtures were irradiated for several hours with a TE laser output of 0.1 joule/pulse at 10 pps with a 300 ns pulse length.

The reaction was noted to proceed with 1 mole of BCl_3 reacting with 2 moles of H_2S to give solids plus 2.5 - 3 moles of HCl. A 1:1 mixture of H_2S and BCl_3 could be totally depleted in H_2S. An initially plausible chemical route appeared to be the following:

$$BCl_3 + nh\nu_{ir} \rightarrow BCl_3^\dagger$$
$$BCl_3^\dagger + H_2S \rightarrow HSBCl_2 + HCl$$
$$HSBCl_2 + H_2S \rightarrow (HS)_2BCl + HCl$$
$$n\{(HS)_2BCl\} \rightarrow products$$

Isotope abundance changes are listed in Table II for irradiations at 10.55 μm and 10.18 μm.

Table II
Isotope Enrichment of Boron Trichloride

	$^{10}BCl_3$	$^{11}BCl_3$
Natural abundance	0.242	0.758
After irradiation of $^{10}BCl_3$	0.169	0.831
After irradiation of $^{11}BCl_3$	0.413	0.587

No isotopic enrichment was observed when pulses were longer than 1 μsec, and negligible reaction occurred under cw irradiation.

The principal studies have been of enrichment as a function of extent of reaction and starting mixture. From these it became apparent that the $HSBCl_2$ was in fact pumped by the laser radiation used. It is now suspected that excited $HSBCl_2$ is quite important in the reaction. (The reaction which forms $HSBCl_2$ goes spontaneously to a small extent.)

The power density threshold observed in this work and the intense visible chemiluminescence obtained at higher pressures are both suggestive of the importance of very high vibrational levels. There is possibly an analogy here to the photo-fragmentation processes recently reported (6,7). The result that no fragmentation occurs at very short pulse (1 ns) (7) durations suggests that the entire excitation process is not carried out in one simultaneous multi-photon event. A picture is gradually emerging of a mixture of simultaneous and sequential events with some possible role of collisions.

On the chemical side, it appears that products and intermediates are being produced which may be different from those reached by thermal routes. Diagnostics are difficult since several species have been formed which have not been previously isolated or characterized.

SUMMARY

Looking back over these three studies with full appreciation for the fact that much more work has to be done, we are led to advance that simple enhancement of first order rates cannot account for the observations.

REFERENCES

1. M.J. Kurylo, W. Braun, C. Nguyen Xuan, and A. Kaldor, J. Chem. Phys. 62, 2065 (1975).

2. R.E. Huie, J.T. Herron, and R.L. Brown, Int. J. Chem. Kinetics (in press).

3. M. Lamotte, H.J. Dewey, R.A. Keller, and J.J. Ritter, Chem. Phys. Lett. 30, 165 (1975).

4. S.M. Freund and J.J. Ritter, Chem. Phys. Lett. 32, 255 (1975).

5. N.V. Karlov, Appl. Opt. 13, 301 (1974).

6. R.V. Ambartzumian, Y.A. Gorokhov, V.S. Letokhov, and G.N. Makarov, JETP Lett. 21, 375 (1975).

7. J.L. Lyman, R.J. Jensen, J.P. Rink, C.P. Robinson, and S.D. Rockwood, Post Deadline Paper, 1975 IEEE/OSA Conference on Laser Engineering and Applications, Washington, D.C.

LASER CHEMISTRY

A.N.Oraevsky, A.V.Pankratov
Lebedev Physical Institute, Acad. of Sciences
Moscow, USSR

The subject of this report opens for the authors limitless possibilities for discussion, while the time for the talk is restricted. Therefore, we shall devote ourselves to the description of kinetic peculiarities in laser triggered chemical reactions. A general "philosophy" of an approach to the problem of laser stimulated chemical reaction has been described by us in /I/, and thus we shall consider this article as a continuation of the noted work.

First of all we should mention some results obtained by various authors during the last few years which permit us to describe laser chemical reactions as a new research trend. Reaction of tetrafluorohydrozine with nitric oxide is a good example of the fact /2/, that under the CO_2-laser radiation, being resonantly absorbed by tetrafluorohydrozine, there occurs a fast-proceeding chemical reaction the activation energy of which exceeds significantly that of the infrared quantum being absorbed.

The introduced in the mixture laser energy turned out to be insufficient for reaction initiation assuming equilibrium distribution of the absorbed energy by the degrees of freedom of the molecules under reaction.

Experimental results made it possible to constitute a concept on the photochemical effect of IR laser radiation which comprises a directed nonthermal stimulation of chemical reactions the activation energy of which may be significantly greater than that of the infrared quantum being absorbed. This situation was confirmed by several investigators /3-8/.

Thus, the analysis of $SF_6 + H_2$ reactions under pulsed CO_2-laser radiation indicated that the period of reaction induction is noticeably less than the time of vibrational-translational relaxation in SF_6 and N_2F_4 molecules /5/.

Under the action of CO_2-laser radiation on the mixture of trans- and cis-isomers of 2-buten is observed mixture enrichment by a cis-isomer, having a lesser absorption coefficient than the trans-isomer; i.e. the radiation caused conversion of the resonantly absorbing component /6/.

The authors of /7/ were lucky to observe conversion of B_2H_6 into the $B_{20}H_{16}$ under the action of CO_2-radiation. Thermal effect of compatible energies does not cause similar transformations of diborane.

The authors of this paper with the collaborators observed BCl_2F and SiF_3Cl formation under the CO_2-laser-action on $BCl_3 + SiF_4$ mixture. Thermal action on this mixture in a broad temperature range does not lead to the formation of the noted molecules.

Up to now we spoke of the action of emission pulses, duration of which was greater than the time of vibrational-vibrational and even vibrational-translational energy exchange between molecules. The activation collisions in those reactions between molecules played an important role. Laser radiation intensity essential to initiate the processes is of the order of several tens Watt/cm^2.

Recently the experiments have been performed on the initiation of chemical reactions by powerful radiation pulses of $10^8 \div 10^9$ W/cm^2 intensity and duration less than 100 nanosecond /7,8/. Under such conditions the CO_2-laser-excitation of BCl_3 and SF_6 high vibrational levels at several torr pressure was shown to proceed without collisions. There was observed a selective SF_6 dissociation and selective chemical reaction of excited BCl_3 molecules with oxygen. In dependence on CO_2--laser radiation frequency there prevailed, among the boron oxides, either $B^{10}O$- or $B^{11}O$-molecules.

Laser-induced reactions proceeding in the nonthermal way, may be called laser chemical. Laser chemical reactions proceeding at the molecular levels can divided into three stages:
1) excitation of molecules by laser radiation;
2) a reaction act conditioned by an excited molecule;
3) secondary reaction acts caused by products of the second stage.

Thermalization of the laser energy absorbed by a molecule is a competing process to the first and second stages. Reaction will have a definitely "laser chemical" character if the rate of vibrationally-excited molecules and the rate of such molecular formation is greater than that of the thermalization of an energy absorbed by a molecule.

The process of V-T (V-R-T) relaxation realizes, primarily, thermalization of the resonantly absorbed energy, as well as energy transfer into the vibrational degrees of freedom in molecules, which do not directly participate in the reaction. Therefore, the molecules with a significant rate of V-T relaxation do not tend to laser chemical processes (for instance, NH_3 and others). In this respect, a steric factor of the reaction rate constant becomes important. The fact is, we can provide a molecule with energy which is close to that of activation , and then the process rate will depend, on the whole, on the steric factor.

II. Another, and presumably most interesting property of laser chemical reactions noted in /2/ is their threshold behavior. It means that at the given content and pressure of chemical mixture there exists such a laser radiation intensity, below which no reaction can be observed. We have studied threshold properties of the reactions listed in Table I. The underlined molecules absorb resonantly the CO_2-laser radiation. The experiments have shown that threshold phenomena were observed both for exothermal and endothermal processes. The difference between them is that with the exothermal processes at the laser beam intensity above threshold we observed a practically complete conversion of the starting reagents, while the product yield of the endothermal reaction is proportional to the deposited energy (number of irradiating acts). Fig. I illustrates the degree of SiF_4 conversion in the reaction with BCl_3 versus CO_2-laser intensity being resonantly absorbed by BCl_3 molecule. The reaction threshold (asterisked) and linear dependence of the reaction output on the laser intensity are well observable. Table 2 shows a sharpness of thresholds within the range of 5÷10% intensity variation. The total conversion of initial reagents, for which a single act of irradiation is quite sufficient, gives place to the lack of conversion even under several thousand shots. Our experiments have demonstrated that threshold laser intensity was not at all a strictly constant value: it depends on interacting molecules, mixture pressure, component concentration ratio, beam diameter, laser pulse duration and so forth.

Therefore, we can speak about an existence of laser chemical reaction in the coordinates of various parameters. Fig. 2 shows teh dependence of threshold intensity on the pressure for I) and 2) reactions from Table I. As an example, the region of existence for N_2F_4 + NO laser-chemical reaction is hatched. Other reactions have a similar behavior.

What are the threshold phenomena in laser chemical reactions associated with ? One may assume that the experimentally observed reac-thresholds appear due to the threshold character of laser-induced thermal or chain combustion. By following the stage division of the process, we introduced above, one can say that the third stage is responsible for the threshold character. Reaction I-4 (Table I) are related to this possibility but threshold behavior of 6-8 reactions cannot be explained by combustion. Moreover, special experiments carried out with N_2F_4 + SO_2 mixture indicated that combustion process was unable to explain threshold character of laser chemical reactions even in the exothermic case. By introducing in the reactor a copper net cylinder we increased significantly the threshold of thermal combustion so that a single laser pulse could cause a full conversion of the mixture. Threshold intensity remained practically unchangeable (see Table 3). From the reactions represented in Table I the process of reaction stimulation 5) seems to have a purely thermal character.
Thermal combustion threshold of this mixture seems to be much less than the threshold for laser chemical reaction.

We believe that the given data are sufficient to reject explanations for the origin of laser-reaction thresholds on the basis of secondary process stage. In this situation it is natural to assume that the activation process of molecule by laser radiation has a threshold behavior. This assumption can be easily checked up. It was shown in a number of papers /9-12/ that visible luminescence appeared due to the population of high energy levels in a molecule produced by the resonant CO_2-laser radiation. We used this circumstance to clarify the above problem. We have studied the dependence of SF_6 and BCl_3 visible luminescence intensity on that of CO_2-laser radiation being resonantly absorbed by these molecules. Irradiation was performed with pulses of $\sim 10^{-2}$ sec and $3 \cdot 10^{-6}$ sec duration. Fig. 3 illustrates the dependence of the BCl_3 visible luminescence excited by 10^{-2} sec duration CO_2-laser pulses. Up to a certain threshold intensity no luminescence was observed. The threshold depends on pressure:

curve I corresponds to 380 torr of initial pressure, curve 2 - to the pressure of 100 torr. Similar dependencies were observed at the study of SF_6 molecules. But in case of SF_6 molecules the threshold intensities are higher.

The study of SF_6 luminescence under the action of CO_2-laser radiation with $3 \cdot 10^{-6}$ sec pulse duration also indicated (Fig. 4) the existence of a threshold intensity (or energy per pulse for a given pulse duration). Threshold depends on SF_6 pressure similarly to the experiments with long pulses. This dependence is demonstrated in Fig. 4 b.

III. How should one interpret the experimentally observed threshold character of luminescence and, along with it, threshold behavior of laser chemical reactions ? For this purpose one has to consider the possible mechanisms of excitation of high vibrational levels in molecules. From a wide variety of mechanisms described in literature the most probable one comprises resonant laser excitation of one or a few low vibrational levels, and a further excitation of high vibrational levels due to vibrational-vibrational energy exchange /13,14/ (Fig. 5). The number of the resonantly excited levels "m" depends on the spectral bandwidth and nonharmonicity factor (Fig.6). Hence for a linear molecule with the rotational constant B the frequency of vibrational-rotational transition from vJ level to $v + 1, J + 1$ level is equal to:

$$\nu_{v+1,v}^{J+1,J'} = \nu_{10} + 2B(J + 1) - v\Delta\nu \qquad (1)$$

The number of resonant lines in the band of vibrational-rotational transitions, by the order of magnitude, is equal to:

$$m \sim \frac{(BkT)^{1/2}}{\Delta\nu} \qquad (2)$$

Theoretical study of this model in various approximations is given in /15-18/. From these works it follows that at laser intensities less than saturation value the population of the v-th vibrational level is the v-th order power function of laser intensity. If one speaks about the high vibrational level population then the order of the power dependence may be very high that causes the growth of population up to reaching a certain value. If the detector sensitivity is a finite value, then such an abrupt dependence on intensity might be considered as the experimentally observed threshold.

To exemplify the above notion we shall consider a simple system consisting of three vibrational levels (Fig.7):

$$\frac{dn_2}{dt} = kn_1^2 - \frac{1}{\tau_2} \cdot n_2$$

$$\frac{dn_1}{dt} = -2kn_1^2 + \sigma I(n_0 - n_1) + \frac{1}{\tau_2} n_2 - \frac{1}{\tau_1} n_1 \quad (3)$$

$$n_0 + n_1 + n_2 = N$$

where I is laser intensity, and σ is cross-section of the resonant photon absorption. The rest notations can be identified from Fig. 7. At small intensities

$$n_1 \simeq \sigma I n_0 \tau_1 \quad ; \quad n_1 = k(\sigma I n_0 \tau_1)^2 \tau_2 \quad (4)$$

The stored vibrational energy per particle is:

$$\varepsilon = \hbar \omega_{10} \left[\sigma I \tau_1 + 2(\sigma I \tau_1)^2 k n_0 \tau_2 \right] \quad (5)$$

However, we cannot be satisfied by such an explanation given to the threshold phenomena. First of all, this is due to a very sharp threshold behavior during laser chemical reaction (Table 2): thousand pulses do not produce a detectable effect in the prethreshold region, while 5÷7% increase in intensity leads to a sufficient effect under the action of a single pulse.

We can propose, as a working hypothesis, a following mechanism for the appearance of threshold phenomena. let us assume that in the process accompaning the resonance molecular excitation (chemical reaction, visible luminescence) there participates not a single but several modes whose frequencies are close to the resonant frequency or its overtones. Those modes can be excited by their parametric interaction with a resonant mode due to nonlinearity (nonharmonicity) of molecular oscillations. The parametric oscillator

$$\ddot{x}_j + 2h\dot{x}_j + \omega_j^2 \cdot (1 + q \cos\Omega t) x_j = 0 \quad (6)$$

has unstable regions as is shown in Fig. 8 /19/. It is expedient to consider the first and second regions, for which $\omega_j/\Omega = 1/2$ and $\omega_j/\Omega = 1$. Value q is defined by the mode interaction and can be estimated in the following way. Energy of the j-th mode can be determined by an equation

$$H = \frac{m_j \dot{x}_j^2}{2} + \frac{k_j x_j^2}{2} + \sum_{k,m} \lambda_{jkm} x_j x_k x_m + \sum_{k,m,n} \mu_{jkmn} x_j x_k x_n \quad (7)$$

If we designate the resonant mode by an index "0" then the terms contributing to the value q, are:

$$\lambda_{jjo} x_j^2 x_o \quad \text{and} \quad \mu_{jjoo} x_j^2 x_o^2 \tag{8}$$

The first term describes a parametric modulation with frequency ω_o. In this case

$$q \cos \Omega t = 2 \frac{\lambda_{jjo}}{2} x_o \cos \omega_o t \tag{9}$$

where x_o is the amplitude of the resonant mode oscillations. For the second term

$$q \cos \Omega t = \frac{\mu_{jjoo}}{2} x_o \cos 2\omega_o t \tag{10}$$

Threshold condition for the first unstable region is:

$$q > 4 h_1 \tag{11}$$

If $h_j \sim 1/\tau_j$, τ_j is time of thermalization of the j-th mode (as, for instance for v-J relaxation) threshold condition will have a form:

$$\frac{\mu_{jjoo}}{2} x_o^2 > 4 \frac{1}{\omega_j \tau_j} \tag{12}$$

x_o^2 is associated with the energy of resonant mode oscillation

$$x_o^2 = 2 \frac{\mathcal{E}_o}{m_o \omega_o^2} \tag{13}$$

Since $\mathcal{E}_o$ depends on laser intensity there follows (from (12)) a threshold condition for intensity. Analogous relation can be obtained for the second region of instability.

Coefficients λ_{jjj} and μ_{jjjj} can be expressed through the frequency factor of nonharmonicity $\Delta \omega = 2\pi \Delta \nu$ /20/. By introducing a similar parameter $\Delta \omega_{jo}$ for λ_{joo} and μ_{jjoo} coefficients the threshold condition (12) can be written in the form:

$$\frac{\Delta \omega_{jo}}{\omega_j^2} \frac{\mathcal{E}_o}{\hbar} > 4 \frac{1}{\omega_j \tau_j} \tag{14}$$

By asubstituting value (5) for $\mathcal{E}_o$ we obtain a threshold condition for laser intensity I:

$$0.25 \cdot \Delta \omega_{jo} \tau_j \left[(\sigma I \tau_o)^2 \frac{\tau_o}{\tau_{vv}} + (\sigma I \tau_o) \right] > 1 \tag{15}$$

Unfortunately $\Delta \omega_{jo}$ coefficients are unknown for majority of molecules. However, even if $\Delta \omega_{jo} \sim 1/\tau_j$, condition (15) can be easily fulfilled. Since the energy of vibrational molecules depends on the parameter $S = k\sigma I n_o \tau$, threshold mechanisms under discussion will, in fact, define threshold value of the parameter S_n. It means that

$$I_n = \frac{S_n}{k \sigma n_o \tau} \qquad (16)$$

The experimentally observed decrease of the threshold intensity with pressure growth demands the assumption of increase of σ with pressure and an independency of τ (or a cery weak dependence) on the initial reagent density.

For pressure used in our experiments

$$\sigma \sim \frac{\Delta \nu}{(\Delta \nu)^2 + (\nu - \nu_o)^2} \qquad (17)$$

where $\Delta \nu$ is linewidth, ν and ν_o are laser and resonant mode frequency, respectively. Coincidence of ν and ν_o is not better than the value of the frequency gap between different rotational levels in lasing or studied molecule. Therefore $(\nu - \nu_o)^2 \gg (\Delta \nu)^2$ for majority cases, and

$$\sigma \sim \frac{\Delta \nu}{(\nu - \nu_o)^2} \sim \text{pressure} \qquad (18)$$

since $\Delta \nu$ is proportional to the pressure. For the independence of τ on pressure one should assume that τ is determined by a unimolecular first order process. All assumptions made for the explanation of the discussed here behavior are of a qualitative character, and a consequent mathematical description is required. Further experiments devoted to the study of intermode energy exchange are needed, and more detailed data on the spectral molecular properties should be available. Hence, a further progress in understanding laser chemical processes is closely associated with advances in a laser spectroscopy of molecules.

TABLE I.

N°	REACTION	ENTHALPY ccal/mole
1.	$N_2F_4 + 4NO = 4FNO + N_2$	-149
2.	$N_2F_4 + 2CO = 2COF_2 + N_2$	-250
3.	$N_2F_4 + 2SO_2 = 2SO_2F_2 + N_2$	-337
4.	$CF_3NF_2 = CF_4 + \tfrac{1}{2}N_2 + \tfrac{1}{2}F_2$	—
5.	$CF_3NF_2 + C_2H_4 \rightarrow CF_3CH=CH_2 + N_2 + \dots$	$\ll 0$
6.	$SiH_4 = Si + 2H_2$	-7.3
7.	$BCl_3 + SiF_4 = BCl_2F + SiF_3Cl$	$+1.3$
8.	$BCl_3 + H_2 = BCl_2H + HCl$	$+14.6$

TABLE 2

N°	COMPOSITION		Laser Beam Diameter (cm)	Pulse Duration (x 10^2 sec)	Laser Intensity (W/cm^2)	Number of Acts	RESULT
	Molecules	Pressure torr					
1	N_2F_4 NO	100 100	0,9	2	17 16	1 $2 \cdot 10^3$	100% Conversion No Reaction
2	N_2F_4 SO_2	200 100	0,8	5	31 29	1 $6 \cdot 10^3$	100% Conversion no reaction
3	SiH_4	35	0,6	3	58 55	70 1	no reaction 1% conversion
4	BCl_3 SiF_4	100 100	0,3	2	210 230	10^2 10^3	no reaction 68% conversion
5	CF_3NF_2	100	0,3	10	45 44	1 $3,5 \cdot 10^3$	100% conversion no reaction

T A B L E 3.

$P_{N_2F_4}$ = 200 torr, P_{SO_2} = 100 torr, p = 5.10^{-2} sec

Diameter of Reaction = 20 mm

Diameter of Beam	Diameter of Net Cylinder (cm)	I_{th} (W/cm²) without Net	I_{th} (W/cm²) with Net	% of Reaction Volume outside Net	Conversion %
0,8	1,8	31	31	90	92
0,8	1,7	31	31	70	77
0,8	1,4	31	31	50	59
0,8	1,1	31	31	30	33
0,3	0,7	43	43	12	20

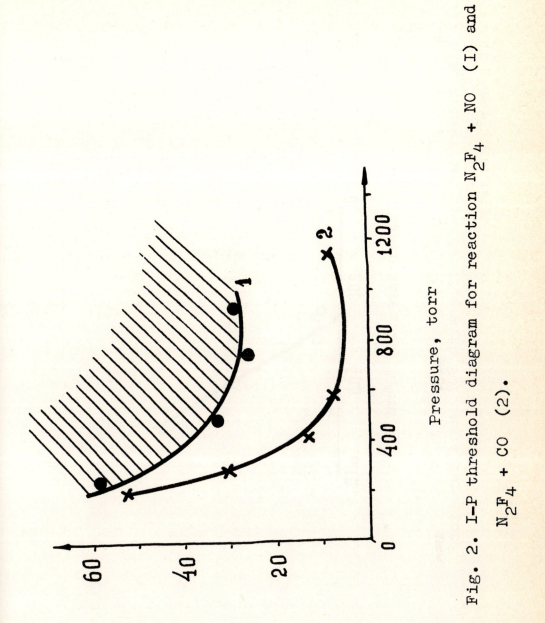

Fig. 2. I-P threshold diagram for reaction $N_2F_4 + NO$ (1) and $N_2F_4 + CO$ (2).

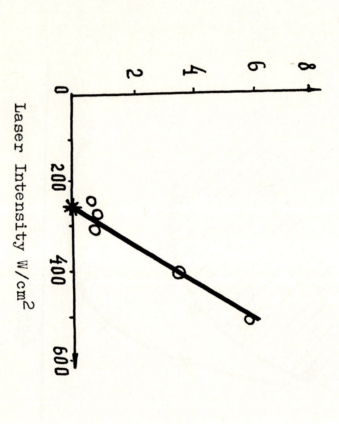

Fig. I. SiF$_4$ conversion in reaction with BCl$_3$ as a function of laser intensity.

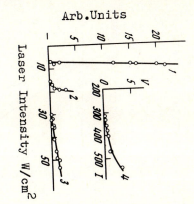
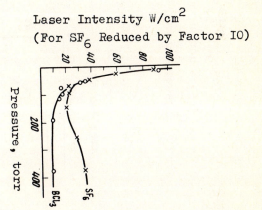

Fig. 3. Intensity of luminescence versus laser intensity (a) and I-P threshold diagram for luminescence (b); long pulse excitation ($\tau_p \simeq 10^{-2}$ sec).

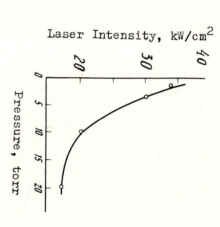

Fig. 4. Intensity of luminescence versus laser intensity (a) and I-P threshold diagram for luminescence (b); short pulse excitation ($\tau_p \simeq 3 \cdot 10^{-6}$ sec).

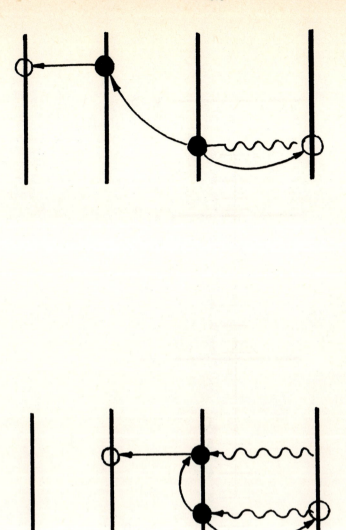

Fig. 5. Diagram of V-V energy exchange.

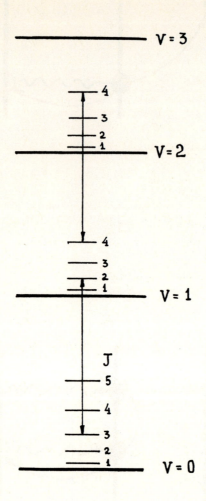

Fig. 6. Compensation of nonharmonic frequency shift by rotational energy.

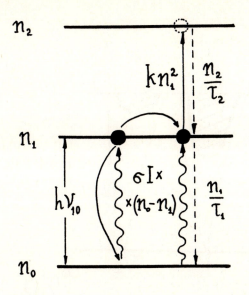

$$\sigma I \tau_1 \ll 1$$

$$n_1 \simeq \sigma I \tau_1 \cdot n_0 \; ; \; n_2 \simeq (\sigma I \tau_1)^2 k \tau_2 n_0^2$$

$$\mathcal{E} = h \nu_{10} (n_1 + 2 n_2)$$

Fig. 7. Three level model of excitation.

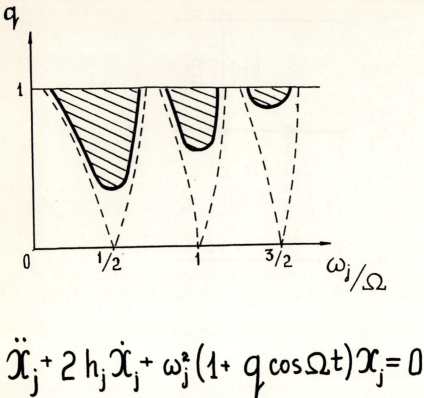

Fig. 8. Regions of Parametric Instability (hatched).

References

1. N.G.Basov, A.N.Oraevsky, A.V.Pankratov. Chapter 7 in the book "Chemical and Biochemical Applications of lasers", ed.by C.B.Moore, Acad. Press, 1974.
2. N.G.Basov, E.P.Markin, A.N.Oraevsky, A.V.Pankratov. DAN USSSR, v. 198, 1043 (1971).
3. N.G.Basov, E.P.Markin, A.N.Oraevsky, A.V.Pankratov, A.N.Skachkov, JETP, Letters, v. 14, 251 (1971).
4. N.G.Basov, E.M.Belenov, E.P.Markin, A.N.Oraevsky, A.V.Pankratov, JETP, Letters, v. 64, 485 (1973).
5. J.L.Lyman, R.J.Jensen, J.Phys.Chem., v. 77, 883 (1973).
6. A.Jogev, R.M.J.Lochwenstein-Benmair, J.Am.Chem.Soc.,v.93,8487(1973).
7. H.R.Bachman, N.Nöth, R.Rink, K.L.Kompa, Chem.Phys.Lett.,v.29,627(1974).
8. R.V.Ambartsumian, V.S.Letokhov, N.V.Chekalin. JETP,Letters,v.21,486((1975).
9. M.C.Borde, A.Henry, M.L.Henry. Comt.Rend.Acad.Sci., B 262, 1389(1966).
10. N.V.Karlov, Yu.N.Petrov, A.M.Prokhorov, O.M.Stel'makh. JETP, Letters, v. 11, 220 (1970).
11. N.R.Jseror, V.Merchand, K.S.Hallsworth, M.S.Richardson, Can.J. of Phys., v.51, 1281 (1973).
12. R.V.Ambartsumian, N.V.Chekalin, V.S.Polyakov, V.S.Letokhov,E.A. Ryabov. Chem.Phys.Lett., v.25, 515 (1974).
13. N.G.Basov, V.T.Galochkin, S.I.Zavorotny, V.N.Kosinov, A.A.Ovchinnikov, A.N.Oraevsky, A.V.Pankratov, A.N.Skachkov, G.V.Shmerling. JETP Letters, v.21m 70 (1975).
14. N.D.Artamonova, V.T.Platonenko, R.V.Khokhlov. JETP, v. 58, 2195 (1970).
15. A.N.Oraevsky, V.A.Savva. Short Commun.in Phys.,FIAN, N 7,50 (1970).
16. N.G.Basov, A.N.Oraevsky, A.A.Stepanov, V.A.Scheglov.JETP, v.65, K837 (1973).
17. B.F.Gordiets, N.S.Mamedov, L.A.Shelepin, Preprint FIAN N 28, 1974.
18. B.F.Gordiets, A.I.Osipov, V.Ya.Panchenko. JETP, v.65, 894 (1973).
19. L.I.Mandel'shtam. Polnoe sobranie trudov, v.IV, Izd.Ac.Sci.USSR, 1955, p.193.
20. L.D.Landau, E.M.Lifshits. Kvantovaya Mekhanika (Nerelyativistskaya teoriya). Fizmatgiz, 1963.

ATOMS IN STRONG RESONANT FIELDS
SPECTRAL DISTRIBUTION OF THE FLUORESCENCE LIGHT

Claude Cohen-Tannoudji
Ecole Normale Supérieure and Collège de France, Paris

1). Introduction. Lowest order Q.E.D. predictions

Resonance fluorescence, i.e. absorption and reemission of resonance radiation by free atoms, is a very important process. By looking at the fluorescence light emitted by these atoms, for example by measuring its polarization, its intensity, its spectral distribution or its time dependance, one gets interesting informations on various important atomic parameters such as g factors, fine or hyperfine structures, radiative lifetimes ...

The physical picture which is usually given for such a process ([1]) can be visualized by the following lowest order diagram :

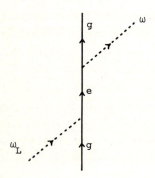

Figure 1 : Lowest order Q.E.D. diagram for resonance fluorescence

One impinging photon, with frequency ω_L, represented by the incoming dotted line is absorbed by the atom which jumps from the ground state g to the excited state e. After a certain amount of time spent in e, the atom falls back to g, spontaneously emitting a photon ω. The scattering amplitude derived from Q.E.D. for such a process obviously contains a $\delta(\omega - \omega_L)$ function which expresses the conservation of energy. Lowest order Q.E.D. therefore predicts that the fluorescence light following a monochromatic excitation must be monochromatic, with the same frequency ω_L as the incident light. In addition, one finds that the scattering amplitude is the greater, the nearer ω_L is to the atomic frequency ω_o. More precisely, if Γ is the radiative width of the excited state, the scattering cross section $\sigma(\omega_L)$ varies as $\left[(\omega_L-\omega_o)^2+(\Gamma/2)^2\right]^{-1}$.

Suppose that the incident light beam contains photons with all frequencies ω forming a white continuous spectrum (or at least a spectrum with a width $\Delta \gg \Gamma$). Every photon ω will be elastically scattered with an efficiency given by $\sigma(\omega)$, so that one predicts from the same lowest order treatment that the spectral distribution of the fluorescence light following a broad line excitation has a lorentzian shape, centred at $\omega = \omega_o$, with a half-width $\Gamma/2$.

All these conclusions are clearly no longer valid in very strong resonant fields. We have now at our disposal laser light sources which can easily saturate an atomic transition : the atom can interact several times with the laser light before emitting spontaneously a photon and a lowest order treatment of the fluorescence process is obviously insufficient. How do atoms behave in strong resonant (or quasi-resonant) light beams ? What kind of light do they emit ? What is the influence of the spectral width of the incident light ? Are the higher order correlation functions of the light important ? These are examples of questions which arise now in connection with laser experiments.

I have already discussed the problem of optical pumping and level crossing experiments performed with lasers at the 1974 Heidelberg conference on Atomic Physics ([2]). So, I will rather discuss in the present talk another problem which is the spectral distribution of the fluorescence light emitted by an atomic beam which is irradiated at right angle by a high intensity laser beam (as the 2 beams are perpendicular, there is no Doppler effect). The first experimental observation of such a spectral distribution has been published last year by Schuda, Stroud and Hercher ([3]). Similar experiments are being performed in other laboratories and will be reported at the present conference in subsequent talks ([4])([5]). Concerning the theory of these effects, several calculations have been published, using different methods and approaches ([6]) $\leftrightarrow$ ([17]). They don't reach all the same quantitative conclusions. Rather than entering into the details of these calculations, I have thought it would be more interesting in this talk to make a few remarks and comments and to try to give some physical feeling about important parameters. Some new theoretical results will be reported at the end of the paper.

2). <u>What does conservation of energy imply</u> ?

The first remark will concern conservation of energy. One could think at first sight that such a principle implies for the fluorescence light following a monochromatic excitation to be always monochromatic, with the same frequency as the incident light. This is not correct at high intensities. Non linear scattering processes can take place in which N impinging photons (with $N > 1$), having all the same energy ω_L (we take $\hbar = c = 1$), disappear and are replaced by N scattered photons with different energies $\omega_1^S, \omega_2^S \ldots \omega_N^S$. Conservation of energy only requires $N\omega_L = \omega_1^S + \omega_2^S + \ldots + \omega_N^S$. As an illustration, we have represented on Fig. 2 such non linear scattering processes corresponding to $N = 2$.

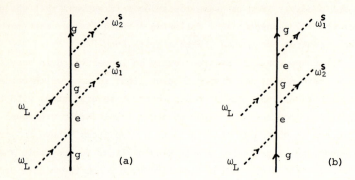

Figure 2 : Example of non linear scattering processes in which 2 impinging photons ω_L, ω_L give rise to 2 scattered photons ω_1^S, ω_2^S with $\omega_1^S + \omega_2^S = 2\omega_L$. The 2 diagrams (a) and (b) differ by the order of emission of the 2 photons ω_1^S and ω_2^S.

I would like also to point out on this example that although $\omega_1^S + \omega_2^S$ is well defined and equal to $2\omega_L$, ω_1^S and ω_2^S are individually spread over finite intervals, which means that inelastic scattering is not monochromatic. Such a finite width of the fluorescence spectrum is due to the energy denominators associated to the intermediate states appearing in diagrams 2a and 2b. When calculating the sum of the 2 scattering amplitudes 2a and 2b, one finds that one of the 2 photons is distributed over an interval of half width $\Gamma/2$ around ω_o [Γ being the natural width of the excited state e] . Consequently, the second photon is distributed over an interval $\Gamma/2$ around $2\omega_L - \omega_o$.

3). The "dressed atom" approach

It would not be a good idea to consider higher and higher order diagrams for understanding the behaviour of an atom in a strong resonant field. For sufficiently large intensities of this field, the perturbation series would not converge, and the situation would be the more difficult, the nearer ω_L would be to ω_o. So we are tempted to try another approach.

Why don't we treat to all orders the coupling between the atom and the incoming photons, neglecting spontaneous emission in a first step ? Let us call "dressed atom" the total isolated system which results from the coupling between the atom and the incoming photons. Such a system has stationary states ψ_α, ψ_β .... with energies E_α, E_β ... which can be calculated easily ([18]). Then we could treat spontaneous emission by using Fermi's golden rule : the dressed atom jumps from ψ_α to a lower level ψ_β, by spontaneously emitting a photon $\omega = E_\alpha - E_\beta$ with a probability per unit time proportional to $|<\psi_\alpha|D|\psi_\beta>|^2$ where D is the atomic electric dipole operator. This process is diagrammatically represented on Fig. 3, where the heavy lines represent the stationary states of the dressed atom.

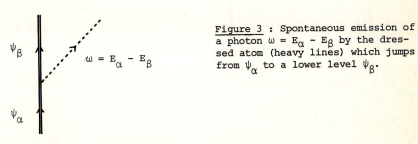

Figure 3 : Spontaneous emission of a photon $\omega = E_\alpha - E_\beta$ by the dressed atom (heavy lines) which jumps from ψ_α to a lower level ψ_β.

4). **The difficulty of dealing with cascades**

The dressed atom approach is very convenient for finding the number and the mean position of the various components of the fluorescence spectrum which correspond to the Bohr frequencies ($E_\alpha - E_\beta$) of the allowed transitions ($<\psi_\alpha|D|\psi_\beta> \neq 0$) of such a system. However, if we want to get more precise informations, concerning for example the widths and the relative amplitudes of these various components, we cannot consider only a single spontaneous emission process as in Fig. 3.

To make this point clear, it will be useful to give some orders of magnitude. An atom, with a thermal velocity $v \sim 10^3$ m.s^{-1}, crossing a laser beam of 10^{-3} m. diameter, spends in this light beam a time $T \sim 10^{-6}$ s, much longer than the radiative lifetime $\tau = \Gamma^{-1}$ of e, which is typically $\tau \sim 10^{-8}$ s. If the light intensity is large enough, the atomic transition is saturated, and the atom spends half of its time in e, so that an average number of $N = \frac{1}{2}\frac{T}{\tau} \sim 50$ spontaneous emission processes can occur during the interaction time T. It follows that the evolution of the dressed atom is more exactly described by the diagrams of Fig. 4 (where, to simplify, we have supposed N to be only equal to 3).

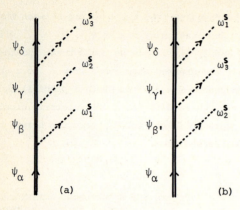

Figure 4 : Sequences of 3 spontaneous emission processes. The 2 sequences (a) and (b) correspond to the same initial and final states ψ_α and ψ_δ, but to a different order of emission of the 3 photons ω_1^s, ω_2^s, ω_3^s.

The dressed atom is "cascading" from ψ_α to ψ_β, then from ψ_β to ψ_γ and finally from ψ_γ to ψ_δ, successively emitting photons with frequencies ω_1^s, ω_2^s, ω_3^s close to $E_\alpha - E_\beta$, $E_\beta - E_\gamma$, $E_\gamma - E_\delta$ (fig. 4-a). But we can imagine other processes, corresponding to the same initial and final states ψ_α and ψ_δ, to the same frequencies ω_1^s, ω_2^s, ω_3^s of the 3 emitted photons, but to a different order of emission of these 3 photons. N! such possibilities exist, of which only 2 are represented on fig. 4.

The difficulty lies in the fact that, being interested in a precise measurement of the frequencies of the photons, we cannot simultaneously determine the time at which they are emitted (time and frequency are complementary physical quantities) and, consequently, we cannot decide what is the quantum path which is followed by the system. We have N! quantum amplitudes which interfere.

One could at least think that one amplitude is much greater than the others because of the energy denominators associated to the intermediate states. This is not true and comes from the periodical structure of the energy diagram of the dressed atom which is itself due to the quantization of the field mode associated to the laser. For any order of emission of the 3 photons, one can find in general intermediate states $\psi_{\beta'}$, $\psi_{\gamma'}$ which introduce small energy denominators by approximately matching the energy of the emitted photon (similar difficulties are encountered when one studies the spontaneous emission from a harmonic oscillator ([19])).

The correct way of pursuing the calculation would be to compute, for all values of N, the N! interfering cascading amplitudes, to deduce from them the N-fold probability distribution $\mathcal{P}^{(N)}(\omega_1^s, \omega_2^s \ldots \omega_N^s)$ for having N spontaneously emitted photons with frequencies ω_1^s, ω_2^s, ..., ω_N^s, finally, after several integrations, to derive from the $\mathcal{P}^{(N)}$ a reduced one photon distribution $\mathcal{J}(\omega)$ giving the probability for any individual photon to have the frequency ω, which is the measured spectral distribution.

5). **Why not calculating directly the spectral distribution $\mathcal{J}(\omega)$?**

Although such an approach is correct, it seems too ambitious. It gives too many informations which are not useful : we are not measuring the $\mathcal{P}^{(N)}$, but $\mathcal{J}(\omega)$. Would not it be possible to calculate directly $\mathcal{J}(\omega)$ without passing as an intermediate step through the $\mathcal{P}^{(N)}$?

We are thus led to the problem of relating directly $\mathcal{J}(\omega)$ to some simple physical quantities characterizing the radiating atoms. Such a problem has been considered in many references [10][20] and I will give here only the results. One finds that the spectral distribution of a given light field is proportional to the Fourier transform (F.T.) of the correlation function (c.f.) of the positive frequency part of the electric field operator. As this electric field is radiated by the atom, it may be related to the atomic electric dipole operator D. So, we find that $\mathcal{J}(\omega)$ is proportional to the F.T. of the c.f. of the atomic dipole moment D. More precisely, let $D_- = d\,|g><e|$ and $D_+ = d\,|e><g|$ be the lowering and raising parts of D, d being equal to the matrix element $<e|D|g>$ (which is assumed to be real). One finds that :

$$\mathcal{J}(\omega) \sim \int_0^T dt \int_0^T dt' \, <D_+(t)\,D_-(t')> \, e^{-i\omega(t-t')} \qquad (1)$$

The integrals over t and t' run over the interval of time $[0, T]$ during which the atom radiates (transit time through the laser beam). The operators $D_+(t)$ and $D_-(t')$ are evaluated in the Heisenberg picture, and the average value is taken within the quantum state of the whole system.

6). **Spin 1/2 representation of the problem**

At this stage of the discussion, and because of lack of time, I will restrict myself to a classical description of the laser field, (but not of the empty modes of the electromagnetic field into which the atom spontaneously emits photons). It would be of course possible to calculate the correlation function written in (1) for the dressed atom introduced above and this has been done [13][17] (such a calculation is considerably simpler than the computation of the whole set of $\mathcal{P}^{(N)}$!). As the number n of impinging photons is very large, we would not make the difference between $\sqrt{n}$ and $\sqrt{n+1}$, and the results would be the same as the ones derived from a classical description of the laser field. Such a classical description will give me the possibility of developping simple geometrical interpretations and fruitful analogies with magnetic resonance experiments.

It is well known that a fictitious spin 1/2 can be associated to any 2 level system, so that our problem can be formulated in the following geometrical terms [21]. We have a spin 1/2 $\vec{S}$, which precesses around a magnetic field $\vec{B}_o$ parallel to Oz with a Larmor frequency equal to the energy separation ω_o between e and g ($\vec{B}_o$ is given by $\omega_o = -\gamma \vec{B}_o$, γ being the gyromagnetic ratio of the spin). To $D_\pm$ are associated the raising and lowering operators $S_\pm = S_x \pm i S_y$, so that we are interested in the c.f. of some transverse components of the spin in the xOy plane.

In this representation, the laser field, of frequency ω_L, is described by an oscillating magnetic field $\vec{B}_1 \cos\omega_L t$ parallel to Ox. We can decompose this oscillating field into 2 left and right circular components, of amplitude $B_1 = \mathcal{B}_1/2$, and keep only the one which precesses around $\vec{B}_o$ in the same sense as the spin $\vec{S}$. Let $\omega_1 = -\gamma B_1$ be the Larmor frequency associated to B_1 (Rabi nutation frequency). ω_1 characterizes the strength of the coupling between the atom and the laser and must be compared to Γ which measures the strength of spontaneous emission. Neglecting the counter-rotating components of $\vec{\mathcal{B}}_1$ is called "rotating wave approximation" (r.w.a) and amounts to ignore Bloch-Siegert's shifts which are much smaller in optical than in RF range. Note that, when doing r.w.a, we don't exclude "light-shifts" [22] which may appear for a quasi-resonant irradiation ($\Gamma < |\omega_L - \omega_o| \ll \omega_o$) and which may be much larger than Bloch-Siegert shifts.

It will be convenient to describe the situation in a reference frame OXYZ rotating around Oz = OZ with the "good" component $\vec{B}_1$ of $\vec{\mathcal{B}}_1$, so that $\vec{B}_1$ is static in this reference frame and parallel to OX (fig. 5). Let $\vec{S}(t)$ be the spin operator in this reference frame. The Larmor precession of $\vec{S}(t)$ around OZ is reduced from ω_o to $\omega_o - \omega_L$, and we can consider that $\vec{S}$ only "sees" 2 static fields $\vec{B}_o$ and $\vec{B}_1$ respectively parallel to OZ and OX and proportional to $\omega_o - \omega_L$ and ω_1 ($\omega_o - \omega_L = -\gamma B_o$, $\omega_1 = -\gamma B_1$). Expressed in terms of $S_\pm(t) = e^{\mp i\omega_L t} \mathcal{S}_\pm(t)$, (1) can be written as:

$$J(\omega) \sim \int_0^T dt \int_0^T dt' <S_+(t) S_-(t')> e^{-i(\omega-\omega_L)(t-t')} \qquad (2)$$

Finally, we have a simple geometrical representation of the internal energy of the atom (precession of $\vec{S}$ around $\vec{B}_o$), of its coupling with the laser (precession around $\vec{B}_1$). The question now is how to describe spontaneous emission in this representation (i.e. the coupling with the empty modes of the electromagnetic field) and how to calculate the c.f. written in (2)?

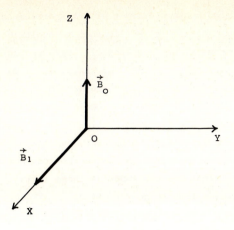

Figure 5 : In the rotating reference frame OXYZ, the fictitious spin $\vec{S}$ associated with the 2-level atom precesses around 2 static fields $\vec{B}_0$ and $\vec{B}_1$ respectively parallel to OZ and OX and proportional to $\omega_0 - \omega_L$ and ω_1.

7). <u>Naïve approach based on "Bloch's equations"</u>

Let's first give a naïve approach of the problem, strongly suggested by the analogy with a magnetic resonance experiment, but which, in the present case, is incorrect. Then, in trying to understand where is the mistake, we will get some physical insight into the problem.

It seems reasonable to describe spontaneous emission by some damping terms in the equations of motion of $< \vec{S}(t) >$. The transfer of atoms from e to g with a rate Γ can be described by $< \dot{S}_z(t) > = \Gamma (S_0 - < S_z(t) >)$, where $S_0 = -1/2$ (after a time large compared to $\tau = \Gamma^{-1}$, all atoms are in g, so that $< S_z > = -1/2$). As g is not affected by spontaneous emission, $< S_{\pm}(t) >$ are damped to zero with a rate $\Gamma/2$
$< \dot{S}_{\pm}(t) > = -(\Gamma/2) < S_{\pm}(t) >$. Adding these damping terms to the ones which describe the precession around $\vec{B}_0$ and $\vec{B}_1$, one gets the following equations which can be considered as the Bloch's equations of the problem :

$$\begin{cases} < \dot{S}_z(t) > = -i(\omega_1/2) < S_+(t) > - \Gamma < S_z(t) > + i(\omega_1/2) < S_-(t) > + \Gamma S_0 & (3\text{-a}) \\ < \dot{S}_{\pm}(t) > = -\left[(\Gamma/2) \pm i(\omega_L - \omega_0) \right] < S_{\pm}(t) > \mp i\omega_1 < S_z(t) > & (3\text{-b}) \end{cases}$$

What is the solution of these equations for an atom flying through the laser beam ? After a transient regim which starts when the atom enters the laser beam at t = 0, and which lasts for a time of the order of $\tau = \Gamma^{-1}$ [damping time of the transient solutions of equations (3)], $< \vec{S}(t) >$ gets a stationary value $< \vec{S} >_{st}$, independant of t, and corresponding to the steady state solution of (3). This situation lasts during all the transit time T through the laser beam (remember that T >> τ). After that, the atom leaves the laser beam at time t = T, and $< \vec{S} >$ damps to zero in a short time, of the order of τ.

At this stage, one is very tempted to consider that the light radiated by the atom corresponds to this evolution of $< \vec{S}(t) >$ (we have to return from the rotating to the laboratory reference frame) and, consequently, that its spectrum is given by the squared modulus of the F.T. of $<S_+(t)> e^{i\omega_L t}$. If such a conclusion were correct, one would get first an elastic component, at frequency ω_L, representing the contribution of the forced steady state motion $<S_+>_{st} e^{i\omega_L t}$ of the dipole moment driven by the laser field and which, as we have seen above, is the main part of the motion of the dipole. Strictly speaking, this elastic component would have a non zero width $1/T$ (corresponding to the finite transit time T), much smaller however than Γ (as $T \gg \tau$). In addition, one would get a small inelastic component, associated with the 2 small transient regims appearing at the 2 small regions where the atom enters or leaves the laser beam. This would suggest that one can suppress these inelastic components just by eliminating the light coming from these 2 regions.

8). What is missing in this approach ? Importance of the fluctuations

The method we have just outlined is not correct. A mathematical argument for showing it is that, when we calculate the squared modulus of the F.T. of $<S_+(t)>$, we find an expression analogous to (2), but where $<S_+(t) S_-(t')>$ is replaced by $<S_+(t)><S_-(t')>$, and these 2 quantities are not equal.

It is perhaps more interesting to try to understand physically where is the mistake. The important point is that the light emitted by the atom is not radiated by its average dipole moment represented by $<S_\pm(t)>$, but by its instantaneous dipole moment $S_\pm(t)$, and, even though the effect of spontaneous emission on $<\vec{S}(t)>$ may be shown to be correctly described by the damping terms of equations (3), such a description is incorrect for $\vec{S}(t)$.

Let's try to visualize the evolution of $\vec{S}(t)$. We can consider the atom as being constantly "shaked" by the "vacuum fluctuations" of the quantized electromagnetic field [23]. These random fluctuations, which have an extremely short correlation time, have a cumulative effect on the atom in the sense that they damp $<\vec{S}(t)>$, but we must not forget that they make the instantaneous dipole moment $S_\pm(t)$ fluctuate permanently around its mean value. The light which comes out is radiated not only by the mean motion of the dipole, but also by its fluctuations around the mean motion.

When we consider the effect of atoms on the incident electromagnetic wave which drives them, i.e. when we study how they absorb or amplify this wave, the average motion $<\vec{S}(t)>$ is very important since it has definite phase relations with the driving field. The fluctuations of $\vec{S}(t)$ act only as a source of noise and can be ignored in a first step. In the problem we are studying here, we cannot ignore the fluctuations since they play an essential role : we are interested in spontaneous emission, not in absorption or induced emission, and the fluctuations of $S_\pm(t)$ entirely determine the inelastic part of the fluorescence spectrum as we will show it now.

9). Elastic and inelastic parts of the fluorescence spectrum

Let us write :

$$S_{\pm}(t) = <S_{\pm}(t)> + \delta S_{\pm}(t) \tag{4}$$

where $\delta S_{\pm}(t)$ is the deviation from the average value and obviously satisfies :

$$<\delta S_{\pm}(t)> = 0 \tag{5}$$

Inserting (4) into (2), and using (5), one gets immediately :

$$<S_{+}(t) S_{-}(t')> = <S_{+}(t)><S_{-}(t')> + <\delta S_{+}(t) \delta S_{-}(t')> \tag{6}$$

One clearly sees from (6) that, in the spectrum of the fluorescence light, there is an elastic component corresponding to the first term of (6) and which is the light radiated by the average motion of the dipole. In addition, we get an inelastic component corresponding to the last term of (6) and which is the light radiated by the fluctuations. The spectrum of this inelastic part is determined by the temporal dependance of these fluctuations, i.e. by their dynamics.

Before studying this problem, let us show how it is possible to derive simple expressions for the total intensity radiated elastically and inelastically, I_{el} and I_{inel}. Integrating (2) over ω, one gets a $\delta(t-t')$ function which gives when using (6) :

$$
\begin{aligned}
I_{el} &\sim \int_0^T dt \ |<S_{+}(t)>|^2 \\
I_{inel} &\sim \int_0^T dt <\delta S_{+}(t) \delta S_{-}(t)> = \int_0^T dt \left[<S_{+}(t)S_{-}(t)> - |<S_{+}(t)>|^2 \right] \\
&= \int_0^T dt \left[\frac{1}{2} + <S_Z(t)> - |<S_{+}(t)>|^2 \right]
\end{aligned} \tag{7}
$$

(We have used the relation $S_{+}S_{-} = \vec{S}^2 - S_Z^2 + S_Z$ and the identities $\vec{S}^2 = 3/4$, $S_Z^2 = 1/4$ valid for a spin 1/2).

A first remark concerning equations (7) is that, when we are interested in a total intensity (integrated over ω), only a knowledge of $<\vec{S}(t)>$ is required. Bloch's equations (2) are sufficient. This justifies the use of such equations (or similar rate equations) for interpreting optical pumping or level crossings experiments where the measured signal is a total intensity integrated over frequencies [2] [27] [28] [29]. Interpreting a spectral distribution is more complicated as it requires the knowledge of 2 times averages such as $<S_{+}(t) S_{-}(t')>$.

Let's come back to equations (7). As the 2 small transient regims near $t = 0$ and $t = T$ have a very small relative contribution (of the order of τ/T), we can replace in (7), $<S_{+}(t)>$ and $<S_Z(t)>$ by the steady state solution of

Bloch's equations $<S_+>_{st}$ and $<S_z>_{st}$. This clearly shows that I_{el} and I_{inel} are proportional to T and that the inelastic part of the fluorescence is radiated uniformly throughout the whole period of time spent by the atom in the laser beam, and not only at the beginning or at the end of this period, as suggested by the naïve approach described above. The calculation of $<\vec{S}>_{st}$ is straightforward and one gets:

$$\frac{I_{el}}{T} \sim \frac{\omega_1^2 \left[\Gamma^2 + 4(\omega_o - \omega_L)^2\right]}{\left[\Gamma^2 + 4(\omega_o - \omega_L)^2 + 2\omega_1^2\right]^2} \qquad \frac{I_{inel}}{T} \sim \frac{2\omega_1^4}{\left[\Gamma^2 + 4(\omega_o - \omega_L)^2 + 2\omega_1^2\right]^2} \qquad (8)$$

For very low intensities of the light beam ($\omega_1 \ll \Gamma, |\omega_L - \omega_o|$), we find that I_{el} varies as ω_1^2, i.e. as the light intensity I, whereas I_{inel} varies as ω_1^4, i.e. as I^2. Most of the light is scattered elastically and we can define a cross section for such a process which is well described by fig. 1. I_{inel} is much smaller and can be considered as due to non linear scattering processes of the type shown in fig. 2.

For very high intensities ($\omega_1 \gg \Gamma, |\omega_L - \omega_o|$), we find on the contrary that I_{el} tends to 0. This is due to the fact that the atomic transition is completely saturated: the 2 populations are equalized ($<S_z>_{st} = 0$) and the dipole moment is reduced to 0 ($<S_\pm>_{st} = 0$). On the other hand, I_{inel} is very large and independant of the light intensity I (this appears clearly on the bracket of the last equation (7) which reduces to 1/2 as $<S_z>_{st} = <S_+>_{st} = 0$). This means that the atom spends half of its time in e and cannot therefore emit more than $\frac{1}{2}\frac{T}{\tau}$ photons. Increasing the incident light intensity cannot change this number.

One therefore concludes that inelastic scattering, which is due to the fluctuations of S_+, is predominant in strong resonant fields. If we ignore these fluctuations, we miss all the physics. One can finally try to understand why these fluctuations are so effective at high intensities ($I_{inel} \gg I_{el}$) whereas they have little influence at low intensities ($I_{inel} \ll I_{el}$). I think this is due to the fact that an atom is the more sensitive to the vacuum fluctuations the greater is the probability to find it in the excited state e. Some components of the vacuum fluctuations are resonant for the atom in e as they can induce it to emit spontaneously a photon whereas they can only produce a level shift of g. At low intensities, most of the atoms are in g and are not very sensitive to the vacuum fluctuations whereas at high intensities half of the atoms are in e and fluctuate appreciably.

10). <u>How to study the dynamics of the fluctuations</u> ?

Let's now discuss the temporal dependance of $<\delta S_+(t) \, \delta S_-(t')>$.

Considering the physical discussion given above, it seems that a good idea would be to try to write down an equation of motion for $\vec{S}(t)$ [and not for $<\vec{S}(t)>$] including the random character of the force exerted by vacuum fluctuations. These fluctuations have a cumulative effect on $\vec{S}(t)$ which we can try to describe by damping terms analogous to those appearing in (2). In addition, $\vec{S}(t)$ fluctuates around its mean value in a way which can be considered as resulting from the action of a random "Langevin force" $\vec{F}(t)$, having an extremely short correlation time and a zero average value [24]. It is clear that some relations must exist between the damping coefficients Γ and the statistical properties of $\vec{F}(t)$ (relations between dissipation and fluctuations) but we will not consider this problem here since, hereafter, we will only use the ultra short memory character of $\vec{F}(t)$. So let's write for example for $S_+(t)$:

$$\dot{S}_+(t) = -\left[(\Gamma/2)+i(\omega_L-\omega_0)\right] S_+(t) - i\omega_1 S_z(t) + F_+(t) \qquad (9)$$

When averaged, (9) reduces to equation (3-b) since $<F_+(t)> = 0$.

Consider now the product $S_+(t) S_-(t')$ with $t > t'$, and let's try to understand how it varies with t. When calculating $\frac{d}{dt} S_+(t) S_-(t')$ and using (9) for $dS_+(t)/dt$, the only difficulty which appears comes from the Langevin term $F_+(t) S_-(t')$, since we know very little about $F_+(t)$. But we only need to calculate $d<S_+(t) S_-(t')>/dt$, so that we only need to calculate the average $<F_+(t) S_-(t')>$. And it is easy to understand that such an average gives 0 since the motion of the dipole at t', $S_-(t')$, cannot be correlated with the Langevin force $F_+(t)$ at a later time t, as a consequence of the ultra short correlation time of $F_+(t)$. It follows that the rate of the t-variation of the 3 correlation functions $<S_i(t) S_-(t')>$ (with $t > t'$, and i = +, -, Z) is described by a set of 3 first order differential equations with the same coefficients as the ones appearing in the Bloch's equations giving the rate of variation of $<S_i(t)>$ [For t' > t, we use the fact that, as $S_+ = (S_-)^+$, $<S_+(t) S_-(t')> = <S_+(t') S_-(t)>^*$]. This important result is a particular case of the "quantum regression theorem" [25]. In the present case, it means that, when the dipole undergoes a fluctuation and is removed from its steady state, the subsequent evolution and the damping of this fluctuation are the same as the transient behaviour of the mean dipole moment starting from a non steady state initial condition.

11). <u>Predicted fluorescence spectrum for an ideal laser light</u>

Once we know how to calculate the dynamics of the fluctuations of $S(t)$, the derivation of $\mathcal{J}(\omega)$ from (2) is simple. We bypass here the corresponding algebra which is straightforward and only give the results for a resonant irradiation ($\omega_L = \omega_0$) and a very high intensity ($\omega_1 \gg \Gamma$). One finds 3 components in the inelastic spectrum : one central component around $\omega = \omega_L$ with a half-width $\Gamma/2$, and 2 equal sidebands around $\omega = \omega_L \pm \omega_1$, with a half-width $3\Gamma/4$ and a height 3 times smaller than the one of the central component.

Such a structure is simple to understand. The 2 sidebands correspond to the modulation of S_Y due to the transient precession of $\vec{S}$ around $\vec{B}_1$ at frequency ω_1 (see fig. 5; as we are at resonance, $B_o = 0$). As the projection of $\vec{S}$ in the plane YOZ perpendicular to $\vec{B}_1$ is alternatively parallel to OY and OZ, and as the 2 damping coefficients associated to S_Z and S_Y are respectively Γ and $\Gamma/2$ (see equations 2), one understands why, when $\omega_1 \gg \Gamma$, the damping of the precession around $\vec{B}_1$ is given by $\left[\Gamma + (\Gamma/2) \right] /2 = 3\Gamma/4$ and this explains the width $3\Gamma/4$ of the 2 sidebands. The central component is associated with the transient behaviour of S_X, which is not modulated by the precession around $\vec{B}_1$ and which has a damping coefficient $\Gamma/2$. This explains the position and the width of the central component.

This result has been derived by several authors using either a classical ([10]) or a quantum ([13])([17]) description of the laser field. Other calculations don't give the same quantitative results ([12])([15])([16]). I think they are based upon too crude approximations (as the one which neglects the interference between different cascading amplitudes in the dressed atom approach described above).

12). Experimental situation

The experiment of Schuda, Stroud and Hercher ([3]) has displayed a 3-peak structure. The precision is perhaps not yet sufficient to allow a quantitative comparison between theory and experiment.

Other experiments are presently being made ([4]). The experimental investigations are rather difficult due to several perturbing effects. One is the spatial inhomogeneity of the laser intensity. As the interval travelled by the atom during its radiative lifetime is short compared to the diameter of the laser beam, each part of the illuminated portion of the atomic beam radiates a 3-peak spectrum with a splitting ω_1 corresponding to the local amplitude of the laser field. A too large spreading of this amplitude would wash out the structure. We must not also forget the elastic component which is not completely negligible when ω_1 is not very large compared to Γ. Let's take for example $\omega_1 = 2\Gamma$, in order to have the 3 peaks just well resolved. From (8), one calculates $I_{el}/I_{inel} = 1/8$. But I_{el} is spread over a very small interval (which is the width $\Delta\nu$ of the laser, or $1/T$), whereas I_{inel} is spread over Γ, or even over $\bar{\omega}_1$, if the spreading of ω_1 is sufficiently large to mask the structure. The ratio between the maxima of the elastic and inelastic components is therefore not $1/8$ but $\bar{\omega}_1/8\Delta\nu$, a number which may be much greater than 1. In such a case, one can get the impression that there is only one elastic component emerging from a broad background. We must have $\omega_1 \gg \Gamma$ in order to have no trouble with the elastic component.

Other possible perturbations of the spectrum calculated above might come from temporal fluctuations of the laser beam. This leads us to the more general problem of the fluorescence light scattered by an atom irradiated by a resonant light which is not an ideal laser light with perfectly well defined phase and amplitude.

13). <u>What happens with a real non ideal laser beam</u> ?

Let's consider a realistic laser light, having a non zero spectral width $\Delta\nu$ and a very large intensity. More precisely, we suppose $\sqrt{\overline{\omega_1^2}} \gg \Gamma, \Delta\nu$ where $\sqrt{\overline{\omega_1^2}}$ is the mean Rabi nutation frequency associated with the probability distribution of the amplitude of the laser. We don't make any hypothesis concerning the relative magnitude of Γ and $\Delta\nu$.

A first important remark is that the knowledge of $\Delta\nu$ is not sufficient for characterizing the light beam. One can imagine different light beams having all the same spectral width $\Delta\nu$, i.e. the same first order correlation function, but completely different microscopic behaviours, corresponding to different higher order correlation functions [20]. One can for example consider a light beam emitted by a laser well above threshold, which has a very well defined amplitude undergoing very small fluctuations, and a phase $\phi(t)$ which, in addition to short time fluctuations, slowly diffuses in the complex plane with a characteristic time $1/\Delta\nu$. At the opposite, we can consider a quasi-monochromatic gaussian field, or a laser just above threshold, for which both phase and amplitude fluctuate appreciably with the same characteristic time $1/\Delta\nu$.

We have done, in collaboration with P. Avan, calculations of the fluorescence spectrum corresponding to different models of laser beams [26]. These calculations show that the shape of this spectrum is very sensitive to the microstructure of the light beam. The 3-peak structure described above is only maintained when the fluctuations of the amplitude are sufficiently small. The 3 components are broadened differently in a way which depends not only on the phase diffusion, but also on the short time fluctuations of this phase $\phi(t)$ [more precisely of $d\phi/dt$]. When the fluctuations of the amplitude are too large, only the central component survives, superposed to a broad background having a width of the order of $\sqrt{\overline{\omega_1^2}}$. This is easy to understand : there is a destructive interference of the various Rabi nutations around $\vec{B}_1$, as a consequence of the too large spreading of the possible values of B_1. To summarize these studies, one can say that they deal with the fluctuations of $\vec{S}$ associated to the fluctuations of the driving field.

We are also investigating the sensitivity of level crossing signals [27] to the fluctuations of the laser beam. The only calculations which have been performed up to now suppose, either a pure coherent field [28][2] or a very broad line excitation ($\Delta\nu \gg \Gamma$, $\sqrt{\overline{\omega_1^2}}$) so that, within the correlation time of the light wave, at most one interaction between the atom and the light can occur [2][29] : in such a case, only the first order correlation function plays a role. It would be interesting to try to fill the gap between these 2 extreme situations.

I would like to conclude with the following remark. The Hanbury-Brown and Twiss experiment has revealed the importance of new experimental methods, such as intensity correlations or photon coincidences, for learning more about light beams [20]. Perhaps, the behaviour of atoms in strong resonant fields could appear as a new interesting probe for exploring such fields.

References

[1] W. Heitler - Quantum Theory of Radiation, 3rd Ed. (1954, London, Oxford Un. Press)
[2] C. Cohen-Tannoudji - Optical Pumping with Lasers. Proceedings of the 4th International Conference on Atomic Physics, in Atomic Physics 4, Plenum Press, p. 589 (1975)
[3] F. Schuda, C.R. Stroud Jr., M. Hercher - J. Phys. B 7, L 198 (1974)
[4] H. Walther - Atomic Fluorescence under Monochromatic Excitation. Proceedings of the 2nd International Laser Spectroscopy Conference, present volume
[5] S. Ezekiel - Private communication
[6] P.A. Apanasevich - Optics and Spectroscopy, 16, 387 (1964)
[7] S.M. Bergmann - J. Math. Phys. 8, 159 (1967)
[8] M.C. Newstein - Phys. Rev. 167, 89 (1968)
[9] V.A. Morozov - Optics and Spectroscopy 26, 62 (1969)
[10] B.R. Mollow - Phys. Rev. 188, 1969 (1969)
[11] M.L. Ter-Mikaelyan and A.O. Melikyan - Soviet Physics JETP, 31, 153 (1970)
[12] C.R. Stroud Jr. - Phys. Rev. A3, 1044 (1971) and Coherence and Quantum Optics ed. L. Mandel and E. Wolf (New York, London, Plenum Press), p. 537 (1972)
[13] G. Oliver, E. Ressayre and A. Tallet - Lettere al Nuovo Cimento, 2, 777 (1971)
[14] R. Gush, and H.P. Gush - Phys. Rev. A6, 129 (1972)
[15] G.S. Agarwal - Quantum Optics, p. 108, Springer Tracts in Modern Physics (1974)
[16] M.E. Smithers and H.S. Freedhoff - J. Phys. B7, L 432 (1974)
[17] H.J. Carmichael and D.F. Walls - J. Phys. B8, L 77 (1975)
[18] C. Cohen-Tannoudji - Cargese Lectures in Physics, vol. 2, p. 347, edited by M. Lévy, Gordon and Breach, New York, 1968
S. Haroche - Ann. de Phys. 6, 189 and 327 (1971)
[19] N. Kroll in Quantum Optics and Electronics, p. 47, Les Houches 1964, edited by C. De Witt, A. Blandin and C. Cohen-Tannoudji, Gordon and Breach, New York (1965)
[20] R.J. Glauber in Quantum Optics and Electronics, p. 63, Les Houches 1964, same reference as **(19)**
[21] C. Cohen-Tannoudji, B. Diu and F. Laloë - Mécanique Quantique, p. 423, Hermann Paris, 1973
[22] J.P. Barrat and C. Cohen-Tannoudji - J. Phys. Rad. 22, 329 and 443 (1961)
C. Cohen-Tannoudji - Ann. de Phys. 7, 423 and 469 (1962)
W. Happer - Rev. Mod. Phys. 44, 169 (1972)
[23] In the Heisenberg picture, spontaneous emission can be described either by radiation reaction or by vacuum field effects.
J.R. Senitzky - Phys. Rev. Letters 31, 955 (1973)
P.W. Milonni, J.R. Ackerhalt and W.A. Smith - ibid. 958 (1973)

(^{24}) A description of Langevin equation approach to damping phenomena may be found in M. Lax, Brandeis University Summer Institute Lectures, Vol. II, ed. by
    M. Chretien, E.P. Gross and S. Deser, Gordon and Breach, New York (1968)
    W.H. Louisell - Quantum Statistical Properties of Radiation, John Wiley and Sons, New York (1973)
See also reference (^{15})
(^{25}) M. Lax - Phys. Rev. 172, 350 (1968)
M. Lax - Reference (^{24})
(^{26}) P. Avan and C. Cohen-Tannoudji - to be published
(^{27}) For recent level crossing experiments done with single mode lasers and atomic beams, see
W. Rasmussen, R. Schieder, H. Walther - Opt. Commun. $\underline{12}$, 315 (1974)
H. Brand, W. Lange, J. Luther, B. Nottbeck and H.W. Schröder - Opt. Commun. $\underline{13}$, 286 (1975)
See also reference (^{4})
(^{28}) P. Avan and C. Cohen-Tannoudji - J. de Phys. Lettres, $\underline{36}$, L 85 (1975)
(^{29}) M. Ducloy - Phys. Rev. $\underline{A8}$, 1844 (1973)
    Phys. Rev. $\underline{A9}$, 1319 (1974)
Level crossing experiments done with broad line sources are described in
M. Ducloy - Ann. de Phys. $\underline{8}$, 403 (1973-74)

PERTURBED FLUORESCENCE SPECTROSCOPY*

W. Happer
Department of Physics
Columbia University
New York, New York 10027

I would like to talk about the application of lasers to an important class of experiments which, for want of a better name, I will call perturbed fluorescence experiments. These experiments are a subgroup of a still more general class of experiments which are illustrated in Figure 1.

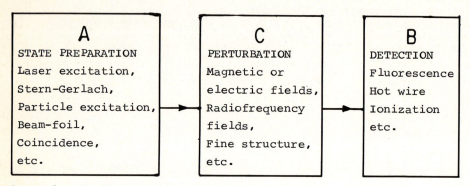

Figure 1. Perturbed fluorescence experiments with lasers are similar to many other 3-step experiments in physics.

In the first part of such an experiment, atoms are prepared in a certain quantum state. These states may be pure states, but most often they are impure quantum states, which are described by a density matrix. Typical examples of state preparation are excitation of atoms with resonant light from a conventional lamp, state selection of an atomic beam with an inhomogeneous magnetic field and <u>laser</u> excitation of atoms. It is the use of lasers for preparing atoms in some quantum state which will be the focus of my talk today.

In the second part of a perturbed fluorescence experiment, the atoms are subject to a perturbation, which is often under the control of the experimenter. For instance, the perturbation could be an externally applied magnetic or electric field, internal couplings of the con-

---

*This work was supported in part by the Joint Services Electronics Program (U. S. Army, U. S. Navy, and U. S. Air Force) under Contract DAAB07-74-C-0341, and in part by the Air Force Office of Scientific Research under Grant AFOSR-74-2685.

stituent angular momenta of the atom, some sort of collisions or a combination of these and other perturbations.

Finally, one detects certain observables which depend on the quantum state of the atom. For instance, one might measure the intensity or polarization of the fluorescent light or the number of atoms in some spatial trajectory in an atomic beam machine. I shall restrict my attention to the detection of polarized or anisotropic fluorescent light.

Perhaps the most important aspect of a perturbed fluorescence experiment is that it allows one to measure the small energy differences between sublevels of an excited state with essentially no limitation of resolution due to Doppler broadening. This is because one is looking directly at beat frequencies or magnetic resonance frequencies in the microwave or radiofrequency range and while these frequencies are shifted by a fractional amount on the order of v/c the shift seldom exceeds 10 KHz in absolute units, and this is usually negligible compared to the natural radiative width of the state. The same fractional shift of an optical wavelength would amount to a few GHz and would be a very serious width to contend with.

I have summarized some of the more common perturbations which are encountered in perturbed fluorescence experiments with lasers in Figure 2. The earliest experiment that I know of is the Hanle effect,[1] i.e. the magnetic depolarization of fluorescent light, which is associated with the crossing of the Zeeman sublevels of an atom at zero field. Only today are we coming to appreciate the full historical significance of Hanle's work (c. 1924). Decoupling experiments in which one analyzes the changes in the polarization of atomic fluorescence associated with the decoupling of internal atomic angular momenta like the nuclear spin I and the electronic spin J were first carried out by Heydenburg[2] and Ellet (c. 1934). Radiofrequency spectroscopy was invented by Rabi[3] (c. 1938) and, of course, Figure 1 is meant to remind one of Rabi's famous atomic beam apparatus. Radiofrequency spectroscopy was first applied to perturbed fluorescence experiments by Brossel and Bitter (c. 1952) who were acting on a suggestion by Kastler. Radiofrequency transitions produced changes in the polarization or anisotropy of the fluorescence. High field level crossing spectroscopy was discovered by Franken and co-workers (c. 1959). When two energy levels cross there can be a sharp change in the transverse polarization of the atom (a change in the coherence) which can be detected by observing appropriately polarized or directional fluorescent light. Finally if an atom is excited by modulated or pulsed light the fluorescence will be modulated or will exhibit a damped ringing which contains the same sort of information about excited state energy level splittings as one obtains from rf or level crossing spectroscopy. This type of experiment was first investigated by Series[6], Dodd and their co-workers (c. 1961) and such experiments have taken on renewed significance now that broadly tunable pulsed lasers are available.

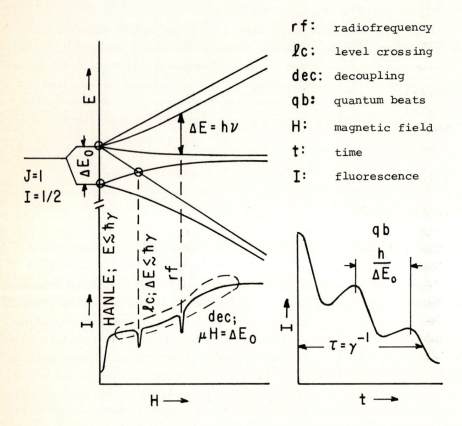

Figure 2. Typical effects used in perturbed fluorescence experiments.

Now let us discuss a few perturbed fluorescence experiments with lasers in more detail. A very abbreviated sketch of a stepwise level crossing experiment of the type developed by Svanberg[7] is shown in Figure 3. Alkali atoms are excited from the ground state to the lowest P state by a powerful conventional lamp. A cw tunable dye laser, pumped by an argon ion laser is used to excite P state atoms to higher D (or S) states. Although it is necessary to stabilize the laser frequency within the absorption linewidth of the optical transition, the width of the level

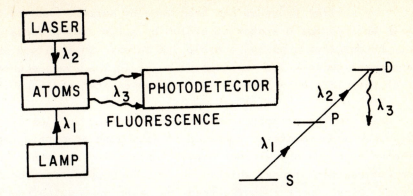

Figure 3. Stepwise perturbed fluorescence experiment.

crossing resonances is basically determined only by the natural lifetime of the excited state and the resolution is essentially independent of the laser frequency stability. Perhaps, the best illustration of the power of this method is the results obtained in the last few years. For example the magnetic dipole coupling constants for Cs^{133} are summarized below:

n $^2D_{3/2}$	A (MHz)
n = 6	16.38 (5)
7	7.4 (2)
8	3.98 (12)
9	2.37 (3)
10	1.52 (3)
11	1.055 (15)
12	0.758 (12)
13	0.556 (8)
14	0.425 (15)
15	0.325 (8)
16	0.255 (12)
17	0.190 (12)
18	0.160 (10)

Except for the n = 6 and n = 8 state which were measured by cascade radiofrequency spectroscopy by Gupta[8] and his collaborators, all of the other states were measured by Svanberg and his co-workers by stepwise spectroscopy with lasers. This is a truly impressive set of measurements, and only four years ago nothing at all was known about the D states of cesium.

Very similar progress has been made by Haroche[9] and co-workers who have excited the D states of sodium by a two-step process with resonant photons from a pulsed tunable dye laser. In a pulsed experiment it is natural to look for the quantum beats in the fluorescence, which occur at the difference frequencies between excited state sublevels.

What are the principal advantages and disadvantages of perturbed fluorescence experiments as compared to saturated absorption spectroscopy, two-photon spectroscopy without Doppler broadening and other high resolution optical techniques, which I shall call Doppler-free optical spectroscopy? First, perturbed fluorescence techniques allow one to measure small energy differences directly in MHz, while it is necessary to take the difference of two optical frequencies in Doppler-free optical spectroscopy. Also, the demands on frequency stability of the laser are much less for perturbed fluorescence experiments, and the signals depend linearly rather than quadratically on the laser intensity. Perturbed fluorescence techniques therefore are ideally suited for measuring hyperfine structure constants, g_J values, tensor polarizabilities and other quantities that determine the splitting of the energy sublevels. However, Doppler-free optical spectroscopy is clearly the only choice for making very precise measurements of isotope shifts scalar polarizabilities and other quantities which are associated with the absolute values of atomic energy levels rather than with the differences in energy between sublevels of the same state. Thus the two techniques are happily complimentary, and each is superior to the other in its natural realm of application.

What are the most promising areas of future application of fluorescence spectroscopy with lasers? Since most of the low lying excited states of atoms have already been thoroughly investigated we must expect that a great deal of attention will be given to high-lying excited states. High-lying excited states with high orbital angular momenta are of particular interest since their magnetic fine structures and hyperfine structures are often strongly influenced or even dominated by core polarization effects. Theoretical capabilities in this area are not very good yet. Measurements of excited state lifetimes electric polarizabilities, quadratic Zeeman effects will also be of considerable interest, since at present it seems that fairly simple semiempirical methods like the Coulomb approximation give quite respectable theoretical predictions. As good experimentalists we hope to find regions where these simple theories fail badly. Finally, it would be very interesting to study the collisional processes which affect some of these highly excited states. Because of the large amount of energy residing in the atom and the close spacing of the energy levels very unusual processes may occur.

References

1. W. Hanle, Z. Physik 30, 93 (1924).
2. A. Ellet and N. P. Heydenburg, Phys. Rev. 46, 583 (1934).
3. I. I. Rabi, et al., Phys. Rev. 53, 318 (1938).
4. J. Brossel and F. Bitter, Phys. Rev. 86, 308 (1952).
5. F. D. Colegrove, et al., Phys. Rev. Letters 3, 420 (1959).
6. J. N. Dodd, et al., Proc. Roy. Soc. Lond. A 273, 41 (1963).
7. S. Svanberg, et al., Phys. Rev. Letters 30, 817 (1973).
8. R. Gupta, et al., Phys. Rev. Letters 29, 695 (1972).
9. S. Haroche, et al., Phys. Rev. Letters 33, 1063 (1974).

LASER SPECTROSCOPY OF SMALL MOLECULES

J.C. Lehmann

Laboratoire de Spectroscopie Hertzienne
de l'Ecole Normale Supérieure
4, Place Jussieu Tour 12 75230 Paris Cedex 05

I - INTRODUCTION

In this paper, we shall describe and comment a few techniques in which a laser excitation followed by an analysis of the subsequent fluorescence light may give numerous informations on the properties of excited molecular states. Although sub Doppler spectroscopy as described in other papers during this conference should in principle, through energy measurements, give the same informations, it will be seen that the "optical pumping" techniques are by far more simple and give more accurate results for many parameters. Moreover, we shall see that optical pumping may introduce changes in the properties of the molecular vapour and therefore permit new types of investigations.

II - EXPERIMENTAL

1) *Selective excitation*

In all the following experiments we investigate separately single rovibronic states of molecules. This means that we must excite with a monochromatic light source a molecular vapour at low density (from a few mtorr to about 1 torr). The required monochromaticity is however only the Doppler width of the lines which is much less severe than for sub Doppler spectroscopy as saturated absorption or multiphoton spectroscopy. Provided that a single level is excited, the first problem is to identify its vibrational v' and rotational J' quantum numbers. This can be done through a spectroscopic analysis of the fluorescence spectrum of the excited vapour. When two or more levels are simultaneously excited, it is generally possible to isolate in the fluorescence spectrum a line issued from a single v', J' level. As an example that we shall use to demonstrate how such methods operate, the iodine molecule I_2 can be excited from the X $^1\Sigma_g^+$ ground electronic state to the B $^3\Pi_{0u}^+$ state by almost any wavelength between 5 000 and 7 000 Å . With dye lasers of spectral width of the order of 1 000 Mhz one can excite selectively most levels up to v' $\sim$ 80 ; and for each value of v', up to J' of the order of 130. Of course all levels cannot be studied, since it would require a physicist's lifetime, but by sampling a large number of levels one can obtain a good idea of how behave molecular parameters.

We shall see now which are these parameters and how they can be measured.

2) *Exponential decay and quantum beats*

If the optical excitation is produced by a short pulse of light as can be produced by a dye laser, excited by a nitrogen laser, that is a pulse which duration is of the order of a few nanoseconds, then the fluorescence observed decreases exponentially following the pulse, with a time constant equal to the inverse of the rate of depopulation Γ of the excited state. If one extrapolates this decay rate to zero vapour pressure, one obtains the natural width $\Gamma_n = 1/\tau$ of the excited level. Moreover the slope of the extrapolation curve gives a measurement of the cross section for depopulation of the excited state due to collisions :

$$\frac{d\Gamma}{dp} = \frac{4\sigma_0}{\sqrt{\pi M k T}} \tag{1}$$

where M is the molecular mass and T the cell temperature. Figure 1 shows an example of a lifetime measurement in iodine.

If now two or more levels are excited <u>coherently</u> by the laser pulse, which requires that their distance in energy ΔE being smaller than the spectral width of the light pulse, and the polarization of the pulse to be adequate, then the exponential decay of the fluorescence is <u>modulated</u> at the frequency $\Delta E/\hbar$. This technique is known as the "quantum beat" method and has been demonstrated by Haroche, Paisner and Schawlow ([1]). It has been used by Paisner and Wallenstein ([2]) to measure the Lande factors of some levels of iodine. Since we have not used this technique we shall not comment on it anymore.

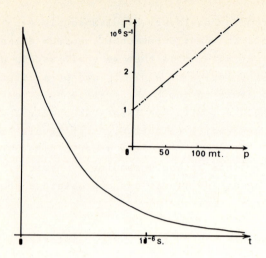

Figure 1 - Lifetime measurement of the B $^3\Pi_{0_u^+}$ v' = 12 J' = 64 state of I_2

3) *Resonances in a modulated light beam*

This technique, due originally to Dodd and Series (^{3}), can be considered as the Fourier transform of the quantum beats : instead of exciting by a short pulse of light several closely spaced level, one uses a cw laser of about the same frequency and spectral width $\Delta\gamma$, but intensity modulated at a frequency ω. One detects in the fluorescence the modulated component at the same frequency ω : $I(\omega)$.

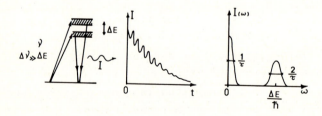

Figure 2 - Quantum beats and resonances in a modulated light beam

Beyond a resonance around $\omega = 0$ whose width give the lifetime of the excited levels, one observes a resonance at $\omega = \Delta E/\hbar$ with the same width $\Delta\omega = 2\pi/\tau$ (In some cases, the width of the two resonances, centered at $\omega = 0$ and $\Delta E/\hbar$ can be different).

Of course if more than two levels are excited coherently, several resonances are observed. This technique gives at first sight the same informations as a double resonance experiment. However it does not require any radiofrequency fields and for example if one studies molecules with Lande factor of the order of nuclear Lande factors, in excited states for which $\tau \sim 10^{-6}$ to 10^{-9} sec, the intensity of the radiofrequency field which would be required for a magnetic resonance experiment is much too high to be obtained (several KGauss in many cases). Moreover the resonances observed with a modulated light beam exhibit no broadening due to a radiofrequency field, and since the polarization required for the exciting light is "coherent", they are much less sensitive to light shifts that in the case of an "incoherent" excitation as is required for double resonance. This technique is therefore especially suitable to study excited molecular states. Figure 3 shows an example of such resonances observed in the $B\ ^3\Sigma_{1u}$ $v' = 0$ $J' = 105$ state of Se_2 at a frequency $\omega = 30$ Mhz ([4]). In this case, the resonances observed correspond to $\Delta m = 2$ Zeeman coherences and the frequency ω is taken fixed while the magnetic field is scanned through the resonance. From the value of H_0, one can deduce the <u>Lande factor</u> for this level.

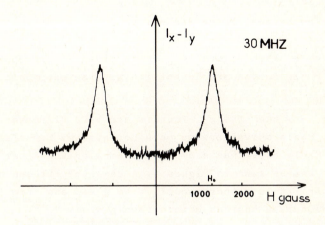

Figure 3 - Resonances of Se_2 at 30 MHz

4) Hanle effect

The simplest method of all to measure lifetimes and Lande factors is the well known Hanle effect or magnetic depolarization of the fluorescence. This technique generally gives Lorentz curves of width $\Delta H = \Gamma/2\gamma$ where γ is the gyromagnetic ratio. The extrapolation of ΔH to zero vapour pressure gives both the cross section σ_2 for destruction of the $\Delta m = 2$ coherences by collision (in the case of linear polarizations) :

$$\frac{d\Delta H}{dp} = \frac{4|\gamma|}{\sqrt{\pi M k T}} \sigma_2 \qquad (2)$$

and the product $g\tau$ of the Lande factor by the lifetime of the level. However it may happen as in the case of iodine, that some Hanle curves are not Lorentzian. This is seen for example in figure 4. It is due to the fact that several hyperfine levels are excited simultaneously and that they have different Lande factors : the two ^{127}I nuclei of a I_2 molecule have each one, a 5/2 nuclear spin, with a Lande factor $g_I = 1.12$ (in nuclear unit). Depending wether the rotational quantum number J' is even or odd, the total nuclear spin I can take the values 1,3,5 (ortho states) or 0,2,4 (para states). This results in 21 or 15 hyperfine sublevels. Some hyperfine structures have been observed by saturated absorption spectroscopy. Due to both the dipole and quadrupole interactions the eigenstates of the molecules must be written

$$|\tau,\epsilon,J',F,M_F> = \sum_I \alpha_I^{\epsilon,F} |\tau,I,J',F,M_F>$$

and the hyperfine Lande factors are equal to

$$g_{\epsilon,F} = \sum_I |\alpha_I^{\epsilon,F}|^2 \{g_J \frac{F(F+1)+J(J+1)-I(I+1)}{2F(F+1)} + g_I \frac{F(F+1)-J(J+1)+I(I+1)}{2F(F+1)}\} \qquad (3)$$

where g_J is the rotational Lande factor.

i/ If $g_J \gg g_I \frac{I}{J}$ then all the $g_{\epsilon F}$ for a given J' are approximately equal, and the Hanle curves are Lorentzian with an extrapolated width equal to $1/2g_J\mu_N\tau$.

ii/ If $g_J \lesssim g_I \frac{I}{J}$, then the $g_{\epsilon F}$ factors are very different from one-another and the Hanle curves are no more Lorentzian. Especially the dispersion shaped Hanle curve has a amplitude smaller than that of the absorption shaped one as seen in Figure 4. A computer fit can be made using g_J as a parameter. It is rather sensitive to g_J/g_I and from the shapes and width of the curves one can obtain both g_J and τ.

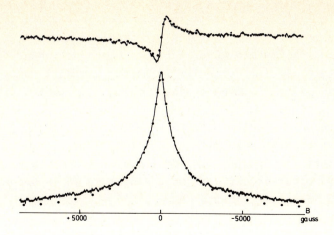

Figure 4 - Hanle effect in the level B $^3\Pi_{0_u^+}$ v'=32 J'=9 and 14 of I_2

III - RESULTS

We shall give here some results obtained in the B $^3\Pi_{0_u^+}$ state of iodine.

1) *Perturbations to the energy*

Let us look for the influence on a state $|B,v,J,\epsilon,F,m_F\rangle$ of the following perturbing terms of the hamiltonian :

$H_Z = \mu_B \vec{H} (\vec{L}+2\vec{S})$ the electronic Zeeman hamiltonian

$H'_R = -2B_R \vec{J} (\vec{L}+\vec{S})$ the off-diagonal part of the rotational hamiltonian $H_R = B_R (\vec{J}-\vec{L}-\vec{S})^2$

$H_D = a \vec{I}.\vec{L}$ the magnetic dipole term of the hyperfine hamiltonian.

A second order perturbation theory gives a correction to the energy :

$$\Delta E = \sum_i \frac{|\langle B,v,J,\epsilon,F,m_F|H_Z + H'_R + H_D|i\rangle|}{E_0 - E_i} \quad (4)$$

where E_0 is the energy of the $|B,v,J,\epsilon,F,m_F\rangle$ state and $|i\rangle$ represents all the other rovibronic states not belonging to the B state. When one develops the squared matrix elements, six terms arise :

1) One in which appear $|\langle B|H_Z|i\rangle|^2$. It gives a quadratic Zeeman effect.

2) One with $|<B|H'_R|i>|^2$. It gives a contribution to the rotational energy due to the moment of inertia of the electronic cloud.
3) One with $|<B|H_D|i>|^2$ which gives a pseudo quadrupole hyperfine term.
4) One crossed term with $<B|H_Z|i><i|H'_R|B>$ + c.c. which is proportional to $\vec{J}.\vec{H}$: it is the linear Zeeman effect, giving rise to the <u>Lande factor</u> g_J.
5) One crossed term in $<B|H'_R|i><i|H_D|B>$ + c.c. which gives $C_I \vec{I}.\vec{J}$, the magnetic dipole hyperfine energy.
6) One crossed term in $<B|H_D|i><i|H_Z|B>$ + c.c. which gives $g_1 \mu_N \vec{I}.\vec{H}$, this term modifies the apparent nuclear Lande factor g_I which becomes $(g_I + g_1)$. It is actually a "chemical shift".

If one considers now that the summations over the $|i>$ states involves approximately the same states in all the 3 latter terms, one should have the approximate relations :

$$\frac{g_J}{2\mu_B B_R} \sim \frac{g_1}{\mu_B a} \sim \frac{C_I}{2B_R a} \qquad (5)$$

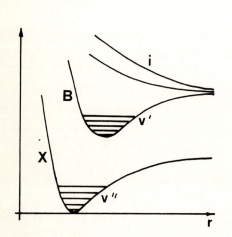

Figure 5 -

Moreover when v become close to the dissociation limit (figure 5) $E_0 - E_i$ becomes very small and all the quantities g_J, g_1 and C_I should increase rather steeply. This is just what has been found for C_I by Levenson and Schawlow ([5]). Figure 6 shows some values of g_J which happen to fit rather nicely with a curve in $\frac{1}{E - E_\ell}$ where E_ℓ is not far from the dissociation limit energy.

Finally g_1 should also increase with g_J and C_I. For example for the v = 62 level excited at 5 017 Å, one should have $g_1 \sim 3$

in nuclear unit. This is an extremely high value since it is even larger than g_I = 1.12. It is no more a small correction but gives rise to a completely new value of the apparent nuclear Lande factor. In the formula

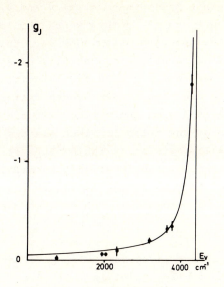

for $g_{\varepsilon F}$ one should therefore replace g_I by $g_I + g_1$. Figure 7 shows resonances in a modulated light beam for this v = 62 level. In the lower part are shown the theoretical position of the resonance for g_1 = 3.4.

If one takes g_I = 0 the resonances should be so close to one another that one would observe a single resonance. A small resonance is also observed close to zero magnetic field which corresponds to a weakly excited v = 71 level. It shows that for such a high vibrational level the Lande factor g_J becomes very high.

Figure 6 -

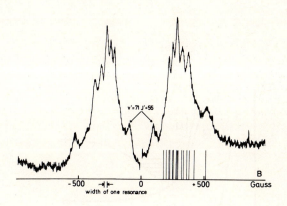

Figure 7 - Resonances in the levels B $^3\Pi_{0_u^+}$ v'=62 J'=27 and v'=71 J'=55 of I_2

2) *Lifetimes and predissociations*

At first sight, it would seem that the lifetime τ of a v,J, level of a definite electronically excited state is independant of v and J. This is due to sum rules over the Franck Condon factors which should give the same total probability of spontaneous emission. However, this is not quite true for several reasons. At first the probability of spontaneous emission depends upon the cube of the frequency of the emitted radiation. This breaks out the sum rule over the Franck Condon factors. Another variation of Γ_{rad} with v and J is due to the mixing of the excited state with other electronic states of the molecule. Such mixings are especially important for long lived excited states as the B $^3\Pi_{0_u}^+$ of iodine and are often dependant on the energy of the level that is mainly on v and to some extent also on J. Also is the matrix element of the electric dipole dependant upon r. This is responsible for another variation of Γ with v.

In the case of iodine another effect gives very strong variations of τ : It is the <u>predissociation</u> of the excited molecule . It is indeed responsible for a leak in the population of the excited state, at a rate Γ_p. Therefore the lifetime τ of a v J excited state is given by

$$1/\tau = \Gamma_{rad} + \Gamma_p \qquad (6)$$

We shall at first only consider natural predissociation.

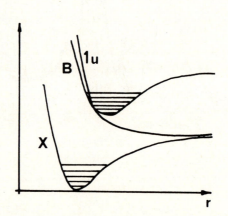

Figure 8 -

Figure 8 shows the potential curves of the X and B levels of iodine together with that of the 1u state responsible for this predissociation. Tellinghuisen suggested ([7]) that the mixing between the B state and the dissociative 1u state is due to the off diagonal part of the rotational hamiltonian : $H'_R = -2B_R \vec{J} (\vec{L}+\vec{S})$. This should give

$$\Gamma_p = k(v) J (J+1) \qquad (7)$$

With the suggested values for k(v) the lifetime τ of any v level should exhibit strong variations with the rotational quantum number J. To check this effect, we measured the lifetimes for a whole set of v J levels. No variations with J was observed for all levels for which 14 < v < 18. However for v = 11, 12 and 13 a dependance of τ with J was indeed observed and seems to fit with the above formula.

The measured radiative lifetimes τ and k(v) are given below.

v	τ_R (μsec)	k(v) (sec^{-1})
11	0.92	66
12	1.15	38
13	1.26	13
14	1.31	≤ 10
15	1.36	"
16	1.23	"
17	1.14	"
18	0.98	"

This seems to indicate that the values of k(v) are much smaller that what was predicted by Tellinghinsen. It was therefore necessary to try to measure directly this parameters. It appeared to be possible through a study of the magnetic predissociation effect :

3) *More about predissociation*

It is known since very long, that a magnetic field induces in the B $^3\Pi_{0_u^+}$ state of iodine a strong predissociation ([8]). It is due to the mixing with a dissociative state through the electronic Zeeman hamiltonian : $H_Z = \mu_B \vec{H} (\vec{L}+2\vec{S})$. Until our works, the magnetic and natural predissociation effects were treated independantly. If one supposes however, that the same 1u state is responsible for both predissociations, then Γ_p is proportional to the squared matrix element :

$$|<B,v,J,M_J|H'_R + H_Z|1u>|^2$$

which develops into three terms, and give for J >> 1 :

$$\Gamma_p = k(v) \; J \; (J+1) + b(v) \; \frac{J^2+M_J^2}{4J^2} \; H^2 + \sqrt{2b(v) \; k(v)} \; M_J \; H \qquad (8)$$

The first two terms are due to the pure natural and magnetic predissociations, the third one is an <u>interference term</u> which involves both types of predissociation.

Due to this third term, the lifetime depend upon the <u>orientation of the molecule</u> and since M_J can be positive or negative, the lifetime is increased by a magnetic field for one orientation, while it is decreased for the other. This effect has indeed be observed ([9]) and results in a strong orientation (up to 30 %) of the excited molecules for intermediate values of the magnetic field.

The second and third terms of formula 8 may also interfere with the Hanle effect in some cases. For example in the v = 6 J = 32 level excited by a 6 328 Å He-Ne laser, a "repolarization" effect is observed due to the fact that when a magnetic field is applied, a desorientation of the excited state starts to occur but is subsequently counterbalanced by the shortening of the lifetime due to the magnetic predissociation. If, for any level, one observes the circularly polarized σ^+ or σ^- fluorescence light, it is more sensitive to the populations of positive or negative values of M_J. It will therefore reflects the rate of orientation of the excited molecules. As can be deduced from formula 8 this rate has a maximum for a non zero value H_m of the magnetic field. From H_m one can deduce the value of k(v) the natural predissociation coefficient. The following table give some measured values of k(v) compared to the values given by Tellinghuisen :

v	k(v) (sec^{-1})	
	our work	Tellinghuisen's values
6	250	2 000
17	5	220
18	13	220
21	31	330

Is is seen that the measured values of k(v) are about 10 times smaller than those predicted by Tellinghüisen. This clearly explains why the dependence of Γ in J(J+1) can be observed only for small values of v. For example, if k(v) = 10 sec^{-1}, k(v) J(J+1) is only of the order of 10^5 sec^{-1} when J = 100 which is still smaller than Γ_{rad} which is of the order of 10^6 sec^{-1}.

IV - CONCLUSION

Optical pumping and related techniques have been widely used since 20 years to study atomic vapours. It is only with the development of tunable lasers that it became possible to extend theses methods to optically excited molecular states. It is undoubtful that the knowledge of series of molecular parameters as lifetimes, Lande factors and others, will be of great help to trace down the exact nature of molecular states.

Besides, it has been seen that predissociation effects may change the nature of the molecular vapour by introducing an orientation. If we go back to formula 8 it can be seen that the natural predissociation alone induces an alignment in the vapour and it has been checked that this alignment indeed exists and has a sign which corresponds to the symetry of the dissociative level involved.

If now one comes back to the ground electronic state, it is repopulated by the radiative decay from the excited state and this repopulation may be used to study some of its highly excited vibrational levels.

Let us also mention that predissociation beeing selectively induced by the optical excitation, it is possible to dissociate only isotopic molecules to produce isotope separation or even or odd molecules to induce an ortho-para pumping as has been recently reported by Lethokov ([10]).

([1]) S. Haroche, J.A. Paisner and A.L. Schawlow : Phys. Rev. Lett., 30 948 (1973)

([2]) J. A. Paisner, I. Wallenstein and A. L. Schawlow : Phys. Rev. Lett., 32 1033 (1974)

([3]) J.N. Dodd and G.W. Series : Proc. Phys. Soc. A 263 353 (1961)

([4]) G. Gouédard and J.C. Lehmann : C.R. : Avril 1975

([5]) M.D. Levenson and A.L.Schawlow : Phys. Rev. A 6 10 (1972)

([6]) M. Broyer, J.C. Lehmann and J. Vigué : J. de Phys., 36 235 (1975)

([7]) J. Tellinghuisen : J. Chem. Phys., 57 2397 (1972)

([8]) J.H. Van Vleck : Phys. Rev., 40, 544 (1932)

([9]) J. Vigué, J. Broyer and J.C. Lehmann : J. Chem. Phys. to be published

([10]) S.A. Bazhutin, V.S. Lethokov, A.M. Makarov and V.A. Semchishen : Zh. E.T.F. Pis. Red. 18 515 (1973)

ATOMIC FLUORESCENCE INDUCED BY MONOCHROMATIC EXCITATION

Herbert Walther
I. Physikalisches Institut
der Universität zu Köln
Köln, Federal Republic of Germany

1.) Introduction

In this review three recent experiments performed with single mode continuous wave dye lasers in our laboratory will be discussed. The first two deal with the atomic fluorescence induced by monochromatic excitation, whereas the third is an investigation of the fine structure splitting of the $3\,^2D$ multiplet of the lithium atom with stepwise excitation.

The problem of monochromatic excitation of free atoms and the frequency distribution of the subsequent spontaneously emitted radiation is an important problem and therefore has received extensive theoretic treatment by many authors (e.g. 1 - 16) using different methods and approaches. The predicted spectra for high power of the exciting radiation include a lorentzian with a hole in the middle (3,7), a three peaked distribution (2,4,6, and 9 - 16), or even more complex structures (5,8). For a detailed theoretical discussion can be referred to the paper of Cohen-Tannoudji in this volume (16).

Another quite important problem in this connection is the observation of the Hanle effect or zero field level crossing under monochromatic laser excitation (17,18,19). To treat this latter problem theoretically the equations which describe the evolution of the atomic density matrix have to be solved. Since the coherence time of the light is long compared to the lifetime of the excited state, along with the nondiagonal elements ρ_{+-}, ρ_{-+} of the density matrix describing the Zeeman coherences in the excited state, also the optical coherences ρ_{eo} (s. Fig. 1) between the excited and the fundamental state have to be considered (17,18). These nondiagonal elements represent the motion of the electric dipole moments driven by the incident laser light.

Density Matrix for Monochromatic Optical Pumping

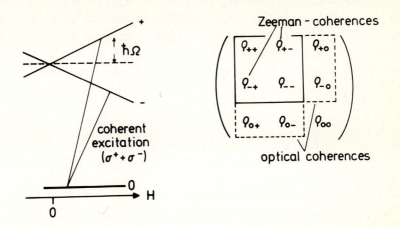

Fig. 1 Density matrix for the level crossing experiments with monochromatic excitation (for details see (17))

The experimental investigation of the problems connected with fluorescence under monochromatic excitation became only possible with the advent of tunable narrow banded lasers. To get rid of the Doppler width of the atomic ensemble well collimated atomic beams have to be used. First experiments to investigate the spectral distribution of the atomic fluorescence have been performed by Schuda et al. (20). Also level crossing experiments with monochromatic excitation have been performed by Rasmussen et al. (19) and later by Brand et al. (21).

In the following two new experiments dealing with the monochromatic fluorescence will be described. The first one is an extension of our level crossing experiment (19) to higher laser powers and the second one was undertaken to investigate the fluorescence spectrum emitted from monochromatically excited atoms with the purpose to obtain the more refined results necessary for a detailed comparison with theory.

2.) Level Crossing Experiment with Monochromatic Excitation

This experiment was performed together with J. Häger, V. Wilke and R. Schieder.

The experimental geometry (s. Fig. 2) was similar to that described in our earlier paper (19). The magnetic field was applied parallel to the direction of the atomic beam. The direction of the exciting light, the direction of the magnetic field and the direction of observation were taken mutually perpendicular. The laser light was linearly polarized perpendicular to the magnetic field. The part of the fluorescent light polarized parallel to the y direction was then observed.

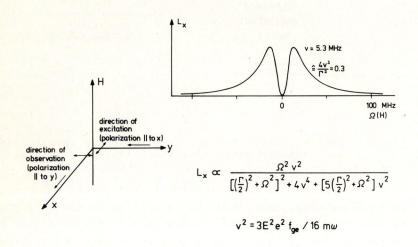

Fig. 2   Experimental geometry and theoretical signal shape (18) for the level crossing experiment. L_x is the fluorescent radiation emitted in x direction. Ω is the Larmor frequency (s. Fig. 1). Γ is the decay constant of the excited state population. E is the field strength of the laser field and ω the resonance frequency of the optical transition at field zero; f_{ge} is the oscillator strength of the optical transition. The signal shape is plotted for $v = 5.3$ MHz.

In the experiment the $(6s^2)\,^1S_0 \rightarrow (6s6p)\,^1P_1$ transition of the the BaI spectrum at $\lambda = 5\,535$ Å was investigated. The laser light was scattered on the free atoms of a well collimated atomic beam (collimation ratio about 1:500). This corresponds to a Doppler width of about 3 MHz. The absorption width of the Ba beam was therefore mainly determined by the natural width of the $^1P_1$ level, which is 20 MHz. For the experiment the laser was tuned to the Ba^{138} transition which has no hyperfine splitting. The dye laser used in the experiment has been described earlier (22). The line width of the frequency distribution of such a free running laser is less than 1 MHz, with a frequency drift of about 1 MHz/min. The signal shape expected from theory for arbitary values of the light intensity (18) is shown in Fig. 2. For vanishing light intensities (v→0), the signal L_x agrees with the expression obtained from the Born amplitude for resonance scattering (19). The decrease of the intensity at higher magnetic fields is due to magnetic scanning, whereas the signal around zero field is mainly determined by the interference of the transition amplitudes. The shape of the signal differs from the lorentzian observed for broad banded excitation. The halfwidth of the dip at zero magnetic field is for low incident intensity 45 % of the usual level crossing signal.

The signal curves obtained in our measurement are displayed on Fig. 3 together with a theoretic fit. There is a very good agreement with the theory for the lowest laser power; the results, however, disagree more and more when the laser power is increased. This disagreement can be explained when the Zeeman splitting of the odd isotopes present in the natural barium mixture is taken into account. This splitting brings some Zeeman components of these isotopes close to the frequency of the laser and the corresponding transitions are therefore excited; in this way the fluorescence intensity is increased and the experiment gives a higher intensity than predicted. Since the high laser intensity broadens the absorption lines, a stronger excitation of the Zeeman components of the odd isotopes occurs; this is the reason that the disagreement increases for larger laser powers. A careful evaluation of this effect results in a very good agreement between theory and experiment.

Avan and Cohen-Tannoudji (18) caculated in their paper also the signal curves which have to be expected when the laser does not agree exactly with atomic resonance. In order to check these results, also measurements with selected detunings of the laser have been performed. Some of these results are shown in Fig. 4. They also indicate a rather good agreement between theory and experiment.

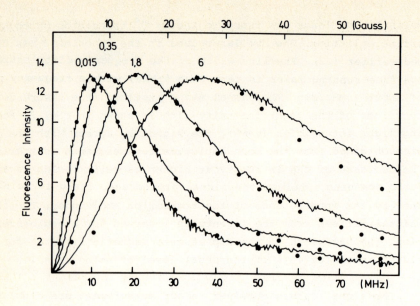

Fig. 3 Level crossing signal for different laser intensities. The parameters given at the signal curves are the $4v^2/\Gamma^2$ values which are proportional to the laser intensity. The signals are normalized so that the maxima of the signals have the same heights. The dots are the theoretical values which follow from the formula shown on Fig. 2.

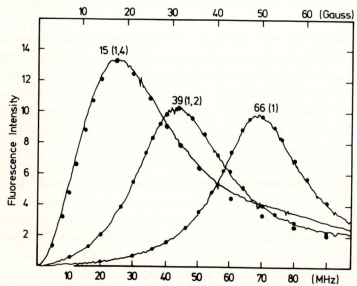

Fig. 4 Level crossing signal for different detunings of the laser frequency with respect to the resonance frequency at zero magnetic field. The parameters give the detuning in MHz. The numbers in brackets represent the $4v^2/\Gamma^2$ values.

3.) Investigation of the Spectral Distribution of Atomic Fluorescence Induced by Strong Monochromatic Excitation.

The experiment described in the following was carried out together with R. Schieder, W. Hartig and V. Wilke.

The experiment we performed was in principle similar to that of Schuda et al. (20); however, some substantial improvements have been made. The main difference was that the atomic beam was placed inside the Fabry-Perot interferometer which analysed the frequency spectrum of the fluorescence. In such an arrangement the observed signal intensity is enhanced by a factor equal to the finesse of the Fabry-Perot (23). We used a spherical Fabry-Perot which was tuned piezoelectrically. The free spectral range was 300 MHz, and the finesse about 50. The dye laser used in the experiment is in principle similar to the one used in the level crossing experiment described above; it provides a much higher frequency stability than the laser which was used by Schuda et al. In the measurements the hyperfine transition $F = 3$, $^2P_{3/2} \rightarrow F' = 2$, $^2S_{1/2}$ was investigated since for this transition no optical pumping of the hyperfine levels of the $^2S_{1/2}$ ground state can occur.

The collimation ratio of the sodium beam was about 1:500 providing a residual Doppler width with respect to the exciting laser beam of about 2 MHz. The direction of the exciting light has been carefully adjusted to be perpendicular to the atomic beam, in order to avoid additional Doppler broadening. The observation of the fluorescent light was performed perpendicular to both the laser and the atomic beams.

The line width observed for the fluorescent light was at low laser intensities about 10 MHz. One contribution for this width results from the residual Doppler width with respect to observation being somewhat larger than 2 MHz as the angle of acceptance for the spherical Fabry-Perot was larger than the divergence of the laser beam. The resolution of the Fabry-Perot was about 3 MHz.

Some results obtained for the frequency distribution of the fluorescent light with a laser power of 1500 mW/cm^2 are shown in Fig. 5. The curves have been measured with the laser on resonance and with selected detunings. For the high laser power used in the experiment

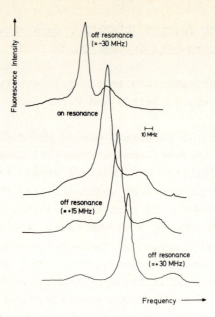

Fig. 5 Fluorescence spectrum for the transition $F=3, {}^2P_{3/2} - F'=2, {}^2S_{1/2}$ of Na. The intensity of the laser was not the same for the measurements off resonance. 1500 W/cm^2 applies for the measurement on resonance.

the power broadened linewidth was so large that besides the F=3 - F'=2 hyperfine component also the F=2 - F'=2 component, which lies 59 MHz towards lower frequencies, was excited with negative detuning. The ratio of the heights of the side maxima to that of the main maximum is about 1:6 instead of 1:3 as predicted by most of the theories for the high power limit (e.g. 6,12,15,16). The deviation may be due to the effects discussed in the paper by Cohen-Tannoudji (16) as e.g. the spatial inhomogeneity of the laser beam or the influence of elastic scattering. Also the polarization of the laser beam has an important influence; this needs to be investigated in more detail. The result obtained is in disagreement with naive predictions based on the conventional phenomenologically damped semiclassical equations, which are usually taken in order to describe coherent transient effects (e.g. 24 - 26).

Similar measurements have also been performed on the $(6s^2)\,{}^1S_0 \rightarrow (6s6p)\,{}^1P_1$ transition of Ba^{138}. In this case the linewidth of the fluorescent light observed for low laser intensities was 10 MHz, too, which is much smaller than the natural width of the transition being 20 MHz. This supports the prediction of Heitler (1) that the spectral width of the fluorescence may be smaller than the natural width.

4.) Determination of the Fine Structure Splitting of the 3^2D Multiplet of Lithium by Stepwise Excitation with cw Dye Lasers.

This experiment was performed together with W. Hartig and V. Wilke (see also (27)).

The fine structure splittings of the low lying states of the ^2P series (n = 2 to 4) in the LiI spectrum have been measured with rather good accuracy by the level crossing method (28 - 30). These distances are about 10 percent smaller than the corresponding values in hydrogen and the deviations increase slightly going from n = 2 to n = 4. Since the effective nuclear charge in lithium is larger than one, one would have expected a fine structure splitting larger than in hydrogen. It was shown that the reason for the smaller splitting may be seen in the fact that spin-other orbit interaction is quite important (30). In consequence of these results it is interesting to see whether the same effect also occurs in the fine structure splitting of the ^2D-series.

Recently the 4^2D multiplet was investigated by means of anti-level-crossing (32). This measurement gives a smaller splitting than the corresponding hydrogen level in agreement with the results for the ^2P series. However, the splitting of 3^2D, which was measured by Meissner et al. (31), gives a larger value than that for hydrogen. In order to clear this discrepancy and to improve the accuracy (the experimental error quoted for the older measurement was 24 MHz) our new experiment was performed. The 3^2D levels were populated by stepwise excitation. To eliminate the Doppler width a highly collimated atomic beam was used.

Another way to study the ^2D multiplet with high resolution would be the use of double quantum transitions (33,34). In the case of lithium, however, the transition probability for double quantum transitions is rather small so that an experiment with single mode cw dye lasers would be rather difficult.

In connection with stepwise excitation the Doppler width in an atomic vapor can also be strongly reduced when the two laser beams are directed in opposite directions. The Doppler width observed is then proportional to the frequency difference between the two beams and not to the transition frequency directly (35).

In our experiment the laser beams were directed collinearly onto an atomic beam. One laser was tuned to the $2^2S_{1/2} - 2^2P_{3/2}$ transition ($\lambda \approx 6708$ Å) and locked to it using standard servo techniques (36). For the excitation of the 3^2D levels two other lasers were used. One of those was locked to the $2^2P_{3/2} - 3^2D_{5/2}$ transition ($\lambda \approx 6103$ Å), the other was used to scan the transition $2^2P_{3/2} - 3^2D_{3/2}$. The fluorescence for the decay $3^2D - 2^2P$ was observed and recorded as a function of the frequency difference of the two lasers exciting the D-levels. This frequency difference was determined by measuring the beat note between the two lasers. A scheme of the experimental set-up is shown in Fig. 6. The three dye lasers are similar to the one described earlier (22). They were pumped by the same Ar^+ laser.

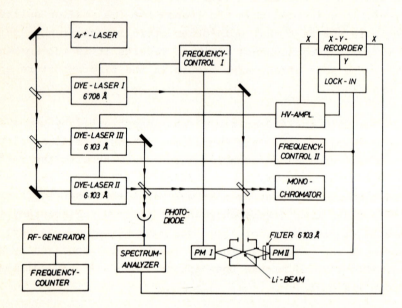

Fig. 6 Experimental set-up for the measurement of the lithium 3^2D fine structure splitting.

The three laser beams excite the atoms of the atomic beam, the collimation ratio of which is 1:800, corresponding to a residual Doppler width of about 4.5 MHz.

In order to frequency stabilize laser I the output was frequency modulated piezoelectrically with 530 Hz. Photomultiplier I (PM I) was used to observe the $\lambda = 6708$ Å fluorescence. The error signal of the servo system was used to change the mean length of the cavity of laser I. The hyperfine splitting of the $2^2P_{3/2}$ level is comparable to the natural line width, which is about 12 MHz (37). Therefore the splitting could not be resolved. However, it can be assumed that our laser was locked close to the frequency of the strongest hyperfine transition $F = 3$, $2^2P_{3/2} - F' = 2$, $2^2S_{1/2}$.

The second multiplier (PM II) was used to lock laser II and to observe the fluorescence induced by laser III. To discriminate the ^2D fluorescence from that of the $^2P_{3/2}$ level an interference filter for $\lambda = 6103$ Å was used for PM II. The fluorescence light resulting from the ^2D levels is amplitude modulated due to the frequency modulation of laser I. This modulation was used to lock laser II to the $2^2P_{3/2} - 3^2D_{5/2}$ transition.

In order that the fluorescence induced by laser III could be separated from the $3^2D_{5/2}$ fluorescence, laser III was frequency modulated with 42.5 Hz and the signal was monitored by a lock-in amplifier. The signal for the modulation of laser III was taken from the reference oscillator of the lock-in.

Scanning laser III across the $^2D_{3/2}$ level, the fluorescence signal was recorded as a function of the difference frequency between laser II and III. A recording of the signal is shown in Fig. 7. The observed line width of about 24 MHz is in good agreement with the expected line width obtained by a convolution of the natural line widths for the ^2P and ^2D states (38) plus residual Doppler width of the atomic beam.

The beat signal between laser II and III was obtained by means of a photodiode and analysed by a radio-wave spectrum analyser. Equidistant frequency marks on the scope of the spectrum analyser were produced by higher harmonics of a radio frequency generator oscillating at 35 MHz. When the laser III was scanned the beat signal was shifted across the frequency marks on the scope. In the case of a coincidence a pulse was given to the xy recorder to provide the frequency scale (Fig. 7) necessary for the evaluation of the measurement. The mean value of the fine structure splitting obtained from fourteen recordings was (1074 ± 3) MHz. The error is the mean square error. The main source of uncertainty results from the line width of the signal and is also to a small amount due to the width of the beat signal, which is about 5 MHz due to the small frequency modulation of laser III.

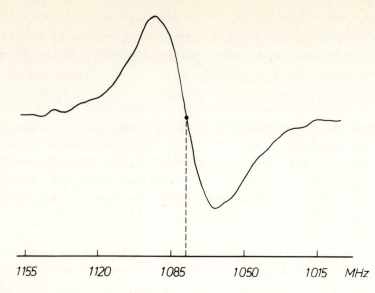

Fig. 7 Fluorescence signal from the transition $3^2D_{3/2} - 2^2P$. Time constant 300 nsec, sweep time 20 sec.

The sign of the fine structure splitting was determined by Meissner et al. (31). This was therefore not checked in our experiment.

The result obtained should be compared with the fine structure splitting for a 3d electron in hydrogen, which is 1082 MHz (15). The corresponding value for lithium is smaller, indicating a similar discrepancy as for the n^2P levels.

In the method used in this paper to determine the fine structure splitting it is not necessary to scan the single mode laser over a larger spectral range. In addition, the measurement of the splitting is reduced to a frequency measurement which can be performed more accurately than a determination of the wavelength difference by means of an interferometer. The photodiodes presently available allow a direct measurement of frequency differences up to about 10 GHz. Using the metal-metal-diodes (40), however, the method can be extended to very large splittings.

Acknowledgement

I would like to thank C. Cohen-Tannoudji for communicating his theoretical results before publication. I also would like to thank W. Rasmussen for valuable discussions and critical comments. The support of the Deutsche Forschungsgemeinschaft is gratefully acknowledged.

References

(1) W. Heitler, Quantum Theory of Radiation, 3rd edition, Oxford University Press, London 1964
(2) P.A. Apanasevich, Optics and Spectroscopy $\underline{16}$, 387 (1964)
(3) S.M. Bergmann, J. Math. Phys. $\underline{8}$, 159 (1967)
(4) M.C. Newstein, Phys. Rev. $\underline{167}$, 89 (1968)
(5) V.A. Morozov, Opt. Spectr. $\underline{26}$, 62 (1969)
(6) B.R. Mollow, Phys. Rev. $\underline{188}$, 1969 (1969)
(7) C.S. Chang and P. Stehle, Phys. Rev. $\underline{A4}$, 641 (1971)
(8) R. Gush and H.P. Gush, Phys. Rev. $\underline{A6}$, 129 (1972)
(9) M.L. Terk-Mikaelyan, A.O. Melikyan, Sov. Phys. JETP $\underline{31}$, 153 (1970)
(10) C.R. Stroud Jr., Phys. Rev. $\underline{A3}$,1044 (1971) and Coherence and Quantum Optics ed. L. Mandel and E. Wolf, New York, London, Plenum Press p 537 (1972)
(11) L. Hahn, I.V. Hertel, J. Phys. $\underline{B5}$, 1995 (1972)
(12) G. Oliver, E. Ressayre, A. Tallet, Lettere al Nuovo Cimento $\underline{2}$, 777 (1971)
(13) G.S. Agarwal, Quantum Optics p. 108, Springer Tracts in Mod. Phys. (1974)
(14) M.E. Smithers, H.S. Freedhoff, J. Phys. B. $\underline{8}$, L432 (1974)
(15) H.J. Carmichael, D.F. Walls, J.Phys. B $\underline{8}$, L77 (1975)
(16) C. Cohen-Tannoudji, Atoms in Strong Resonant Fields, Spectral Distribution of the Fluorescent Light. Proceedings of the 2nd International Laser Spectroscopy Conference, present volume
(17) C. Cohen-Tannoudji, Optical Pumping with Lasers, Proceedings of the 4th International Conference on Atomic Physics in Atomic Physics 4, Plenum Press p 589 (1975)
(18) P. Avan, C. Cohen-Tannoudji, Journ. de Physique Lettres $\underline{36}$, L85 (1975)
(19) W. Rasmussen, R. Schieder, H. Walther, Opt. Comm. $\underline{12}$, 315 (1974)
(20) F. Schuda, C.R. Stroud Jr., M. Hercher, J. Phys. $\underline{B7}$, L198 (1974)
(21) H. Brand, W. Lange, J. Luther, B. Nottbeck, H.W. Schröder, Opt. Comm. $\underline{13}$,286 (1975)
(22) H. Walther, Physica Scripta $\underline{9}$,297 (1974)
(23) A. Kastler,Appl.Opt. $\underline{1}$, 17 (1962) see also L.D. Vil'ner, S.G. Rautian, S.A. Khaikin, Opt. a. Spectr. $\underline{12}$, 240 (1962)
(24) G.B. Hocker, C.L. Tang, Phys. Rev. Lett. $\underline{21}$, 591 (1968)
(25) P.W. Hoff, H.A. Haus, T.J. Bridges, Phys. Rev. Lett. $\underline{25}$,82 (1970)
(26) R.G. Brewer, R.L. Shoemaker, Phys. Rev. Lett. $\underline{27}$, 631 (1973)
(27) W. Hartig, V. Wilke, H. Walther, Opt. Comm. in press
(28) K.C. Brog, T.G. Eck, H. Wieder, Phys. Rev. $\underline{153}$, 91 (1967) and B. Budik, H. Bucka, R.J. Goshen, A. Landmann, R. Novick, Phys. Rev. $\underline{147}$, 146 (1966)
(29) R.C. Isler, S. Marcus, R. Novick, Phys. Rev. $\underline{187}$, 66 (1969)
(30) K.L. Bell, A.L. Steward, Proc. Phys. Soc. (London) $\underline{83}$, 1039 (1964)
(31) K.W. Meissner, L.G. Mundie, P.H. Stelson, Phys. Rev. $\underline{74}$, 932 (1948)

ON THE $2P_{3/2}-2S_{1/2}$ ENERGY DIFFERENCE IN VERY LIGHT MUONIC SYSTEMS

E. Zavattini
CERN, Geneva, Switzerland

I will talk about some measurements that we hope it will be possible to perform in a near future, and present the results[1] of an experiment undertaken recently at the CERN Synchro-cyclotron by a group-collaboration from the Pisa-Bologna-Saclay-CERN laboratories.

This experimental line started from the realization that by performing, on a simple muonic system (and in particular let us consider the cases of μP, μD, μHe, and μHe), an accurate measurement of the energy difference between the 2S and the 2P levels, we can measure a quantity which is essentially determined by the electronic vacuum polarization of the electromagnetic field acting between the muon and the nucleus[2].

Let us spend a little time on treating this general fact in more detail. If we write

$$V(r) = V_C(r)[1 + \beta(r)], \qquad (1)$$

the potential between two (distinct and spinless) particles of unitary (and point-like) charges, and assume that their masses are much bigger than the electron mass, then if $V_C(r)$ is the familiar Coulomb potential (r being the distance between the two charged particles), for values of $r \geq 10^{-13}$ cm, according to QED $\beta(r)$ is a small correction which depends, apart from r, only on the two quantities α (fine structure constant = = 1/137) and λ_e (the reduced Compton wavelength of the electron = 3.8×10^{-11} cm).

Generally, under these particular conditions, $\beta(r)$ is a correction entirely due to the electronic vacuum polarization of the electromagnetic field acting between the two point-like charges (Uheling term). The correction $\beta(r)$ can be calculated by developing it in a power series of the parameter α and using the general prescription of quantum electrodynamics; the dependence of $\beta(r)$ on r, considering only terms up to α^2 (inclusive), is represented in Fig. 1 [3]. Looking at Fig. 1 we can make the following remarks:

i) $\beta(r)$ in the region considered here ($r \geq 10^{-13}$ cm) is not more than 8‰;

ii) $\beta(r)$ is a "short-range" correction which for $r > \lambda_e$ goes very quickly to zero.

Therefore considering that the Bohr radius of a light muonic atom is in value very near to λ_e (in Fig. 1, a_μ^B and a_e^B are the Bohr radii of the μP and eP systems, respectively), it is clear that the *energy difference* S_μ between the 2S and 2P levels in a muonic atom will be quite sensitive to the magnitude of the "short-range" electron vacuum polarization correction $\beta(r)$. Moreover, since the muon mass is so large compared to the electron mass, corrections introduced by the muon radiation reaction

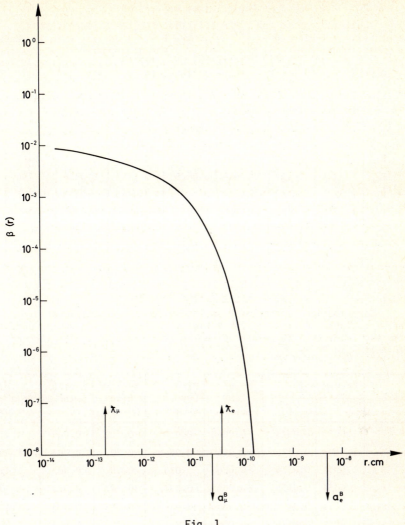

Fig. 1

(Lamb shift terms) are greatly reduced and, in fact, for the case of a muonic atom, they contribute to S_μ for a very small quantity.

Concluding a precise measurement of S_μ gives an almost unique opportunity to check the correction $\beta(r)$ directly.

Let us now go to the real cases. The expected energy differences S_μ between the various 2S and 2P levels for the four above-mentioned simpler muonic systems μP, μD, $(\mu^3 He)^+$ and $(\mu^4 He)^+$ are shown in Tables 1, 2, 3, and 4, respectively; in all cases considered here the 2P levels are always less bound than the 2S ones.

Table 1 (see Ref. 4)

Contribution to the 2S-2P splitting for the μP muonic atom.
The states are indicated as $^{2F+1}L_j$, where $F = j + $ nuclear spin I. Energies are given in units of $\alpha^2 Ry = 0.13461$ eV. The disappearance rate of the 2P state is 1.2×10^{11} sec^{-1}: in the given units the width of the 2P level is 0.0006 $\alpha^2 Ry$ (which for λ causes a linewidth of 20 Å).

Transition	Vacuum polar. α	Vacuum polar. α^2	Muonic Lamb shift	Fine struct.	Hyperfine struct.	Nuclear size ±0.0006	Total in $\alpha^2 Ry$ ±0.0006	λ in Å ±20 Å
$^1S_{1/2} \to {}^3P_{1/2}$	1.5225	0.0112	-0.0049	0	0.1417	-0.0258	1.6447	56000
$^1S_{1/2} \to {}^3P_{3/2}$	1.5225	0.0112	-0.0049	0.0625	0.1135	-0.0258	1.6790	54860
$^3S_{1/2} \to {}^1P_{1/2}$	1.5225	0.0112	-0.0049	0	-0.0845	-0.0258	1.4185	64930
$^3S_{1/2} \to {}^3P_{1/2}$	1.5225	0.0112	-0.0049	0	-0.0280	-0.0258	1.4750	62450
$^3S_{1/2} \to {}^3P_{3/2}$	1.5225	0.0112	-0.0049	0.0625	-0.0563	-0.0258	1.5092	61030
$^3S_{1/2} \to {}^5P_{3/2}$	1.5225	0.0112	-0.0049	0.0625	-0.0337	-0.0258	1.5318	60130

Table 2 (see Ref. 5)

Contribution to the 2S-2P splitting for the μD muonic atom.
The states are indicated as for Table 1. Energies are given in units of $\alpha^2 Ry = 0.14182$ eV. The width of the 2P level is 0.0006 $\alpha^2 Ry$ (natural linewidth of λ is 20 Å).

Transition	Vacuum polar. α	Vacuum polar. α^2	Muonic Lamb shift	Fine and hyperf. struct.	Nuclear size ±0.009	Nuclear polar. ±0.002	Total in $\alpha^2 Ry$ ±0.009	λ in Å ±300
$^2S_{1/2} \to {}^2P_{1/2}$	1.6051	0.0175	-0.0060	0.0192	-0.202	0.003	1.437	60960
$^2S_{1/2} \to {}^2P_{3/2}$	1.6051	0.0175	-0.0060	0.0846	-0.202	0.003	1.502	58300
$^2S_{1/2} \to {}^4P_{1/2}$	1.6051	0.0175	-0.0060	0.0336	-0.202	0.003	1.451	60350
$^2S_{1/2} \to {}^4P_{3/2}$	1.6051	0.0175	-0.0060	0.0902	-0.202	0.003	1.508	58080
$^4S_{1/2} \to {}^2P_{1/2}$	1.6051	0.0175	-0.0060	-0.0240	-0.202	0.003	1.394	62850
$^4S_{1/2} \to {}^2P_{3/2}$	1.6051	0.0175	-0.0060	-0.0413	-0.202	0.003	1.459	60030
$^4S_{1/2} \to {}^4P_{1/2}$	1.6051	0.0175	-0.0060	-0.0096	-0.202	0.003	1.408	62210
$^4S_{1/2} \to {}^4P_{3/2}$	1.6051	0.0175	-0.0060	0.0467	-0.202	0.003	1.464	59800
$^4S_{1/2} \to {}^6P_{3/2}$	1.6051	0.0175	-0.0060	0.0506	-0.202	0.003	1.468	59650

With reference to these tables the following remarks have to be made:

a) The different errors quoted include only the uncertainty introduced by the errors with which the various form factors are experimentally known. No errors due to the inaccuracy or approximations of the theoretical calculations are included; according to Di Giacomo[4] these are on the level of the natural linewidth Γ.

For the cases of Tables 2, 3, and 4 the quoted uncertainties[5-7] are always much bigger than the respective natural linewidths; for the case of Table 1 ($\bar{\mu}P$ system)

Table 3 (see Ref. 6)

Contribution to the 2S-2P splitting for the $(\mu^3He)^+$ muonic ion. The states are indicated as for Table 1. Energies are given in units of $\alpha^2 Ry = 0.14438$ eV. The disappearance rate of the 2P state is 2×10^{12} sec^{-1}; in the given units the width of the 2P level is 0.0096 (which for λ causes a natural linewidth of 8 Å).

Transition	Vacuum polar.		Muonic Lamb shift	Fine and hyperf. struct.	Finite size ±0.27	Total in α^2Ry ±0.27	λ in Å ±300
	α	α^2					
$^1S_{1/2} \to {}^3P_{1/2}$	11.372	0.079	-0.096	-0.673	-2.76	7.922	10840
$^1S_{1/2} \to {}^3P_{3/2}$	11.372	0.079	-0.096	0.468	-2.76	9.063	9475
$^3S_{1/2} \to {}^1P_{1/2}$	11.372	0.079	-0.096	0.399	-2.76	8.994	9548
$^3S_{1/2} \to {}^3P_{1/2}$	11.372	0.079	-0.096	0.133	-2.76	8.728	3839
$^3S_{1/2} \to {}^3P_{3/2}$	11.372	0.079	-0.096	1.266	-2.76	9.861	8709
$^3S_{1/2} \to {}^5P_{3/2}$	11.372	0.079	-0.096	1.160	-2.76	9.755	8803

Table 4 (see Ref. 7)

Contribution to the 2S-2P splitting for the $(\mu^4He)^+$ muonic ion. The states are indicated as L_j. Energies are given in units of $\alpha^2 Ry = 0.145687$ eV. In these units the width of the 2P level is 0,0096 (which for λ causes a natural linewidth of 8 Å).

Transition	Vacuum polar.		Muonic Lamb shift	Fine struct.	Nuclear polar.	Finite size ±0.0597	Total in α^2Ry ±0.0597	λ in Å ±47
	α	α^2						
$S_{1/2} \to P_{1/2}$	11.4347	0.0789	-0.0982	0	0.0213	-1.9678	1.3795	7359
$S_{1/2} \to P_{3/2}$	11.4347	0.0789	-0.0982	1.0000	0.0213	-1.9678	1.5251	8136

the quoted inaccuracy, due to the r.m.s. experimental error, is comparable to the linewidth Γ.

b) The uncertainty introduced by the experimental errors on the r.m.s. are smaller than the vacuum polarization term in α^2 for the cases of the μP and $(\mu^4He)^+$ muonic systems.

c) As expected, the contribution due to the muonic Lamb shift is in all cases a very small part of the total, which is mainly determined by the vacuum polarization term in α.

d) The fine and hyperfine structure contributions are also small compared to the total.

Let us now go on to some experimental considerations. So far, the method envisaged for measuring the 2P → 2S energies level difference with good accuracy is the following: it consists of sending short pulses of almost monochromatic electromagnetic radiation on a muonic system previously prepared in a 2S state, and varying the radiation wavelength until a 2S → 2P transition takes place; since the 2P level is highly unstable (for instance the 2P level lifetime in a muonic helium system is $\tau_{2P} \simeq 5 \times 10^{-13}$ sec), the 2S → 2P induced transition will quickly be followed by a fast muonic X-ray (∼ 8 keV) emission which is therefore used to signal the transition and identify the "resonance" wavelength.

In order to perform this type of measurement, various different problems have first to be solved; in the following we will first state and then discuss the two most relevant ones, and give the answers when these are available.

1) Formation of a muonic system in a 2S state (metastable) and in a medium so that it lives for a long enough time to perform an experiment on it. Let us first talk about the formation of a muonic metastable 2S state.

At the CERN muon channel of the 600 MeV Synchro-cyclotron, measurements have been done of the fraction $F(^4He)$ of metastable muonic systems $(\mu^{-4}He)^+_{2S}$ initially formed per negative muon stopped in a gaseous ^4He target (7-10 atm, 300°K); it has been found experimentally that $F(^4He) \simeq 4\%$. This value agrees well with the result of a theoretical calculation. No experiment has been done using a gaseous ^3He target, but one expects that in this case the situation will be similar to that obtained with a ^4He target.

There are no measurements on the fraction $F(H_2)$ or $F(D_2)$ of metastable muonic systems $(\mu^-P)_{2S}$ and $(\mu^-D)_{2S}$ initially formed when negative muons are stopped in a gaseous hydrogen or deuterium target, respectively. It has to be said, however, that simple theoretical estimates indicate that most probably $F(H_2)$ and $F(D_2)$ will at least be as big as $F(^4He)$.

As regards the lifetime of these metastable muonic systems, once they have been formed, let us first look at those channels and their rates through which the metastable 2S muonic state will spontaneously disappear, i.e. at the limiting conditions of zero density; these are given in Table 5 for only the two most important cases of $(\mu^-P)_{2S}$ and $(\mu^{-4}He)^+_{2S}$ muonic systems.

It is, however, clear that to obtain the total disappearance rate $\lambda^{2S}(p)$ -- p being the pressure of the gas target at 300°K -- we have to add to the rates of Table 5 the disappearance rates of the metastable 2S state via Stark mixing and Auger transitions, which will take place owing to the inevitable collisions experienced by the metastable muonic systems against the neighbouring atoms of the medium (at pressure p) in which they have been formed.

The value of the total disappearance rate $\lambda^{2S}(p)$ for the case of the muonic metastable state $(\mu^{-4}He)^+_{2S}$ has been experimentally determined at CERN[8] for the cases in which the 2S systems were formed in a gaseous target at densities corresponding to pressures within 7-50 atm, 300°K. The results indicate that with this

Table 5

Decay channels (present and corresponding rates for $(\mu P)_{2S}$ and $(\mu^4 He)^+_{2S}$ muonic systems in eV^{-1}; M1 transitions from 2S to 1S are neglected[8]).

Process	$(\mu P)_{2S}$	$(\mu^4 He)^+_{2S}$
Muon decay	4.54×10^5	4.54×10^5
Muon capture	80	45
Two-quantum decay to the 1S level	1.66×10^3	1.06×10^5

range of pressure the 2S-state lifetime is at least 1.4 μsec long. Also here one expects that for the $(\mu^{-3}He)^+_{2S}$ the situation will be very similar to the one found for the $(\mu^{-4}He)^+_{2S}$ system.

No data exist for the total disappearance rate of muonic metastable atoms $(\mu^- P)$ or $(\mu^- D)$ formed in a gaseous target; estimates show that it is conceivable that already at a pressure of 1 atm the 2S-state lifetime could be as short as a fraction of a nanosecond.

2) The almost monochromatic electromagnetic radiation pulses must be adjustable in their wavelength values around the values shown in Tables 1 to 4, at least within the errors indicated there; and moreover, each pulse must contain sufficient energy to be able to induce the selected 2S → 2P transition at an experimental detectable level.

For the cases of the muonic system $(\mu^{-4}He)^+_{2S}$ and $(\mu^{-3}He)^+_{2S}$ given in Tables 3 and 4, the range of wavelength needed is easily covered by dye lasers.

As a result of a compromise between the existing experimental and physics interests, and using the general method outlined above, a Pisa-Bologna-Saclay-CERN Collaboration performed this type of experiment at CERN (as part of a first step) searching for the $2S_{1/2} \to 2P_{3/2}$ transition on the muonic metastable 2S system $(\mu^{-4}He)^+_{2S}$ [9,1]. In what follows we will briefly mention some details of this experiment and give the first result, together with some conclusions drawn from it.

In Fig. 2 a simplified view of the set-up is shown. The negative muon beam entering the target (filled with ^{4}He has) was bunched in pulses, a few milliseconds wide, at the same repetition rate as that of the ruby laser (0.25 Hz). The electromagnetic pulses were produced by an infrared dye laser (excited by a Q-switched ruby laser) which could be tuned using a diffracting grading inserted in the infrared dye laser cavity[9].

The main characteristics of the lasers are given in Table 6. The energy per pulse ensured that for the experimental conditions chosen we would have a probability for the $2S_{1/2} \to 2P_{3/2}$ transition, at the resonance wavelength, of at least 15%. The target

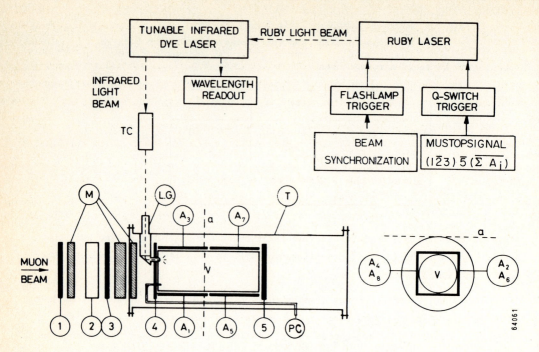

Fig. 2 Very simplified view of the whole apparatus. M = CH_2 moderators; 1,3,4,5 = plastic scintillators; 2 = anticoincidence Čerenkov counter; T = invar steel tank; V = useful volume for stopping muons; A_1-A_8 = Na(Tl) counters; L.G. = light-guide used to inject the infrared radiation into the target; TC = optical telescope; PC = optical fibre supplying the energy-monitoring signal.

Table 6

Characteristics of the dye laser used in the present apparatus

Dye	HITC [a]
Solvent	DMSO [b]
Molar concentration (moles/litre)	$\sim 5 \times 10^{-5}$
Average ruby pumping energy (J)	1.2
Average pulsed infrared output energy (mJ)	300
Infrared power at 8150 Å (MW)	~ 15
Radiation pulse duration (nsec)	~ 20
Bandwidth of radiation (Å)	~ 6

a) 1,3,3,1',3',3'-hexamethyl-2,2'-indotricarbocyanine iodide: purchase from K. & K. Labs., Plainview, N.Y., USA.

b) Dimethylsulfoxide: purchased from Carl Bittman A.G., Basel, Switzerland.

is surrounded by eight NaI(Tl) crystals in order to detect, after a muon stopped in the target (with or without infrared light entering in it), any X-rays of energy between 4 and 14 keV and at any time t (from the muon stop signal initial time) contained in an interval of about 2 μsec. The main difficulty in performing this experiment was represented by the high rate of accidental X-ray background.

In Fig. 3 is shown the results[1] for a series of exploratory runs in which the wavelength interval 8090-8160 Å was covered (about 8×10^5 flashes to the ruby laser): a peak appears in the distribution, and this can be well fitted by a Lorentzian curve with $\Gamma = 8$ Å. A fit of the data with a simple straight line (case of no effect) gives a χ^2 corresponding to a confidence level of about 4.5%. Taking the peak as evidence that the $2S_{1/2} \to 2P_{3/2}$ transition has been induced by the infrared light we obtain for the resonance wavelength

$$\lambda = 8117 \pm 5 \text{ Å} . \tag{2}$$

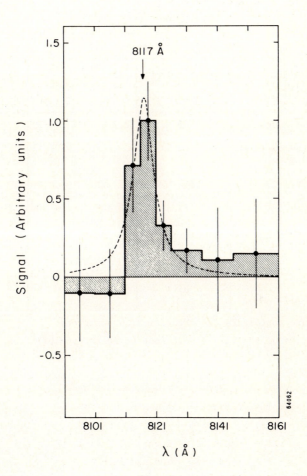

Fig. 3

Comparing the value (2) with the theoretical predictions, we conclude the following:

i) The difference DS between the theoretical and the experimental transition energy (see Table 4) for the $2S_{1/2} \to 2P_{3/2}$ line in the muonic $(\mu^- He)^+$ ionic system is $DS_\mu = (0.016 \pm 0.060)\alpha^2 Ry$, and is compatible with zero within the errors. This indicates that in this case, within the quoted accuracy, there is agreement between the QED prediction as far as concerns the electronic vacuum polarization and the experiment. A better understanding of this result is obained by comparing it to other results which are most relevant with respect to tests of QED; this is done in Table 7 [10].

ii) On the other hand, if we assume QED to be valid comparing the result (2) with theoretical values (Table 4), we obtain for the r.m.s. charge radius of the ^{4}He nucleus the value $\langle r^2 \rangle^{1/2} = 1.644 \pm 0.005$ fm, where the error [3] contains also the uncertainty in the theoretical calculation [11].

Table 7 (see Ref. 10)

Tests of the vacuum polarization effects S_{vp}

Experiment	Exp. precision $\Delta S/S$ (ppm)	$\frac{S_{vp}}{S}$ (Th.)	$\frac{S_{vp}}{S_{vp}}$ (1‰)	Mean momentum transfer
Lamb shift in hydrogen	60	2.5×10^{-2}	2-3	αm_e
Hyperfine structures in hydrogen and muonium	4 a)	α^2	100	αm_e
$(\mu^4 He)^+$ $\Delta E(2P_{3/2} - 2S_{1/2})$	6000	1.2	5	αm_μ
$g_e - 2$	3	10^{-4}	30	m_e
Hyperfine structure in positronium	25	$\frac{1}{3}\frac{\alpha}{\pi}$ b)	40	$2m_e$
High-Z μ-mesic atoms	50	5×10^{-3}	10	$\frac{Z\alpha m_\mu}{n}$
$g_\mu - 2$	27	$2\frac{\alpha}{\pi}$	7	m_μ

a) This figure represents the theoretical uncertainty.
b) Contribution of the vacuum polarization in the annihilation channel.

REFERENCES

1) A. Bertin et al., Phys. Letters 55B, 411 (1975).

2) See, for instance, A.I. Akhiezer and V.B. Berestetskii, Quantum Electrodynamics (Interscience Publishers, NY, 1969), Section 53.2 and references quoted therein.

3) E.A. Uehling, Phys. Rev. 48, 55 (1935).
J. Schwinger, Phys. Rev. 75, 651 (1949).
L.L. Foldy and E. Eriksen, Phys. Rev. 95, 1048 (1954).
G. Källèn and A. Sabry, K. Danske Vidensk. Selsk. Mat.-Fys. Medd. 29, 17 (1955).
See also, E. Campani, Thesis, University of Pisa (1970).

4) A. Di Giacomo, Nuclear Phys. B11, 411 (1969) and Erratum B23, 671 (1970).
E. Campani, Nuovo Cimento Letters 4, 512 (1970) and Thesis, University of Pisa (1970).

5) G. Carboni, Nuovo Cimento Letters 7, 160 (1973). In this calculation the measured values of the r.m.s. of the deuteron and the electric quadrupole moment of the deuteron have been taken respectively from Landot-Börenstein, Neue Serie Gruppe 1, 2 Kernzadien (Berlin, 1967) and H.G. Kolsky et al., Phys. Rev. 87, 395 (1952).

6) E. Campani, Nuovo Cimento Letters 4, 982 (1970). In this calculation the value of the r.m.s. of the ^{3}He nucleus has been taken from H. Callard et al., Phys. Rev. Letters 11, 132 (1963). Moreover, in this case the contribution due to the ^{3}He nuclear polarizability (which could be not negligible) has not been included.

7) E. Campani, Nuovo Cimento Letters 4, 982 (1970); J. Bernabeu et al., Nuclear Phys. B75, 59 (1974). The values presented here are those taken from Ref. 1.

8) A. Bertin et al., Nuovo Cimento 26B, 433 (1975).

9) A. Bertin et al., Nuovo Cimento 23B, 490 (1974).

10) R. Barbieri, CERN TH-1963 (1975). (This is a partial reproduction of Table 1.)

11) Taking for the r.m.s. charge radius of ^{4}He the weighted average of the existing experimental results (see Ref. 1), one obtains $\langle r^2 \rangle^{1/2} = 1.650 \pm 0.025$ fm.

ULTRAFAST VIBRATIONAL RELAXATION AND ENERGY TRANSFER IN LIQUIDS

W. KAISER AND A. LAUBEREAU
Technische Universität München, München, West Germany

Stimulated Raman scattering is a convenient tool for coherent excitation of well-defined molecular vibrations (in the electronic ground state). Coherent probe scattering in a phase matched geometry allows the determination of the dephasing time τ of the excited vibrational mode. Experiments of this sort have been performed on a number of vibrational modes of different molecules. For relatively long pump pulses with $t_p \gg \tau$ the excitation process is highly selective. Only one molecular vibration with the largest scattering cross-section is excited. The situation is different for the transient stimulated Raman process where $t_p \lesssim \tau$. In this case it is possible to excite vibrational modes of small frequency differences $\delta\omega$ (e.g. different isotope species) simultaneously. It can be readily shown that neighboring vibrational modes are excited with equal amplitude and phase when the duration of the incident light pulse is sufficiently short i.e. $t_p \ll 2\pi/\delta\omega$. After the short pump pulse has passed the medium, the collectively vibrating (isotope) components relax freely oscillating with their individual resonance frequencies. A beating between the vibrating systems is predicted where maxima and minima occur after a time interval of $1/\delta\omega$.

Recently we have observed the beating phenomenon of isotope components by measuring the coherent probe scattering of a second short light pulse as a function of delay time t_D. The incident probe pulse generates scattered signals at slightly different directions. The scattering directions result from the different vibrational frequencies of the isotope components. There are two possibilities to investigate the collectively vibrating molecular systems. 1) A beam direction is selected where one vibrational component has a predominant scattering signal. 2) A direction is chosen where two (or more) beams have scattering intensities of comparable magnitude. In this case we observe as a function of delay time t_D strong beats of the coherent scattering signal. This result was expected since each coherent probe beam reflects the phase information of the respective vibrational component. The high selectivity of the method results from the critical phase matching of the coherent probe scattering.

One can show that for two isotope components a and b the scattered signal $S^{coh}(t_D)$ has the form (for $t_D > t_p$):

$$S^{coh}(t_D) \simeq \text{const} \left(1 + \frac{2N_b}{N_a} \cos \delta\omega t_D \right) e^{-t_D/\tau}$$

In addition to the exponential decay with the dephasing time τ, the scattering signal is modulated with the beat frequency $\delta\omega$ of the coherently excited isotope components.

We have extended our calculations to several isotope species and to longer light pulses, $t_p \lesssim 2\pi/\delta\omega$ and find the same beating phenomenon.

In Fig. 1 we present a first example of isotope beats. Common CCl_4 was investigated where spontaneous Raman data show line splitting ($\delta\omega \simeq 3$ cm^{-1}) of the totally symmetric vibration around 459 cm^{-1} on account of the natural abundance of the chlorine isotopes. The three most abundant molecular components $C^{37}Cl^{35}Cl_3$, $C^{35}Cl_4$ and $C^{35}Cl_2{}^{37}Cl_2$ have a concentration ratio of 1 : 0.77 : 0.49. Single pulses of approximately 3.5 psec duration were used for the excitation and the probing pulse.

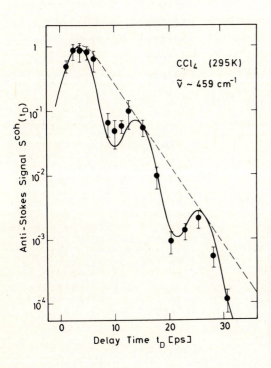

Fig.1. Coherent scattering signal of common CCl_4 representing the beating of the isotope components separated by $\delta\omega/2\pi c = 2.9$ cm^{-1}. The broken line indicates the dephasing of one isotope component.

The collective beating of the CCl_4 isotope species is readily seen from Figure 1. The beat frequency is deduced to be $\delta\omega/2\pi c$ = 2.9 ± 0.15 cm^{-1} in good agreement with spontaneous data. The curve is calculated using known parameters of CCl_4 and a dephasing time τ = 3.6 ± 0.5 psec which was determined in an independent experiment where one isotope component was investigated separately. The good agreement between the calculated curve and our experimental points should be noted.

We now turn to the discussion of a series of experiments to study the population lifetime τ' of the v = 1 state of a normal mode of vibration in the electronic ground state. In a first group of investigations, the molecules were excited via stimulated Raman scattering and the degree of excitation was monitored by spontaneous anti-Stokes scattering of a delayed probe pulse. In a second experimental system ultrashort infrared pulses excite directly the vibrational mode of interest and a second properly delayed probe pulse monitors the excitation using a transition to a fluorescent electronic state. The second technique is described below.

The population relaxation time τ' was studied in a variety of molecular systems. Three examples are briefly discussed here in order to demonstrate the potential of the method. The first important question is concerned with possible energy transfer between vibrational states of different molecules. The most direct experiment was performed in a mixture of CH_3CCl_3 and CD_3OD (see Fig. 2).

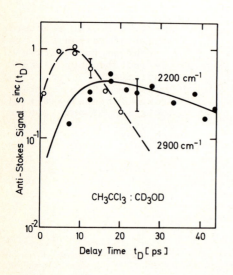

Fig.2. Measurement of the energy relaxation time τ' and of the intermolecular energy transfer in the mixture CH_3CCl_3 : CD_3OD. Open circles denote probe scattering from the CH_3-valence bond vibration of CH_3CCl_3 which is excited by the pump pulse. Full points represent scattering observed from the CD_3-valence bond vibration of CD_3OD directly indicating intermolecular transfer of vibrational energy.

First the ν_H-mode of CH_3CCl_3 was excited by an intense pump pulse via stimulated Raman scattering. The rise and decay of the excess population of the $v = 1$ state at a frequency shift of $\nu \sim 3000$ cm^{-1} was observed with the help of spontaneous anti-Stokes scattering of delayed probe pulses. Then, the spontaneous anti-Stokes signal was measured at a frequency shift of $\nu \sim 2200$ cm^{-1} where the ν_D-mode of CH_3OD is located. Indeed, the subsequent rise and decay of the population of the ν_D-mode of 2200 cm^{-1} was clearly observed. We feel that this experiment gives convincing proof for vibrational energy transfer from the CH_3CCl_3 molecule to CD_3OD. It should be noted that this efficient vibrational transfer process is quasi-resonant since the energies of a CCl_3-chlorine vibration plus the ν_D-mode add up with good accuracy to the energy of the primarily excited ν_H-vibration. In a mixture of CH_3CCl_3 and CD_3OD (60 : 40) the first excited ν_H-mode has two decay routes: 53% of the decay processes proceeds via $\nu_H \rightarrow \nu_D$ transfer while the rest decays via the bending mode $\nu_H \rightarrow 2\delta_H$. These numbers are estimated from the relative intensities of the various scattering signals. The transfer rate $\nu_H \rightarrow \nu_D$ of the vibrational energy in this mixture is measured to be $k \simeq (13\ psec)^{-1}$.

The concentration dependence of the energy (population) relaxation time τ' of the ν_H-stretching mode was studied in several $CH_3Cl_3 : CCl_4$ mixtures with mole fractions $x = 1, 0.8, 0.6$ and 0.4 of CH_3CCl_3.

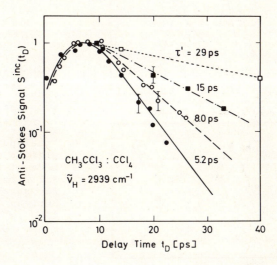

Fig.3. Incoherent probe signals of the ν_H-vibration (2939 cm^{-1}) versus delay time t_D in various mixtures $CH_3CCl_3 : CCl_4$. Mole fractions of CH_3CCl_3, $x = 0.4(\ )$, $0.6(\ )$, $0.8(o)$ and $1.0(o)$.

It was found that the energy relaxation time of the ν_H-mode of CH_3CCl_3 at $\nu \sim 3000$ cm^{-1} increases strongly from 5.2 psec to 29 psec when x decreases from 1.0 to 0.4 (see Fig. 3). Since the vibrational (v = 1) levels of CCl_4 are small ($\nu \leq 800$ cm^{-1}) compared to the excited ν_H-vibration ($\nu \approx 30000$ cm^{-1}), the solvent CCl_4 is considered to be an energetically "inert" molecule in the mixture. Our data indicate that $\tau'^{-1} \propto x^2$ i.e. the excited molecule appears to interact with more than one molecule of its kind during the decay. It is interesting to note that the Raman line width of the ν_H-mode of CH_2CCl_3 (a measure of the dephasing time in this molecule) does not change by more than 10% in the same mixtures studied. i.e. line width measurements do not allow the determination of the strong concentration dependence of τ' found in these investigations.

Energy transfer between neighboring vibrational energy levels was observed in ethanol. This molecule has 21 molecular vibrations, five of which are at energies between 2972 and 2877 cm^{-1}. The spontaneous Stokes Raman spectrum is depicted in Fig. 4 indicating three distinct vibrational modes.

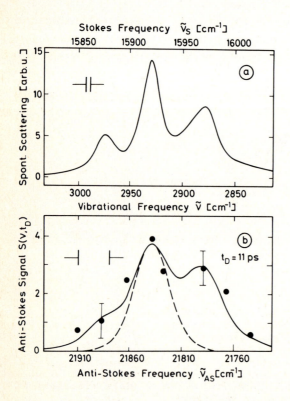

Fig.4. a) Spontaneous Stokes spectrum of the CH-stretching vibrations. b) Anti-Stokes probe scattering of CH-stretching vibrations at t_D = 11 ps; broken line indicates spectral profile of the pumped mode at 2928 cm^{-1}. Solid curve calculated for quasi-equilibrium between levels around 2900 cm^{-1}.

The most intense Raman line at $\nu = 2928$ cm^{-1} participates in the stimulated Raman process. As a result this energy level is primarily excited by the incident pump pulse. Extending the argument of resonant energy transfer to ethanol, rapid energy exchange is expected between the five vibrational levels around 3000 cm^{-1}. The spontaneous anti-Stokes signal of the probe pulse was measured for frequency shifts of 2830 to 3000 cm^{-1} with a time delay of 11 psec. Since the anti-Stokes spectrum is a direct measure of the population of various energy states, we are able to determine which energy states are populated immediately after the pump pulse has left the sample. The experimentally observed spectrum is depicted in Fig. 3b. The spectrum extends over 150 cm^{-1} while the instrumental resolution is 30 cm^{-1}. This finding strongly suggests rapid energy transfer between the ν_H-energy states around 3000 cm^{-1}.

Very recently we have devised an experimental system for the study of molecular vibrations in highly diluted solutions. This method is particularly well suited for investigations of the electronic ground state of fluorescent molecules. Fig. 5 illustrates schematically the relevant transitions involved in our measurements.

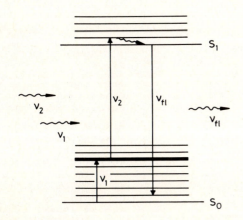

Fig.5. Schematic of the molecular energy levels and transitions during the vibrational excitation and probing process.

A first powerful short infrared pulse excited the molecular system via infrared absorption. The frequency ν_1 of the infrared pulse is properly selected in order to interact only with one well-defined vibrational mode. Our pump pulse has a duration of approximately 3 psec and contains 10^{15} infrared quanta at ~3000 cm^{-1}. The molecular system is subsequently interrogated by a second pulse of frequency ν_2. The second probe pulse promotes molecules which are vibrationally excited to a level close to the vibrational ground state of the first excited singlet state S_1. The fluorescence originating from this singlet state is experimentally observed. It should be noted that energy levels smaller than $h\nu_1$ (in the electronic ground state) do not interact with the probing pulse of frequency ν_2. The time integrated fluorescent radiation serves as a direct measure of the instantaneous vibrational excitation of the energy level $h\nu_1$. The fluorescent signal is measured as a function of delay time t_D between the infrared excitation pulse and the probing light pulse. With this technique it is possible to study well-defined vibrational excitations even in large molecules with complicated vibrational spectra. Of special interest is the high sensitivity of this method for molecules with high fluorescent quantum efficiency.

We have applied this technique to dynamic investigations of dye solutions in a large concentration range of 10^{-6} M to 10^{-3} M. As an example we discuss results on the molecule Coumarin 6 in the solvent CCl_4. The Coumarin 6 molecule is depicted on the top of Fig. 6.

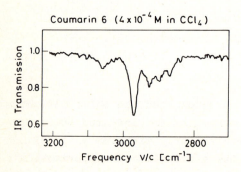

Fig.6.
Top: The Coumarin 6 molecule.

Bottom: Infrared absorption spectrum of Coumarin 6 in CCl_4 (10^{-3} M) around 3000 cm^{-1}. The line at 2970 cm^{-1} is excited in the picosecond experiment.

Two C_2H_5-groups are bonded to the conjugated ring system. The infrared transmission spectrum of Coumarin 6 around 3000 cm^{-1} is presented in the lower part of Fig. 6. Several absorption peaks are clearly resolved in the spectrum. The bands between 2865 cm^{-1} and 2970 cm^{-1} are readily interpreted as normal vibrational modes of the two ethyl groups of Coumarin 6. The CH-valence bond vibrations of the ring system produce the absorption peak at 3055 cm^{-1}. Most important for our investigations is the absorption maximum at 2970 cm^{-1} which corresponds to the asymmetric CH_3-mode. It is this vibrational mode which is populated by the resonant short infrared pulse. The degree of vibrational excitation is subsequently monitored as a function of time by our fluorescence technique. As frequency of the probing pulse we choose $\nu_2 = 2\nu_L = 18910$ cm^{-1} (0.53 µm) i.e. the second harmonic of the Nd-glass laser.

Experimental results for Coumarin 6 in CCl_4 are presented in Fig. 7 for a concentration of 3×10^{-5} M.

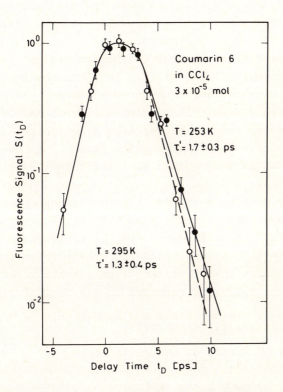

Fig.7. Ultrafast vibrational relaxation of Coumarin 6 in CCl_4 at 295 K (open circles) and 253 K (full points). The asymmetric CH_3-mode at 2970 cm^{-1} in the electronic ground state is excited and the vibrational excitation is observed as a function of time with a novel fluorescence probing technique.

The observed fluorescence signal, $S(t_D)$, initiated by the probe pulse, is plotted as a function of delay time t_D for two temperatures, 253 K and 295 K. The time scale of a few psec should be noted. The fluorescence signal increases sharply within approximately 1 psec to a maximum value at $t_D \simeq 2$ psec. The build-up of excess population in the vibration spectrum at $\nu_1 = 2970$ cm^{-1} is directly seen from the rise of the signal curve. Of special interest are the decaying parts of the signal curves which extend over two orders of ten. Fast relaxation of the vibrational excitation with slightly different slopes for the two temperatures is clearly indicated by the data. At room temperature a time constant of 1.3 ± 0.4 psec is directly obtained from the exponential slope (open circle). At $-20°$C the energy relaxation time is measured to be $\tau' = 1.7 \pm 0.3$ psec. We have ascertained that these time constants represent true relaxation times. The liquid cell containing the Coumarin molecules was replaced by a nonlinear crystal. The intensity of the sum frequency at $\nu_1 + \nu_2$ was measured as a function of the delay time between the two pulses of frequency ν_1 and ν_2. In this way the time resolution of our system was tested and found to be better than one psec.

At the present time the energy relaxation process is not yet definitely determined. There are two possible physical mechanisms which lead to a decay of our fluorescence signal: 1) energy transfer to vibrational states of approximately equal energy with smaller absorption cross-sections to the excited electronic state; 2) energy decay to lower vibrational states which cannot interact with the probe pulse. Work is in progress to illucidate the relaxation process in more detail.

In summary we wish to emphasize that different experimental techniques are now available for the study of ultrafast dynamical processes in liquids. In this short note, investigations are reviewed of collective vibrational processes, of energy relaxation times and intermolecular vibrational transfer of polyatomic molecules.

For details the reader is referred to the following papers:

A. Laubereau, D. von der Linde and W. Kaiser,
  Phys. Rev. Letters **28**, 1162 (1972).

A. Laubereau, L. Kirschner and W. Kaiser,
  Opt. Commun. **9**, 182 (1973).

A. Laubereau, G. Kehl and W. Kaiser
  Opt. Commun. **11**, 74 (1974).

A. Seilmeier, A. Laubereau and W. Kaiser
  to be published.

A. Laubereau, G. Wochner and W. Kaiser,
 to be published.

STUDIES OF CHEMICAL AND PHYSICAL PROCESSES WITH PICOSECOND LASERS

Kenneth B. Eisenthal

One of the basic questions in chemistry today is concerned with the degradation of energy in a molecular system. The time dependent redistribution of energy between the various degrees of freedom within a molecule on excitation to some excited state and the interactions and energy exchange of the excited molecule with surrounding molecules and external fields is of fundamental importance to a description of molecular phenomena. It is the completion between the various dissipative pathways which determines whether light is emitted or nonradiative physical and chemical processes dominate in the degradation of energy by the molecules of interest. Studies of these processes in the picosecond time domain brings new insight of these decay mechanisms since the competitive channeling of energy is often determined on this time scale.

In this talk I will present some of our work on the cage effect and electron transfer processes in liquids.

Photodissociation and the Cage Effect

The chemistry following the photodissociation of a molecule AB into the fragments A and B is strongly dependent on the frequency and nature of the collisions of the fragments with surrounding molecules. In the gas phase for example the probability of the original fragments re-encountering one another and then recombining to give the parent molecule AB is close to zero. It is far more likely in a low density gas that the fragments will collide and react with other A and B fragments generated by photodissociation of other AB molecules elsewhere in the sample or react with other species present in the system. On the other hand we would find in carrying out this same experiment in the liquid state that the original fragments A and B cannot so readily escape from one another due to their collisions with the surrounding solvent molecules of the liquid. The chemistry following a dissociation event is therefore dependent on the probabilities of 1) recombination of the original fragments which results in no net chemical change, 2) escape of the fragments from the dissociation of other AB molecules or their reactions with other species present in the system. In a liquid this enhanced probability of the original fragments re-encountering one another and reacting to yield the parent molecule was first postulated by Franck and Rabinowitch[1] in 1934 and is called the cage effect.

Since the cage effect is critically dependent on the interactions of the fragments with the surrounding solvent molecules it can serve as a probe of the liquid state itself, i.e. the nature of translational motions and energy and momentum exchanging processes. The nature of the chemical processes in liquids is intimately connected with the nature of the liquid state itself.

Although there has been discussion of primary and secondary cage effects, it is perhaps more useful to consider all recombinations of original partners as due to the cage effect. To arbitrarily separate the cage effect into primary and secondary processes is questionable since the cage is not, as originally postulated, a static structure with fragments bouncing around in a "rigid" solvent environment. As we shall demonstrate, the description of the cage effect must be of a dynamic nature dependent on the translational motions of both the fragment and solvent molecules.

Because the A and B fragments are generated by photodissociation, their distribution is not initially spatially random. Since A and B in the early time domain are more likely to be near each other, there are local concentration gradients in the solution, and one cannot use a conventional kinetic treatment to describe the dynamics of the geminate and nongeminate recombination processes. If we use a concentration-diffusion or random-flights description at least two problems come to mind. One is that the diffusion coefficient in the usual description is assumed to be independent of the separation of A and B, which may be incorrect for the processes considered here, since A and B are less than a few molecular diameters apart in the early stages of the reaction. Second, the motions of A and B may be correlated and not describable by a random walk since the motion of one fragment influences the motion of the solvent molecule which can effect a drag on the other fragment. Furthermore, in these processes it may not be accurate to describe the solvent as a continuous and isotropic medium; motions in certain directions with certain displacement sizes may be favored.

Since the processes involved in "cage" effect reactions are in a time domain beyond the scope of conventional dynamic methods, the extensive studies of these phenomena have heretofore been investigated by indirect and time-independent methods. For example, information on the quantum yields for dissociation obtained from scavenger experiments have provided valuable insights into the nature of the "cage" processes.[2] However, with these methods no measurements of the dynamics of the geminate recombinations were possible, and thus we had no idea of the time scale for the geminate processes, i.e., if it was of 10^{-13} sec or 10^{-11} sec or 10^{-9} sec duration. To determine the nature of "cage" effect reactions, it is necessary to obtain information on the early time motions of the fragments since this is the key to the partitioning between geminate and nongeminate recombinations.

The system selected for this study was I_2 since it is a simple molecule of great interest and a great deal is known about its spectroscopic properties and chemistry. Two different solvents, namely hexadecane and carbon tetrachloride, were used in these experiments.[3] The system is pulsed with an intense 5300-Å picosecond light pulse (half-width ~5 psec). At this frequency I_2 is excited to the $^3\Pi_{0^+_u}(v' \approx 33)$ state. (A small fraction of I_2 molecules are excited to the $^1\Pi_u$ state and directly dissociate.)

I_2 molecules in the $^3\Pi_{0^+_u}$ state undergo a collisionally induced predissociation leading to a pair of ground state, $^2P_{3/2}$, iodine atoms. The iodine atoms can geminately recombine or can escape and subsequently react with iodine atoms produced elsewhere in the liquid. The population of I_2 molecules is monitored with a weak 5300-Å picosecond light pulse from times prior to the strong excitation pulse up through 800 psec after the excitation pulse. The strong excitation pulse depopulates a good fraction of the groundstate I_2 molecules and thus yields an increase in the transmission of the probe pulse. As the iodine atoms recombine, the population of absorbers (iodine molecules) increases and therefore the transmission of the probe pulse decreases. In this way we can follow the recombination dynamics of the iodine atoms by monitoring the time-dependent population of I_2 molecules. The picosecond time resolution is readily obtained by spatially delaying the probe pulse with respect to the excitation pulse; a 1-mm path difference in air is equal to a 3.3-psec time delay.

In Figures 1 and 2 we see the transmission of the probe light increases to a peak value at about 25 psec after the strong excitation pulse in both the CCl_4 and hexadecane solvents. The transmission reaches a stable value in both solvents at about 800 psec. The residual difference in absorption between the long time values (800 psec) and the initial absorption ($t < 0$) is due to those iodine atoms which have escaped their original partners. The iodine atoms which have escaped will recombine at much later times ($>10^{-8}$ sec) with iodine atoms from other dissociation events, i.e., the nongeminate recombination. We thus have observed the dynamics of the geminate recombination (the cage effect) and the escape of fragments leading to the nongeminate recombination processes. The geminate recombination times (decrease by e^{-1} from peak values) are about 70 psec in hexadecane and 140 psec in carbon tetrachloride. It seems unlikely that a description of the cage effect as the collisions of the iodine atoms inside a static solvent cage would be consistent with the time scale of these geminate recombinations.

To describe the dynamics of the recombination, Noyes's treatment based on a random flight model was used.[4] The theoretical curves, the solid and dashed lines, are shown in Figures 1 and 2. In comparing the theory to our experimental results we find that, if we adjust the theoretical curve to fit the long time behavior, where one would expect the random walk description to be most valid, the early time behavior is too rapidly decaying. The lack of agreement between theory and our experimental results can be due to the crudeness of the theory, e.g., assuming one distance between the iodine atoms on thermalization rather than a distribution of distances, or to the more fundamental issues mentioned earlier, namely the correctness of a random walk description for atoms within several Ångstroms of each other. We are presently investigating the issue *via* a simulated molecular dynamic calculation of the dissociation and recombination to better theoretically describe the cage effect in liquids.

Figures 1 and 2 also provide information on the dynamics of the collision-induced predissociation process which generates the iodine atoms from the laser excited bound $^3\Pi_0^+{}_u$ state. The observed peak in the probe transmission occurs at a time (25 psec) significantly after the decay of the excitation pulse (full width less than 8-10 psec) in both the CCl_4 and hexadecane solvents. Therefore, the continual rise in the transmission of the probe light after the excitation pulse cannot be due to the further depopulation of the ground-state iodine molecules by the excitation pulse. The rise time of the transmission can be explained by assuming that the probe light can be absorbed not only by ground-state but also by the $^3\Pi_0^+{}_u$ excited iodine molecules. The photodissociative recoil studies of Busch, et al.,[5] on I_2 show that I_2 in the $^3\Pi_0^+{}_u$ state does absorb at 0.53 μ. Thus, the probe pulse is monitoring the change in both the ground and excited iodine molecule populations. We therefore conclude that after the excitation pulse reduced the ground-state population the subsequently observed increase in the probe transmission is due to the decay of the excited iodine molecules into iodine atoms. We have thus obtained from this experiment what we believe to be the first direct observation of the dynamics of a collision-induced predissociation in the liquid state and find a pseudo-first-order rate constant of about 10^{11} sec^{-1}. This is about 10^5 larger than the spontaneous predissociation process observed in I_2 at low pressures in the gas phase.

We have considered up to this point the nature of the photodissociation of I_2 and the recombination of the iodine atoms in inert liquids. However, in some liquids the I_2 molecule can have a weak interaction in the ground state and in addition it is known that iodine atoms in some liquids can form transient complexes ($\sim 10^{-6}$ sec) with the solvent molecules. If we excited I_2 in such a liquid we can ask the questions 1) how do the interactions in the liquid alter the predissociation of excited I_2 and 2) what is the dynamics of the iodine atom molecule reaction in the liquid of interest.

The system selected for study was 10^{-2} M I_2 in benzene. There is some evidence for the formation of a weak ground state I_2-Benzene complex at room temperature but there remains some controversy as to the existence of this weak complex.[6] The existence of an iodine atom benzene complex has been established[7] and the absorption spectrum of this transient species has been observed and is found to peak around 5000 Å with a spectral bandwidth of several hundred Angstroms. On excitation of this system with an intense 5300 Å picosecond pulse and probing with an attenuated picosecond pulse at the same frequence at times before and after excitation yields the results shown in Fig. (3, 4)[8]. The absorption is seen to increase and reach a peak value at about 15-20 psec after the excitation pulse and then remains constant to the longest time of our measurements at 1.2 nsec. This behavior is considerably different from that of I_2 in CCl_4 where a decrease in absorption is observed to reach a peak value at about 25 psec after the excitation pulse. In the latter case the processes involved are excitation of I_2 followed by a collision induced predissociation leading to the generation of ground state iodine atoms. In the I_2-benzene system the sharp increase in absorption with time indicates that the collision induced predissociation is far more rapid than in the I_2 - CCl_4 system and also that the absorption coefficient of the iodine atom-benzene complex is larger than that of ground state I_2 in benzene at the probe wavelength. The enhancement of the collision induced predissociation process can be due to the charge transfer interactions between ground state I_2 with benzene or perhaps more likely to the interactions between excited I_2 and benzene since the electron affinity of excited I_2 is greater than ground state I_2 by about 2 eV.

The decrease of the probe transmission levelling off at about 15-20 psec after the excitation pulse, can be initially due to the absorption of the excited I_2 "complexed" with benzene as well as the absorption of the iodine atom-benzene complex. The absorption of the species generated by the excitation pulse must be greater than that of ground state iodine to yield the observed decrease in transmission. However at longer times the absorption is due to the iodine atom benzene complexes since the benzene-excited iodine complexes dissociates to produce the iodine atoms which then react with benzene to form the long lived iodine atom-benzene complex. Unless the benzene-excited iodine molecule complex and the iodine atom-benzene complex absorption at the probe frequency are accidentally the same we conclude that the observed time dependence cannot be due to the excited iodine molecule absorption alone and must therefore reflect the time dependence of the iodine atom benzene reaction.

Excited State Electron Transfer Processes

One of the key processes by which organic molecules in excited electronic states degrade their electronic energy is by charge transfer interactions between the excited molecule and surrounding ground state molecules. The transfer of an electron from a ground state donor molecule D to excited accpetor molecule A* quenches the normal fluorescence of A*, leads to the appearance of a new emission in low dielectric solvents, can produce ion radicals in polar solvents, and can change the chemistry of the system. The physical and chemical natures of these diverse processes have been extensively studied since the discovery of excited-state charge-transfer complexes by Leonhardt and Weller,[9] but heretofore not in the subnano-second time region which is of key importance to an understanding of these events. In addition to our interest in the charge-transfer process and the subsequent energy dissipation, the electron-transfer reaction between A* and D provides an excellent vehicle for testing the theories of diffusion-controlled chemical reactions.

In conventional kinetic treatments it is assumed that the reactivity of a molecule does not change in any interval of time subsequent to the formation of the reactive molecule. In other words, it is assumed that the reactivity at time t is the same as it was at earlier times and will be the same at future times. This is equivalent to stating that there is an equilibrium distribution of reacting molecules at all times and thus the reaction can be described by a time independent rate constant. However, in a highly reactive system this description is incorrect.

The time dependence of the chemical reaction can be viewed in the following way. At time $t = 0$, the molecules A* and D are randomly distributed, but as time proceeds those distributions in which an A* is near to a D are preferentially depleted since there is a higher probability for reaction than for those distributions in which A* and D are far apart. This produces a spatially nonuniform distribution of molecules leading to a flux of molecules from the more concentrated to the less concentrated regions of the liquid. Since the distribution of molecules is changing with time, the rate "constant" for the reaction is also changing with time. To test theory and its limits adequately, it is necessary to determine the full time bahavior of the chemical reaction. Picosecond laser studies were therefore initiated on the anthracene (acceptor) and diethylaniline (donor) system to determine the key parameters of the electron-transfer process and to test the theories of diffusion-controlled chemical reactions.[10]

A single laser pulse was extracted from the train of pulses generated by a mode-locked ruby laser and was frequency doubled from the fundamental at 0.6943μ to 0.3472μ. The 0.3472μ phase was then used to excite the ^1L_a state of anthracene. The fundamental frequency of the laser at 0.6943μ is resonant with a transition of the excited charge-transfer complex[11] $(A^--D^+)^*$. The sequence of steps in the experiment are

Formation of acceptor A*

$$A + 2\hbar\omega(0.3472\mu) \longrightarrow A^*$$

Electron transfer

$$A^* + D \xrightarrow{k(t)} (A^--D^+)^*$$

Detection of $(A^--D^+)^*$ formation

$$(A^--D+)^* + \hbar\omega(0.6943\mu) \longrightarrow (A^--D^+)^{**}$$

To separate changes in absorption by $(A^--D^+)^*$ in time due to the $(A^--D^+)^*$

population growth on the one hand from absorption changes due to changes in the orientational distribution of $(A^--D^+)^*$, the absorptions of probe pulses polarized in the directions both parallel and perpendicular to the excitation pulse polarization are carried out. The sum of the absorbances for the three directions is independent of any changes in the orientational distribution of $(A^--D^+)^*$ and gives the charge-transfer population at the time of the measurement.

From Figure 5 we note that our experimental results are in good agreement with the theoretical curve which includes all transient terms. This full transient term behavior, or equivalently the time changing character of the rate "constant," is thus established. There are two parameters which can now be legitimately extracted from the fitting of the theoretical expression to the experimental curve. One is the distance of separation between D and A* at which electron transfer occurs, R, and the second is the rate constant, k, for the reaction between D and A* at an equilibrium separation of R. The values obtained are $R = 8$ Å and $k = 10^{11}$ l./(mol sec). On varying the diethylaniline concentration from 0.1 to 1 M and maintaining a constant anthracene concentration with hexane as the solvent, good agreement with theory is obtained. In all cases the same values of R and k are obtained, namely 8 Å and 10^{11} l./(mol sec), respectively. However, in systems containing 3 M diethylaniline or in the neat diethylaniline liquid the formation of the excited charge-transfer complex $(A^--D^+)^*$ follows an exponential time dependence characteristic of a bimolecular process with a single time-independent rate constant of 10^{11} sec^{-1}. At these high diethylaniline concentrations it therefore appears that translational motions as contained in the diffusion treatment are not of key importance.

To determine the nature of any geometrical effects on the dynamics of the electron transfer process the acceptor anthracene was linked to the donor dimethylaniline via three methylene groups, A - $(CH_2)_3$ -D, as was initially done by Chandross and Thomas, Mataga and coworkers and the Weller group.[12]. From our previous studies on the free anthracene and diethylaniline system the rate of formation of the charge transfer complex in the hooked together molecule should be of the order of $10"$ sec^{-1} if there are no orientational requirements for the electron transfer. As in our previous studies the anthracene part of the hooked together molecule is excited at 3472Å and the electron transfer is monitored by the appearance of the absorption at 6943Å.[13] (The absorption sepctrum of the hooked together molecule is a superposition of the A and D molecules thus indicating that there is no significant ground state interactions between A and D). In the solvent hexane we find that the electron transfer from D to A* is initially rapid, to about 40 psec, and then levels off or increases very slowly thereafter, Fig. 6. This is contrary to our previous results which we attribute to degradation of the sample used in our earlier work.[14] The number of charge transfer complexes formed in the hooked together molecule is significantly less, by roughly a factor of four, than for the unhooked system at a high concentration of D. A possible explanation of the diminished complex formation for the hooked together molecule would be the requirement for the D and A* moieties to rotate into some overlapping sandwich configuration for electron transfer to occur. If the time required for the rotation about the methylene groups into a favorable geometry is longer than the lifetime of A* then only some fraction of the A*-$(CH_2)_3$-D molecules will achieve this configuration before the A* decays back to the ground state. If rotation of the D and A* during the lifetime of the A* is important then the dynamics and number of charge transfer complexes formed should be viscosity dependent. Thus in a more viscous solvent such as hexadecane the charge transfer process should be impeded relative to hexane due the increased time required for the rotation of the D and A* groups into the favorable geometry. (At room temperature the viscosity of hexadecane (3.34 cp) is ten times greater than that of hexane (.33 cp) and their dielectric constants are about the same, $\varepsilon \sim 2$). As shown in Fig. 7 the rate of formation of the charge transfer complex and the number of charge transfer complexes formed, roughly less by a factor of four than the free system, are the same in hexadecane and hexane.

These results suggest that in the ground state roughly one fourth of the A-$(CH_2)_3$-D molecules in both hexane and hexadecane are in a configuration with the A and D moieties in a roughly sandwiched configuration as limited by the methylene groups with the remaining molecules having the A and D in the non-overlapping extended configuration. On excitation of the A portion an electron is transferred from D to A* only for those molecules in the overlapping configuration. The rapid rise in about 40 psec is attributed to the electron transfer occurring in those molecules having this favorable configuration. For those molecules in the extended configuration there is negligible electron transfer since the A* decays (5 nsec.) before the D and A* portions can rotate about the methylene groups to the configuration favorable for electron transfer. The observation of the fluorescence from the anthracene portion of the hooked together molecule in hexane (T. Okada, et al)[12] with a decay time of 5.5 nsec. is consistent with this interpretation of our picosecond measurements.

We therefore conclude from these preliminary studies that there are strong geometrical requirements for electron transfer in the non-polar (low dielectric) solvents hexane and hexadecane, and that the rotation of D and A* about the methylene group to a favorable geometry has a time constant in excess of 5nsec. in these solvents. Before more general conclusions can be drawn concerning the geometrical restrictions for the electron transfer process in non-polar solvents, it will be necessary to study the temperature and viscosity dependence over a wider range than reported in these initial studies.

Measurements of the time evolution of the electron transfer process in the

A-$(CH_2)_3$-D in the high dielectric solvent acetonitrile is shown in Fig. 8. A similar curve and amplitude is found in the solvent methanol to the longest times of the measurement in methanol, i.e. 125 psec. As with the low dielectric solvents the initial rise is rapid and peaks in about 40-50 psec after the excitation pulse. However, the amplitude of the effect is larger in the high vs. low dielectric solvents and the magnitude in the polar solvents is comparable to that of the unconnected donor and acceptor at a high donor concentration. In addition we note that there is a marked decay with time probably due to a recombination of the $D^{\cdot +}$ and $A^{\cdot -}$ moieties leading either to the ground states of A-$(CH_2)_3$-D molecule or the triplet state of the anthracene portion ^3A*-$(CH_2)_3$-D since this latter process is energetically feasible. The higher amplitude of the electron transfer process in the high dielectric solvents can be due to substantially all of the ground state A-$(CH_2)_3$-D molecules being in the overlapping sandwich configuration favorable for electron transfer on excitation of A to A*. It is not clear why a larger fraction of the ground state molecules in the polar solvents should be in this configuration relative to the low dielectric solvents. In fact from simple arguments of the dipolar interactions of the A and D moieties with the solvent molecules one would expect that more of the A-$(CH_2)_3$-D molecules would be in the extended configuration in polar vs. non-polar solvents. Another possibility for the larger effect in the polar solvents might be the relative shifting of the A*-$(CH_2)_3$-D and $A^{\cdot -}$-$(CH_2)_3$-$D^{\cdot +}$ energy surfaces in the high dielectric solvents leading to a substantial increase of the electron transfer probability in the extended configuration. Further experiments are clearly required to resolve this issue of the role of the solvent (whether an active or an inactive role) in electron transfer processes.

REFERENCES

(1) J. Franck and E. Rabinowitch, Trans. Faraday Soc., $\underline{30}$, 120 (1934)
(2) F. W. Lampe and R. M. Noyes, J. Amer. Chem. Soc., 76, 2140 (1954).
(3) T. J. Chuang, G. W. Hoffman, and K. B. Eisenthal, Chem. Phys. Lett., 25, 201 (1974).
(4) R. M. Noyes, J. Chem. Phys. 22, 1349 (1954); J. Amer. Chem. Soc., 78, 5486 (1956).
(5) G. E. Busch, R. T. Mahoney, R. I. Morse, and K. R. Wilson, J. Chem. Phys., 51, 837 (1969).
(6) M. Tamres and J. Yarwood, Spectroscopy and Structure of Molecular Complexes, edited by J. Yarwood, (Plenum Press, London, 1973).
(7) S. J. Rand and R. L. Strong, JACS $\underline{82}$, 5 (1960); T. A. Gover and G. Porter, Proc. Royal Soc. $\underline{a262}$, 476 (1961); N. Yamamoto, T. Kajikawa, H. Sato, and H. Tsubomura, JACS $\underline{91}$, 265 (1969).
(8) K. Gnadig and K. B. Eisenthal, unpublished results
(9) H. Leonhardt and A. Weller, Ber. Bunsenges, Phys. Chem., 67, 791 (1963); A. Weller, "5th Nobel Symposium," S. Claesson, Ed., Interscience, New York, N. Y., 1967, p 413; M. Ottolenghi, Accounts Chem. Res., 6, 153 (1973); T. Okada, T. Fujita, M. Kubota, S. Masaki, N. Mataga, R. Ide, Y. Sakata, and S. Misumi, Chem. Phys. Lett. 14, 563 (1972); E. A. Chandross and H. T. Thomas, ibid., 9, 393 (1971).
(10) T. J. Chuang and K. B. Eisenthal, J. Chem. Phys. $\underline{62}$, 2213 (1975); W. R. Ware and J. S. Novros, J. Phys. Chem., 70, 3246 (1966); R. M. Noyes, Progr. Reaction Kinetics, 1, 129 (1961).
(11) R. Potashnik, C. R. Goldschmidt, M. Ottolenghi, and A. Weller, J. Chem. Phys. 55, 5344 (1971).
(12) E. A. Chandross and H. T. Thomas, Chem. Phys. Lett. $\underline{9}$, 393, 397 (1971); T. Okada, T. Fujita, M. Kubota, S. Masaki, and N. Mataga, Chem. Phys. Lett. $\underline{14}$, 563 (1972); A. Weller, private communication.
(13) K. Gnadig and K. B. Eisenthal, unpublished results.
(14) T. J. Chuang, R. J. Cox, and K. B. Eisenthal, JACS $\underline{96}$, 6828 (1974).

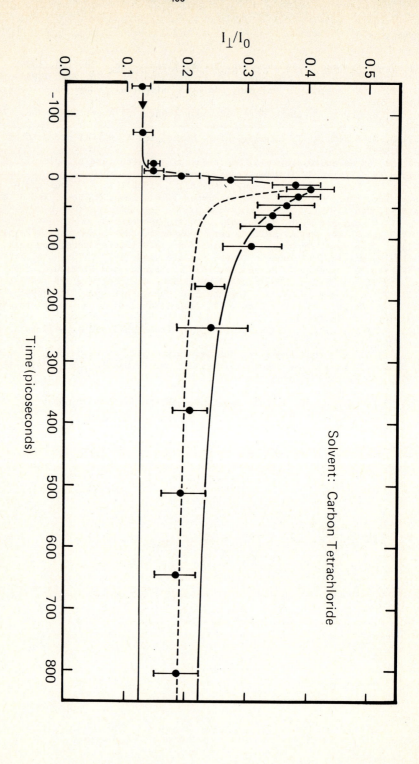

Figure 1. Probe transmission $I_\perp/I_0$ as a function of time after excitation for iodine in CCl_4. Dashed and solid curves are the decay functions calculated from Noyes random flight model.

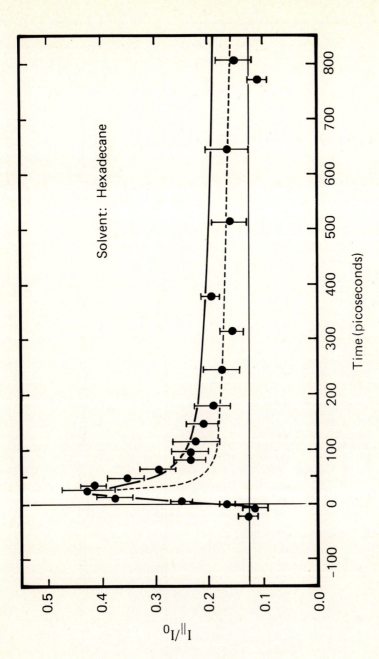

Figure 2. Probe transmission $I_\parallel/I_0$ as a function of time after excitation of iodine in hexadecane. Dashed and solid curves are the decay functions calculated from Noyes random flight model.

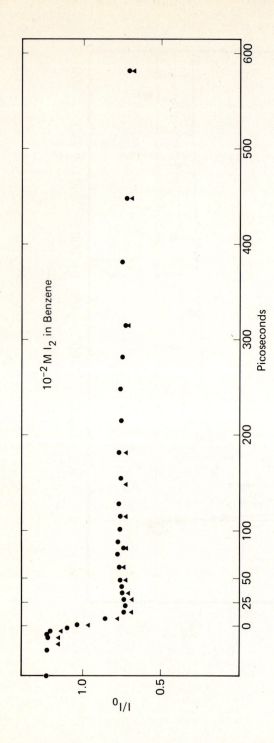

Figure 3. Probe transmission I/I_0 as a function of time on excitation of 10^{-2} M I_2 in benzene.

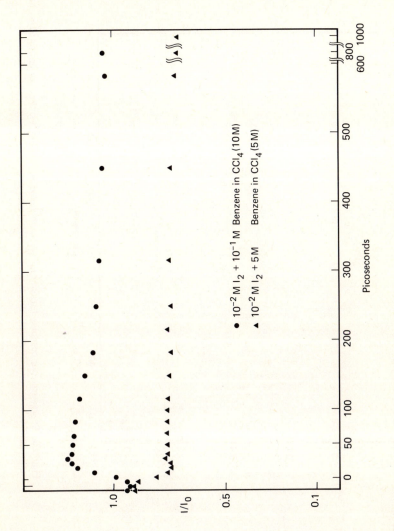

Figure 4. Probe transmission I/I_0 as a function of time on excitation of, · 10^{-2}M I_2 + 10^{-1}M benzene in CCl$_4$ and ▲10^{-2}M I_2 + 5M benzene in CCl$_4$.

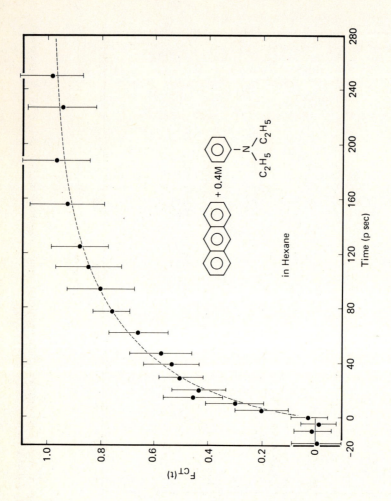

Figure 5. Charge-transfer complex formation, F_{CT} (normalized population) vs. time for anthracene plus diethylaniline in hexane. The solid line is the theoretical curve.

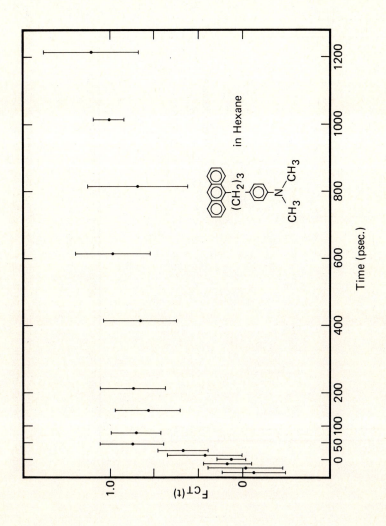

Figure 6. Normalized population, F_{CT}, of charge transfer species as a function of time on excitation of A-$(CH_2)_3$-D in hexane.

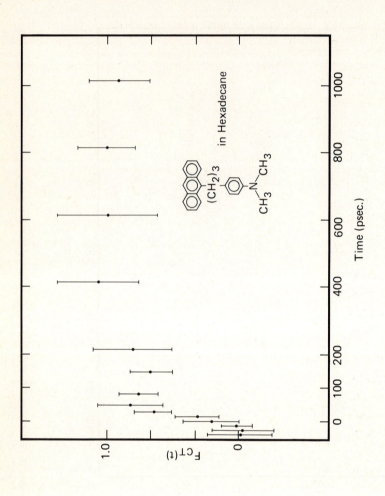

Figure 7. Normalized population, F_{CT}, of charge transfer species as a function of time on excitation of A-$(CH_2)_3$-D in hexadecane.

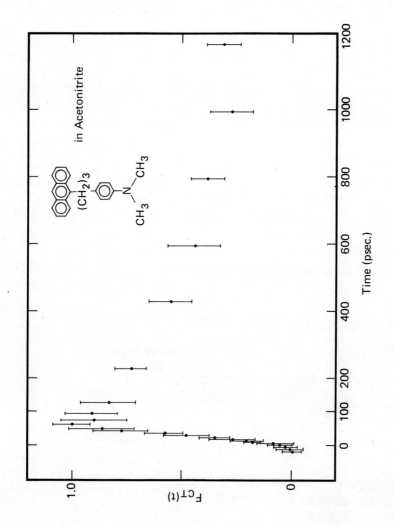

Figure 8. Normalized population, F_{CT}, of charge transfer species as a function of time on excitation of A-$(CH_2)_3$-D in acetonitrile.

TIME RESOLVED SPECTROSCOPY WITH SUB-PICOSECOND OPTICAL PULSES

C. V. Shank and E. P. Ippen
Bell Laboratories
Holmdel, New Jersey 07733

The passively modelocked CW dye laser is a unique source of optical pulses for investigating ultrafast processes on picosecond or even subpicosecond time scale. In this discussion we will describe the generation of optical pulses as short as a few tenths of a picosecond at high repetition rates. Techniques will be described for the application of these pulses to perform time resolved spectroscopy on a subpicosecond time scale.

The optical cavity we have used in passive modelocking the CW dye laser[1,2] is shown below in Figure 1.

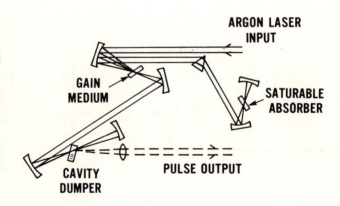

Figure 1
Modelocked CW Dye Laser Cavity

The cavity is a six mirror optical cavity with three focal spots. The center focal region contains a free-flowing stream of Rhodamine 6G in ethylene glycol. This region is optically pumped with a focused Argon laser beam creating optical gain. At one end of the cavity an acoustic cell is used to select single output pulses. At the other end of the cavity the focal region contains a second free-flowing stream with the saturable absorber. The saturable absorber consists of a mixture of two dyes. The first saturable absorber, DODCI, with a lifetime of 1.2ns, is normally used in passively modelocking the dye laser. The second absorber is malachite green with a lifetime in the picosecond range. The slow absorber provides a means of modelocking the laser while the fast absorber acts to stabilize the short pulse operation above threshold. The fast absorber alone will not modelock the dye laser. The shortest pulses are obtained near $\lambda = 6150 \overset{\circ}{A}$ which is the peak of the malachite green absorption.

To determine the optical pulse duration we measure the pulse autocorrelation function by generating second harmonic in KDP. A convenient way of doing this is shown below in Figure 2.

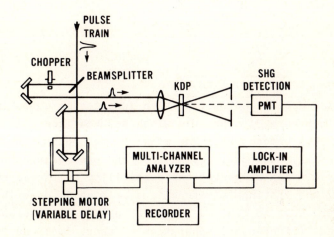

Figure 2
Experimental Apparatus for Autocorrelation Measurement

The beam from the laser is split into two parallel beams having a variable relative time delay. These beams are then focused onto the KDP crystal and second harmonic is generated at an angle bisecting the two beams. This technique provides a measurement of the autocorrelation function with no background.

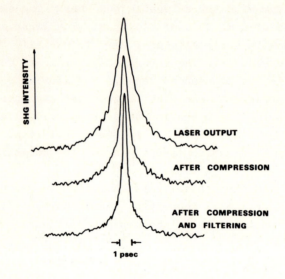

Figure 3
Measured Pulse Autocorrelation Functions

A typical pulse optical autocorrelation function is shown in the upper trace in Figure 3. The pulsewidth deduced from this trace is approximately 0.9 picosecond. The wings of the pulse fit very closely to an exponential. The oscillating spectrum was measured to be approximately 20Å and suggested the possibility of excess bandwidth.
Using the grating-pair technique originally described by Treacy[3] we were able to achieve direct temporal compression of the pulse. A slant (light path) separation of about 10cm was found to be optimum for compression with a pair of 1800 line/mm gratings. The autocorrelation of the compressed pulses is shown in the middle trace in Figure 3. Compression in the center of the trace is better than a factor of two. A smaller compression ratio in the wings indicates the frequency sweep is not linear. The pulses were further shortened by spectral filtering and selecting only the linear portion of the frequency sweep. The result is shown in lower trace of Figure 3 where

the low frequency components of the spectrum were selected. The resulting trace has a width (FWHM) of .5 psec and implies a pulse width of about .3 psec.

The short compressed pulses have been correlated with the original uncompressed pulses from the modelocked laser in KDP to measure the pulse shape. By filtering the generated second harmonic to detect particular sum frequencies, dynamic spectra were also obtained.

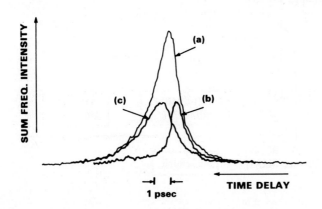

Figure 4
Dynamic Spectragram of a Modelocked Pulse (a) All Frequencies, (b) low Frequency Components of the Pulse (c) High Frequency Components of the Pulse

The predicted asymmetry[4] of the laser output pulses is clearly demonstrated in Figure 4(a). These pulses are characterized in time by a sharply rising leading edge and a longer trailing edge. The dynamic spectral behavior is illustrated by traces 4(b) and 4(c). Trace (b) is obtained when only the low frequency components of the generated UV are detected and for trace (c) only the high frequency components are monitored. These results show both the frequency sweep and its nonlinearity. The pulses are chirped "up," sweeping from red to blue in time, and the rate of sweep decreases significantly in the more slowly varying trailing edge.

The experimental set up shown in Figure 2 is easily adapted as shown in Figure 5 to perform time resolved spectroscopy. The grating pair and birefringent filter are used to provide the shortest compressed pulse. A simple pump probe type experiment is possible with the relative delay provided.

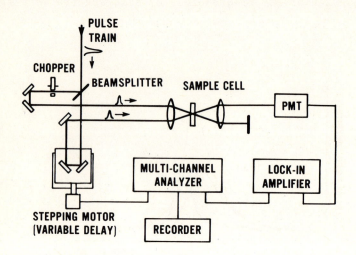

Figure 5
Experimental Arrangement for Time Resolved Spectroscopy Measurements

We have investigated the absorption recovery of the triphenylmethane dye malachite green. The result for malachite green in methanol is shown in Figure 6.

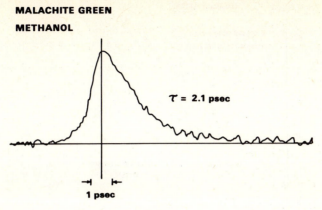

Figure 6
Absorption Recovery of Malachite Green

The recovery is very close to exponential in shape with a time constant of 2.1 psec. This is one of the shortest absorption recovery times ever observed. Even so, from the quantum efficiency measurements of Förster et al[5], one would predict that the lifetime of the excited state would be at least an order of magnitude shorter. The longer time measured here is for complete recovery of the ground state indicating that some additional process takes place after relaxation of the excited state. With higher solvent viscocities the absorption recovery time was observed to increase as did the quantum efficiency.

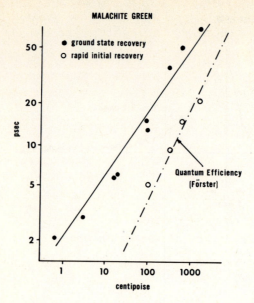

Figure 7
Measured Recovery Time Versus Viscocity

The experimental results (the solid circles) and an excited state lifetime calculated from the quantum efficiency (dashed line) are plotted in Figure 7.

At higher solvent viscocities we noted that the decay began to appear more like a two component exponential decay. A semilog plot of the results for glycerol is shown in Figure 8.

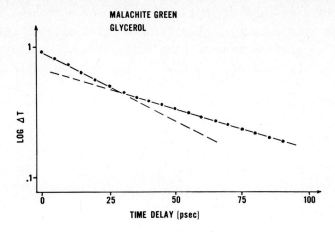

Figure 8
Absorption Recovery of Malachite Green in Glycerol

These results suggest a simple model which is consistent with the experimental observations.

TWO COMPONENT ABSORPTION RECOVERY

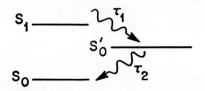

Figure 9
Simple Model Illustrating a two Component Absorption Recovery

In Figure 9 we have a simple model whereby molecules are excited from the ground state to the first excited singlet level, undergo a rapid relaxation to some high lying vibrational level of the ground state, So, in a time τ_1 and then relax to the ground state in a time τ_2.

Under these conditions the decay of the absorption to a short pulse excitation is given by:

$$\Delta\alpha \sim \frac{1}{1-\tau_1/\tau_2}\left\{e^{-t/\tau_2}+\left(1-\frac{2\tau_1}{\tau_2}\right)e^{-t/\tau_1}\right\}$$

Fitting the data to this expression we were able to determine τ_1 and τ_2. In Figure 7 we have plotted τ_1 (the open circles) and τ_2 (the closed circles). Note that the τ_1 points fall very close to the excited state lifetime predicted by Förster's quantum efficiency measurements. This time is of course the lifetime of the S_1 state in our simple model.

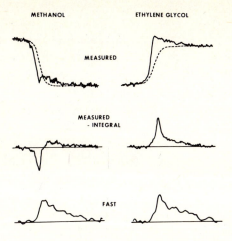

Figure 10
Absorption Recovery in Chloroaluminum Pthalocyanine in
Methanol and Ethylene Glycol

The thermalization of vibrational levels in organic molecules can also be investigated with an absorption recovery experiment. In Figure 10 we show some experimental measurements for chloroaluminum pthalocyanine in two solvents. In the solvent methanol the dye acts as a saturable absorber. In ethylene glycol, a slight spectral shift due to the solvent causes excited state absorption to be the dominant effect. Thus the phase of the pump induced absorption changes are opposite for the two cases. The measured curves are actually a sum of two responses, a slow response corresponding to recovery of the electronic transition and fast response determined by the vibrational relaxation. Since the electronic recovery is so long (1nsec.), this contribution amounts to an integral of the pulse intensity. The pulse integrating effect can be directly computed from the pulse autocorrelation measurements. The result of subtracting the integral response from the measured response are shown in the two middle curves in Figure 10. An additional unavoidable coherent beam coupling artifact must be subtracted from these curves to produce the "fast" response shown in the lower curves. For the methanol solvent case we measure a 5ps exponential decay constant and for ethylene glycol a 6ps decay time. These times give a measure of the vibrational relaxation in chloroaluminum pthalocyanine but do not

in themselves provide details of the relaxation process. The
measurement includes a sum of contributions from ground and excited
states.

      The experimental system we describe here is very versatile
and has provided a means of measuring a variety of phenomena. With a
small adaption we were able to induce a transient birefringence and
measure a relaxation time of 2.1 psec for CS_2.[6] By inducing transient
dichroism we were able to measure molecular reorientation in organic
dye molecules.[7]

REFERENCES

1. C. V. Shank and E. P. Ippen, Appl. Phys. Lett. $\underline{24}$, 373 (1974).
2. E. P. Ippen, C. V. Shank and A. Dienes, Appl. Phys. Lett. $\underline{21}$, 348 (1972).
3. E. B. Treacy, IEEE J. Quant. Electron., $\underline{QE-5}$, 454 (1969).
4. H. A. Haus, C. V. Shank and E. P. Ippen, Opt. Commun. (1975).
5. T. H. Förster and G. Hoffman, Z. Physik. Chem. NF75 (1971) 63.
6. E. P. Ippen and C. V. Shank, Appl. Phys. Lett. $\underline{26}$, 92 (1975).
7. C. V. Shank and E. P. Ippen, Appl. Phys. Lett. $\underline{26}$, 62 (1975).

QUANTUM ELECTRODYNAMIC CALCULATION OF QUANTUM BEATS IN A SPONTANEOUSLY RADIATING THREE LEVEL SYSTEM[*]

A. Schenzle[†] and Richard G. Brewer
IBM Research Laboratory
San Jose, California 95193

ABSTRACT: The coherence properties of a spontaneously radiating three-level atomic system are investigated in a first principle calculation that utilizes either coherent or incoherent preparation. Quantum beat phenomena are explained for the first time in a manner that reveals the statistical nature of the preparative step as well as a possible radiative correction to the beat frequency.

The spontaneous emission of light in a suitably prepared three level quantum system can give rise to the well-known "quantum beat" effect.[1] The usual atomic or molecular level configuration, shown in Fig. 1(a), involves two neighboring upper states that can spontaneously radiate to a common lower level. It is now realized that the nature of the emission depends in a critical way on how the preparation is performed. For example, in the earlier work a pulsed incoherent optical field that simultaneously excites both transitions, places the upper two levels in quantum mechanical superposition. The spontaneous emission that follows occurs isotropically, it exhibits a beat or modulation, the quantum beat, at a frequency corresponding to the splitting of the upper two levels, and the intensity depends linearly on the molecular number density N. Equivalent preparations that produce the same beat phenomenon have been achieved with a pulsed laser source[2] and also in beam foil experiments.[3]

A different manifestation of the quantum beat effect has arisen in recent photon echo measurements[4,5] where laser radiation places all three levels in superposition. For this case, the emission is directional and intense, the intensity varies as N^2, and the two closely spaced levels can appear either in the upper (Fig. 1a) or in the lower state (Fig. 1b).

For a single atom, neoclassical theory (NCT) predicts beats for either level configuration. On the other hand, quantum electrodynamic (QED) arguments yield a beat only for the Fig. 1(a) level structure.[6,7] Regarding Fig. 1(b), Breit[8] argued long ago using QED that an excited atom in level 3 can spontaneously emit one photon by passing either to level 1 or 2, but not to both states; of course, this situation excludes the possibility of a quantum beat. For the case of many atoms, each in three-level superposition, both NCT and QED predict beats for either level structure.[6,7]

Our purpose is not to compare NCT with QED, but to present a more unified and quantitative treatment of the general problem of spontaneous emission of a three-level system subject to different types of preparation. We present a QED treatment which reveals new features of quantum beats. Our results appear to describe observations to date and includes for the first time the statistical properties of the optical source which prepares the sample. A possible radative correction to the beat frequency is also indicated in the solution.

Spontaneous emission of an ensemble of atoms can be described by two stages: (1) an external perturbation that prepares atoms in an initial state and (2) the spontaneous emission that follows. We characterize the exciting field by F and $F^\dagger$ and the emitted light field of wavelength λ by its creation and annihilation operators b_λ, $b_\lambda^\dagger$. The correlation of the emitted field has to be determined by the characteristics of the perturbation field. If we assume the perturbing force to be weak, we can show that

$$<b_\lambda^\dagger(t+\tau)b_{\lambda'}(t)> = \sum_{\nu\nu'}\int_0^T\int_0^T G_{\lambda\lambda'}^{\nu\nu'}(t,\tau,t',t'')<F_\nu^\dagger(t')F_{\nu'}(t'')>dt'dt'' \qquad (1)$$

where now all the basic physics is buried in the integral kernel G, and T is the finite time of interaction with the external field. To evaluate the time evolution of G, it is necessary to account for the atomic degrees of freedom. The atoms can be characterized by a number of atomic transitions, described by the dipole operators

$S_{i\ell}^\dagger$ where i labels the transition and ℓ the individual atom. In the Heisenberg picture, the dipoles, being driven by an external force, lead to the following equation of motion

$$\frac{d}{dt} S_{j\ell}^\dagger(t) = i\omega_{j\ell} S_{j\ell}^\dagger(t) + \sum_\nu F_\nu^\dagger(t) K_\nu \qquad (2)$$

which formally integrated leads to the dipole-dipole correlation, when the external field terminates, namely,

$$\langle S_{j\ell}^\dagger(T) S_{i\ell'}^-(T)\rangle = \sum_{\nu\nu'} \int_0^T \int_0^T K_\nu K_{\nu'} e^{i\omega_{j\ell}(T-t')-i\omega_{i\ell'}(T-t'')} \langle F_\nu^\dagger(t') F_{\nu'}(t'')\rangle dt' dt'' \qquad (3)$$

If $F_\nu(t)$ represents a stationary field, we have to assume that its single time average vanishes, i.e. $\langle F_\nu(t)\rangle = 0$ for all t, which might be interpreted to mean that $F_\nu(t)$ experiences random fluctuations of its phase. Therefore, such an external field does not introduce a dipole moment in the classical sense, since $\langle S_{j\ell}^\dagger(t)\rangle = 0$. It is remarkable that a randomly fluctuating light field can introduce coherence in an atomic three-level system eventhough the expectation value of the two dipole moments vanish. The reason is that the two transitions being driven by the same field experience random phases whereas their relative phase remains fixed. This corresponds to a nonvanishing value of $\langle S_1^\dagger(t) S_2^-(t')\rangle$. On the other hand, the oscillating atomic dipoles are the source of an electromagnetic field, emitted into the vacuum

$$\frac{d}{dt} b_\lambda^\dagger(t) = i\omega_\lambda b_\lambda^\dagger + i\sum_{j\ell} g_{j\lambda} u_\lambda(\ell) S_{j\ell}^\dagger(t) \qquad (4)$$

where g is the dipole matrix element and u contains the spatial dependence. After integration, we can relate the correlations of the emitted light field to those of the dipoles

$$\langle b_\lambda^\dagger(t_1) b_{\lambda'}(t_2)\rangle = \sum_{\ell\ell' ij} \int_0^{t_1}\int_0^{t_2} g_{i\lambda} g_{j\lambda'}^* u_\lambda(\ell) u_{\lambda'}^*(\ell')$$

$$\times e^{i\omega_\lambda(t_1-t')-i\omega_{\lambda'}(t_2-t'')} \langle S_{i\ell}^\dagger(t') S_{j\ell'}^-(t'')\rangle dt' dt'' . \qquad (5)$$

In order to combine expressions ( ) and ( ) to find the integral kernel G explicitly, we have to relate the dipole correlation at an arbitrary pair of times (t',t") to their initial value at $t = T$.

We describe the atom dynamics in a master equation approach[9] by defining a reduced density matrix

$$W(t) = tr_{light} \rho(t) \tag{6}$$

where $\rho(t)$ is the total density matrix in the product space (field $\otimes$ atom). To be more specific, we consider the three-level systems of Fig. 1 which we will refer to as cases A and B. Here, dipole transitions are allowed between the levels $1 \leftrightarrow 3$ and $2 \leftrightarrow 3$. The total Hamiltonian of these structures reads

$$H = H_A + H_F + H_I, \quad H_A = \sum_{\ell,j} \hbar\omega_{j\ell} a^\dagger_{j\ell} a_{j\ell}, \quad H_F = \sum_\lambda \hbar\omega_\lambda b^\dagger_\lambda b_\lambda$$

$$H_I = \sum_{\ell,\lambda} (g_{13} s^\dagger_{1\ell} + g_{23} s^\dagger_{2\ell}) u_\lambda(\ell) b^\dagger_\lambda + h.c. \tag{7}$$

where

$$s^\dagger_{1\ell} = a^\dagger_{1\ell} a_{3\ell} \text{ in A and } s^\dagger_{1\ell} = a^\dagger_{3\ell} a_{1\ell} \text{ in B.}$$

With these definitions, we can write the master equation in Born-Markoff approximation

$$\frac{dW}{dt} = -i\sum(\omega_{1\ell}-\omega_{3\ell})[a^\dagger_{1\ell}a_{1\ell},W] - i\sum(\omega_{2\ell}-\omega_{3\ell})[a^\dagger_{2\ell}a_{2\ell},W]$$

$$-i\sum_{\ell\ell'ij} \Delta^{ij}_{\ell\ell'}[s^\dagger_{i\ell},s^-_{j\ell'},W] + \sum_{\ell\ell'ij} K^{ij}_{\ell\ell'}([s^\dagger_{i\ell}W,s^-_{j\ell'}]+[s^\dagger_{i\ell},Ws^-_{j\ell'}]) \tag{8}$$

with $K^{ij}_{\ell\ell'} = \pi\sum_\lambda g_{13\lambda}g_{3j\lambda}u_\lambda(\ell)u^*_\lambda(\ell')\delta(\omega_{j\ell}-\omega_\lambda)$ and $\Delta^{ij}_{\ell\ell'} = P\int d^3j D(j)g_{13\lambda}g_{3j\lambda}u_\lambda(\ell)u^*_\lambda(\ell') \times (\omega_\lambda-\omega_{j\ell})^{-1}$ where D is the field density of states. Comparing Eq. (9) with the well known master equation for the two-level system,[9] we identify familiar terms for the two individual transitions $1 \leftrightarrow 3$ and $2 \leftrightarrow 3$. In addition, explicit interaction terms appear. The interaction terms result from the fact that both transitions are coupled to the same ensemble of light modes.

With Eq. (8), we are able to calculate the time evolution of ensemble averages in single particle approximation. For example, in case A we find for the single particle averages

$$\frac{d}{dt}\langle S_{1\ell}^{\dagger}(t)\rangle = (i\omega_{1\ell}-K^{11})\langle S_{1\ell}^{\dagger}\rangle - (i\Delta^{12}+K^{21})\langle S_{2\ell}^{\dagger}(t)\rangle , \tag{9}$$

and for two-particle averages

$$\frac{d}{dt}\langle S_{1\ell}^{\dagger}(t)S_{2\ell'}^{-}(t)\rangle = [i(\omega_{1\ell}-\omega_{2\ell'})-K^{11}-K^{22}]\langle S_{1\ell}^{\dagger}S_{2\ell'}^{-}\rangle$$

$$- (K^{12}+i\Delta^{12})\langle S_{1\ell}^{\dagger}S_{1\ell'}^{-}\rangle - (K^{21}-i\Delta^{21})\langle S_{2\ell}^{\dagger}S_{2\ell'}^{-}\rangle . \tag{10}$$

The solution of Eq. (10) has the eigenvalues

$$\Omega_{1,2} = -(K^{11}+K^{22})$$

$$\Omega_{3,4} = -(K^{11}+K^{22}) \pm i[(\omega_1-\omega_2)^2-4(K^{12})^2+4(\Delta^{12})^2]^{\frac{1}{2}}$$

In the Markoff approximation, the two-time correlation needed to evaluate (5) follows from (9) and (10). The same calculation for case B leads to

$$\frac{d}{dt}\langle S_{1\ell}^{\dagger}(t)\rangle = (i\omega_{1\ell}-K^{11}-K^{22})\langle S_{1\ell}^{\dagger}(t)\rangle , \tag{11}$$

$$\frac{d}{dt}\langle S_{1\ell}^{\dagger}(t)S_{2\ell'}^{-}(t)\rangle = [i(\omega_{1\ell}-\omega_{2\ell'})-2K^{11}-2K^{22}]\langle S_{1\ell}^{\dagger}(t)S_{2\ell'}^{-}(t)\rangle . \tag{12}$$

The existence of a nonvanishing cross correlation $\langle S_{1\ell}^{\dagger}(t)S_{2\ell'}^{-}(t)\rangle$ produces an intensity modulation in the emission field

$$\langle b_{\lambda}^{\dagger}(t)b_{\lambda}(t)\rangle = \sum_{ij\ell\ell'} Q_{ij} e^{\alpha_{ij}(t-T)} \langle S_{i\ell}^{\dagger}(T)S_{j\ell'}^{-}(T)\rangle \tag{13}$$

where the quantum beat frequency is now given by the imaginary part of

$$\alpha_{12} = -(K^{11}+K^{22}) + i\sqrt{(\omega_1-\omega_2)^2-4(K^{12})^2+4(\Delta^{12})^2} \quad \text{for case A,} \tag{14a}$$

$$\alpha_{12} = -2(K^{11}+K^{22})+i(\omega_1-\omega_2) \quad \text{for case B.} \tag{14b}$$

Here, it is understood that ω_i includes the traditional two-level Lamb shift calculation. We note that the level structure, case A, exhibits a frequency shift in the term $[4(\Delta^{12})^2-4(K^{12})^2]$ that also contributes to the Lamb shift and has not been introduced in quantum beat calculations previously. It corresponds to the process where an atom emits a photon in going from state 1→3 and absorbs another in going from state 3→2 (or vice versa). However, this shift is not expected to be significant for most level structures because the angular momentum quantum numbers for states 1 and 2 must be the same. For case B, an atom initially in state 3 cannot exhibit an emission-absorption process involving all three levels, and therefore, the beat frequency $\omega_1 - \omega_2$ will be unshifted as (14b) indicates.

Quantum beats will not be observed for all types of preparation, however. The single atom diagonal terms ($\ell = \ell'$) of Eq. (13), for example, are determined by the two-time correlation function

$$\langle S^\dagger_{1\ell}(T) S^-_{2\ell}(t) \rangle \propto \begin{cases} \langle a^\dagger_{1\ell} a_{3\ell} a^\dagger_{3\ell} a_{2\ell} \rangle = \langle a^\dagger_1 a_2 \rangle & \text{for case A}, \quad (15a) \\ \langle a^\dagger_{3\ell} a_{1\ell} a^\dagger_{2\ell} a_{3\ell} \rangle = 0 & \text{for case B}. \quad (15b) \end{cases}$$

Equation (15b) agrees with earlier discussions[6-8] that a beat effect is not predicted in a single atom for case B. This is a pure quantum effect due to the Fermion nature of electrons. For case A, a beat will be observed where we see from (15a) that the preparation requires that only the upper levels 1 and 2 be in superposition. This is the usual experimental situation where pulsed incoherent light excites both transitions during preparation, thereby correlating the upper two levels but not the lower one. In addition, from (13) it follows that the emitted light intensity will vary linearly with the number of molecules N since there are as many diagonal terms. Furthermore, since these terms are $\vec{k}$ independent, the radiation will be isotropic.

Consider now the off-diagonal terms ($\ell \ne \ell'$) of (13). These become important when a coherent optical source, a laser beam, places all three levels in superposition. Both case A and B yield beats in this circumstance. The beat frequency is still given by (14) but

$$\langle S^\dagger_{1\ell}(T) S^-_{2\ell'}(t) \rangle = \langle (a^\dagger_3 a_1)_\ell (a^\dagger_2 a_3)_{\ell'} \rangle \text{ for case A and B}. \quad (16)$$

Since (13) contains N^2 off-diagonal terms of the form (16), the radiated intensity will vary as N^2. When the system is prepared by a laser beam, the emission will be in the same direction. Consideration of (13) reveals factors of the form $e^{i\vec{k}\cdot\vec{z}}$ which when summed over the sample length yield the familiar antenna pattern, an intense forward lobe.

We see that the off-diagonal terms of (13) correspond to a two-atom correlation, namely, one transition of one atom being correlated with a second transition of another atom - the two coherent radiation fields giving rise to a beat. On the other hand, the diagonal elements of (13) correspond to correlation of two levels within each atom.

It seems clear that inhomogeneous dephasing arising from the Doppler effect will quench the emission of coherently prepared samples in a time the order of T_2^*. Therefore, for times $t > T_2^*$, beats from two-atom correlations disappear whereas the one atom beats of case A level structure survive. On the other hand, since the inhomogeneous dephasing is reversible, beats in the two-atom case can be recovered in a photon echo experiment.[4,5]

We have derived the coherence properties of a spontaneously radiating three-level system in a first principle calculation and have thereby avoided the previous wave function approach[6,7] which cannot take into account statistical properties in general.

The comments of Professors S. Stenholm and P.R. Berman are appreciated.

REFERENCES

* Work supported in part by the U. S. Office of Naval Research under Contract No. N00014-72-C-0153.

† On leave from the Institute of Theoretical Physics, University of Stuttgart, Stuttgart, Germany.

1. J. N. Dodd, R. D. Kaul and D. M. Warrington, Proc. Phys. Soc. (London) $\underline{84}$, 176 (1964); J. N. Dodd, W. J. Sandle, and D. Zissermann, Proc. Phys. Soc. (London) $\underline{92}$, 497 (1967); A. Corney and G. W. Series, Phys. Rev. $\underline{121}$, 508 (1961).
2. S. Haroche, J. A. Paisner and A. L. Schawlow, Phys. Rev. Letters $\underline{30}$, 948 (1973).
3. K. Tillman, H. J. Andrä, and W. Wittman, Phys. Rev. Letters $\underline{30}$, 155 (1973) and references therein.
4. L. Q. Lambert, A. Compaan and I. B. Abella, Phys. Lett. $\underline{30A}$, 153 (1969); L. Q. Lambert, Phys. Rev. $\underline{B7}$, 1834 (1973).
5. R. L. Shoemaker and F. A. Hopf, Phys. Rev. Letters, $\underline{33}$, 1527 (1974).
6. W. W. Chow, M. O. Scully and J. O. Stoner, Jr., Phys. Rev. $\underline{A11}$, 1380 (1975).
7. R. M. Herman, H. Grotch, R. Kornblith and J. H. Eberly, Phys. Rev. $\underline{A11}$, 1389 (1975).
8. G. Breit, Revs. Modern Physics $\underline{5}$, 91 (1933).
9. H. Haken, <u>Handbuch der Physik</u>, (Springer-Verlag, Berlin 1970), Vol. XXV/2c; G. S. Agarwal, Proceedings of the Third Rochester Conference on Coherence and Quantum Optics (Plenum Press, N.Y., 1973), p. 157.

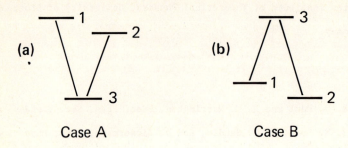

FIG. 1. Two possible level structures that give rise to a quantum beat effect during spontaneous one-photon emission.

COLLISION INDUCED OPTICAL DOUBLE RESONANCE

Stig Stenholm
Department of Theoretical Physics
University of Helsinki
SF-00170 Helsinki 17, Finland

1. Introduction

Nonlinear spectroscopy using lasers has already proved itself a powerful tool for the investigations of atomic and molecular spectra. In gaseous samples at low pressures the constituents are essentially independent of each other and the free particle theory applies. When the pressure is increased, the particles start to collide and the idealized picture breaks down.

In most works the effects of collisions are a nuisance, that one tries to avoid as much as possible. Recently it has, however, been realized that collisions give rise to entirely new phenomena, which provide a qualitative manifestation of their presence in contrast to their quantitative perturbation of other effects. These new phenomena can be used to investigate the collision process itself and it is expected that they will be able to provide much more detailed information than earlier methods. Some progress has been made, and it is my conviction that more will be forthcoming.

In this talk I will try to summarize the effects of collisions on atoms (and mutatis mutandis on molecules) in terms of a density matrix description. This is found in Sec. 2. In Sec. 3 I will describe some experiments, which have confirmed the existence of population transfer in molecular collisions, and finally in Sec. 4 I will briefly discuss the possibility to observe the transfer of coherence in a collision.

2. Collision effects on the density matrix

As most spectroscopic measurements are best described by the use of a density matrix, it is convenient to give the effects of collisions in terms of its change in one such event. This assumes that the collisions are of brief duration and that their separation is sufficiently long, i.e. we consider the impact-approximation limit.

Because of the linearity of the quantum-mechanical time-evolution equation, the change in the density matrix will be a linear function of its elements before the collision

$$\langle\mu|\frac{d\rho}{dt}|\mu'\rangle = \sum_{\nu\nu'} \Gamma_{\nu\nu'} \left[\Pi_{\mu\mu'}{}^{\nu\nu'} - \delta(\nu,\mu)\delta(\nu',\mu') \right] \langle\nu|\rho|\nu'\rangle . \quad (1)$$

(Only the collision-dependent part of the time evolution is written out.) The quantum numbers ν,μ include the labels for the atomic states and the translational degrees of freedom. Such equations have been derived by Sobel'man et al. [1] and Berman [2]. The function $\Gamma_{\nu\nu'}$ is the collision-induced decay rate of $\langle\nu|\rho|\nu'\rangle$, and the scattering function Π is normalized

$$\sum_{\mu\mu'} \Pi_{\mu\mu'}{}^{\nu\nu'} = 1 . \quad (2)$$

The functions of (1) can be expressed in terms of the S-matrix (cf. Ref. [1]) in the form

$$\Gamma_{\nu\nu'} \Pi_{\mu\mu'}{}^{\nu\nu'} = \overline{S_{\mu\nu} S^+_{\nu'\mu'}} , \quad (3)$$

where the bar denotes an average over all possible collisions.

It has been pointed out by Berman and Lamb [3] and Berman [4] that for different scattering cross-sections in the different atomic states, one cannot assign a classical trajectory to the atomic off-diagonal density matrix elements. The diagonal elements, on the other hand, can be labelled by definite values for the velocity parameters. With state-independent scattering this holds for all elements. It is then possible to describe the velocity classically and Eq. (1) becomes

$$\frac{d}{dt} \rho_{\alpha\beta} = \sum_{\alpha'\beta'} \int \Gamma_{\alpha'\beta'}(v') \Pi_{\alpha\beta}{}^{\alpha'\beta'}(v,v') \rho_{\alpha'\beta'}(v')dv' - \Gamma_{\alpha\beta}(v)\rho_{\alpha\beta}(v) . \quad (4)$$

Here the indices α,β refer to atomic states only. This approach to the theory of collisions is discussed in detail in Ref. [5].

When we consider the collisional coupling of matrix elements to themselves, $\Pi_{\alpha\beta}{}^{\alpha\beta}$, we obtain the ordinary shifts and broadenings of spectral lines, caused by phase perturbation during the collision.

As an integral-kernel, the operators $\Pi(v,v')$ smear the velocity over an interval Δu. If this is of the order of the natural line width γ, it affects the results only little, but if it is of the order of the Doppler width ku, it essentially destroys all memory of the velocity before the collision.

In spectroscopy and laser physics one has mostly been interested in the collisional effects on one transition a-b only. Here we want to discuss the effects of collisions coupling two different transitions a-b and c-d, see Figure 1. Here we have two different consequencies.

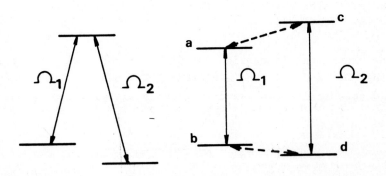

Figure 1

The elements $\Pi_{aa}{}^{cc}$ and $\Pi_{bb}{}^{dd}$ transfer populations between the levels. If the accompanying velocity smearing Δu is small, collisions can lead to velocity selective resonances, which have been reported by Shoemaker et al. [6] . These experiments will be discussed in the next section.

The elements $\Pi_{ab}{}^{cd}$ transfer coherens between the transitions. Such collisional transfer of coherence has been observed in the radio frequency range (see e.g. Series [7]). It is probable that the effect exists at optical frequencies too, and in the final section we will discuss the possibilities to design an experiment which could detect its presence.

3. Transfer of population: collision-induced double resonance

In the molecule $^{13}CH_3F$ the ν_3-band line $(J,K)=(4,3)\rightarrow(5,3)$ can be tuned to resonance with the 10 μm line of a CO_2 laser by a static Stark field ϵ. In this field the levels are split into $(2J+1)$ components, with different M quantum numbers. The splitting between the sublevels of the upper state is denoted by

$$(E(M+1)-E(M))_u = \Delta_u \epsilon \qquad (5)$$

and for the lower levels

$$(E(M+1)-E(M))_l = \Delta_l \epsilon . \qquad (6)$$

Measuring the tuning rates Δ_u, Δ_l one can determine the dipole moment of the levels (see [6]).

The experiment is carried out with two parallel laser beams of frequencies Ω_1 and Ω_2 and amplitudes E_1 and E_2 respectively. Each field selects its own velocity group satisfying the resonance conditions

$$\Omega_1 = \omega_{ab} + kv_1$$
$$\Omega_2 = \omega_{cd} + kv_2 . \qquad (7)$$

When we require these two velocity groups to coincide we obtain the resonance condition for collision-induced double resonance

$$\Omega_1 - \Omega_2 = \omega_{ab} - \omega_{cd} . \qquad (8)$$

The frequency difference $\Omega_1-\Omega_2$ was locked in the experiments and could be determined with very high accuracy. For details of the experiment see Ref. [6].

The observed double resonance can be explained by a straightforward transfer of population between the levels $|J,K,M\rangle$ with different values of M. It is shown by Schmidt et al. [8], that phase-changing collisions have a negligible influence on this molecule and that the average velocity jump $\Delta u \sim 1 \text{m/s}$, which corresponds to a velocity smearing of $k\Delta u \sim 10^5$ Hz. This causes no appreciable broadening as it is of the order of the experimental line width, which is determined mainly by the molecular transit time across the laser beam.

The transfer of population take the forms

$$\rho_{aa} = \kappa_{ac} \gamma \rho_{cc} + \ldots \tag{9}$$

and correspondingly for levels b-d. The resulting equations have been solved in Ref. [6] and the observed quantity becomes after velocity and Doppler averaging

$$S \propto \kappa \frac{N E_1^2 E_2^2}{\{[(\omega_{ab}-\omega_{cd})-(\Omega_1-\Omega_2)]^2 + 4\gamma^2\}}. \tag{10}$$

The result is similar to the ordinary three-level double resonance (described in Ref. [9]) but is proportional to κ, which denotes the ratio of M-changing line broadening processes to the total line width. The resonance condition agrees with the one derived in Eq. (8), and the line width is 2γ, i.e. twice that of the three-level resonance. This derives from the fact that population transfer, even without a velocity smearing is essentially an incoherent process and the widths of the two transitions involved add directly.

An example of the observed double resonances is presented in Table I. All the observed lines can be described in accordance with Eqs. (5), (6) and (8) in the form

$$\Omega_1 - \Omega_2 = (m \Delta_1 - n \Delta_u)\varepsilon, \tag{11}$$

where m and n are integers.

TABLE I

Line	Tuning rate (kHz V^{-1}cm)	Line Center (V cm^{-1}) Observed	Line Center (V cm^{-1}) Predicted	Line width (kHz, FWHM)
a	370.9 ($4\Delta_1 - 2\Delta_u$)	80.90	80.78	960
b	326.2 ($3\Delta_1 - \Delta_u$)	92.00	91.85	640
c	281.99 ($2\Delta_1$)	106.42	106.42	410
d	237.00 ($\Delta_1 + \Delta_u$)	126.58	126.51	810
e	192.48 ($2\Delta_u$)	155.90	155.90	400
f	147.8 ($3\Delta_u - \Delta_1$)	203.08	203.17	750

The lines c and e correspond to three-level resonances and have been used to fix the parameters Δ_u and Δ_1. It is seen that their width is roughly half the width of the collision-induced resonances in agreement with the theoretical result (10).

The collisions in CH_3F can easily change the quantum number M because the permanent dipole moments of these molecules have a long range anisotropic interaction which readily dips the angular momentum vectors without much affecting the other degrees of freedom of the molecule. The probability of this process can be determined from the ratio of the three-level resonances to the collision-induced ones, and the cross section is found to be $\sigma \sim 10^2$ Å^2.

The same resonances have also been observed in Lamb-dip experiments Here only one laser beam is used, but it is reflected back into the sample The two counter propagating waves interact with the velocity groups

$$\Omega_1 = \omega_{ab} + kv_1$$
$$\Omega_1 = \omega_{cd} - kv_2 \quad . \qquad (12)$$

Setting these velocities equal we obtain the resonance condition

$$\Omega_1 = \frac{1}{2}(\omega_{ab} + \omega_{cd}) \quad . \qquad (13)$$

Because the ordinary Lamb dip resonances are found at $\Omega_1 = \omega_{\alpha\beta}$ ($\alpha\beta$ = ab or cd) we observe the collision-induced dips midway between the ordinary ones. Here the two oppositely propagating waves play the same roles as the two unidirectional waves of different frequency in the previous experiment, and the theoretical treatments almost coincide.

Lamb dip spectroscopy has been used by Johns et al. [10] to observe collision-induced four-level resonances in the molecules CH_3F, NH_3 and H_2CO. The general conclusions agree with those of the work in Ref. [6].

It may be expected that collision-induced double resonance may provide a valuable supplement to the transient methods (Ref. [8]) when one wants to investigate collisional processes. The state dependence of the cross section, the validity of selection rules and the collisional velocity changes can all be inferred from the experiments. Their usefulness can possibly be extended by the use of foreign buffer gases as the collisional agent.

4. Transfer of coherence: transient measurements

In this section we will consider the possibility of coherence transfer between a pair of transitions a-b and c-d. Again we assume velocity smearing to be negligible and postulate that if $\rho_{cd}=0$ and $\rho_{ab} \neq 0$ before the collision, we have

$$\rho_{cd} = \zeta \, \rho_{ab} \qquad (14)$$

just after. This assumes the impact approximation. (The parameter ζ plays a role similar to κ in Eq. (9).)

Because the newly created coherence ρ_{cd} developes with the frequency ω_{cd} different from ω_{ab}, the two off-diagonal elements dephase at the frequency

$$\Delta\omega = \omega_{ab} - \omega_{cd} \, . \qquad (15)$$

In steady state measurements the observation involves an average over a very long time interval. The collision can take place at any instant within this, and summing contribution from all possible collision times we will average the collisional coherence to zero. Consequently we expect the coherence transfer to be most easily observed in transient experiments where the dephasing is a less severe restriction.

Here we will describe an idealized experiment where, in principle, it should be possible to detect collisional transfer of the type (14) (see [11] for details). The experiment is a double-resonance echo experiment. We assume two laser beams acting on the two transitions of Figure 1, and we write

$$\rho_{ab} = \exp\left[-i(\Omega_1 t - kz)\right] \rho_1$$
$$\rho_{cd} = \exp\left[-i(\Omega_2 t - kz)\right] \rho_2 \qquad (16)$$

At time $t=0$ we apply a $\pi/2$-pulse to transition a-b only. For later times $t>0$ we have

$$\rho_1(t) = \exp\left[-(i\Delta\omega_1 + \gamma_{ab})t\right] \rho_0 \quad , \qquad (17)$$

where

$$\Delta\omega_1 = \omega_{ab} - \Omega_1 + kv \quad . \qquad (18)$$

We assume a collision to take place at t_1 and from (14)-(17) we obtain

$$\rho_2(t_1) = \zeta \exp\left[-i(\Delta\Omega + \Delta\omega_1)t_1 - \gamma_{ab} t_1\right] \rho_0 \qquad (19)$$

with

$$\Delta\Omega = \Omega_1 - \Omega_2 \quad . \qquad (20)$$

After the collision ρ_2 will develop in time with the exponent $i\Delta\omega_2 + \gamma_{cd}$, where

$$\Delta\omega_2 = \omega_{cd} - \Omega_2 + kv \quad . \qquad (21)$$

At the time $\tau > t_1$ we apply a π-pulse to transition c-d only. The effect of an idealized π-pulse is to take the complex conjugate of the off-diagonal element of the density matrix. We obtain for

$$\rho_2(\tau) = \zeta^* \exp\left[(i\Delta\omega_2 - \gamma_{cd})(\tau - t_1)\right] \exp\left[i(\Delta\Omega + \Delta\omega_1)t_1 - \gamma_{ab}t_1\right] \rho_0^* \, . \tag{21}$$

The time development of ρ_2 continues and we obtain for $t > \tau$

$$\rho_2(t) = \exp\left[-i\Delta\omega_2(t - 2\tau) - \gamma_{cd} t\right] \exp\left[(i\Delta\omega - \gamma_{ab} + \gamma_{cd})t_1\right] \zeta^* \rho_0^* \, . \tag{22}$$

The first term gives the echo at $t = 2\tau$ when integrated over the inhomogeneous frequency distribution. It is seen to decay with the rate γ_{cd}. This echo should be totally absent without the collisional coupling of the levels.

In a real rxperiment we observe signals from all collisions $t_1 < \tau$. If there is no π-pulse on the transition a-b it is easily seen that collisions with $t_1 > \tau$ do not give rise to an echo. We have to sum over an ensemble of collision events with $0 \leq t_1 < \tau$ with the probability of a collision in $(t_1, t_1 + dt_1)$ being (dt_1/T), where T is the average time between collisions. For simplicity we assume $\gamma_{ab} = \gamma_{cd}$ and obtain

$$\int_0^\tau e^{i\Delta\omega t} \, dt/T = \frac{e^{i\Delta\omega\tau} - 1}{i\Delta\omega T} \tag{23}$$

Hence we see that the echo signal is modulated at the frequency $\Delta\omega$. This can be used to identify an observed echo as deriving from two transitions instead of one.

The presentation above is, of course, too idealized. The effect is, however, present, even in more realistic treatments and should be observable. In an experiment an CH_3F the decay time $\gamma^{-1} \sim 1\mu s$ and $\Delta\omega/2\pi \sim 10^7 Hz$. The latter corresponds to a modulation of the echo with period $0.1 \, \mu s$. The width of the echo signal is about $0.2 \, \mu s$ and hence it should be possible to observe the modulation. In order to optimise the transfer of coherence one should choose a line pair like d of Table I. Here $|\Delta m| = 1$ for both the upper and lower states and this ought to favour simultaneous transitions in both states. The experiment may, however, offer considerable difficulties, but if successful it does provide an interesting new collision-induced spectroscopic effect.

References

[1] V. A. Alekseev, T. L. Andreeva and I. I. Sobel'man, Soviet Phys. JETP, 35, 325 (1972); ibid 37, 413 (1973).
[2] P. R. Berman, Phys. Rev. A5, 927 (1972); ibid A6, 2157 (1972).
[3] P. R. Berman and W. E. Lamb, Jr., Phys. Rev. A2, 2435 (1970).
[4] P. R. Berman, Appl. Phys. 6, 283 (1975).
[5] S. Stenholm, Phys. Rev. A2, 2089 (1970).
[6] R. L. Shoemaker, R. G. Brewer and S. Stenholm, Phys. Rev. Lett. 33, 63 (1974) and Phys. Rev. A10, 2037 (1974).
[7] G. W. Series, Physica, 33, 138 (1967).
[8] J. Schmidt, P. R. Berman and R. G. Brewer, Phys. Rev. Lett., 31, 1103 (1973) and P. R. Berman, J. M. Levy and R. G. Brewer, Phys. Rev., A. to appear (1975).
[9] M. S. Feld and A. Javan, Phys. Rev. 177, 540 (1969).
[10] J. W. C. Johns, A. R. W. McKellar, T. Oka and M. Römheld, to be published.
[11] S. Stenholm, to be published.

HIGH RESOLUTION STUDIES WITH DOPPLER FREE
RESONANCES; RECENT WORKS AT MIT

Ali Javan
Department of Physics
Massachusetts Institute of Technology
Cambridge, Massachusetts 02139

ABSTRACT

This paper gives a summary of recent MIT works in high resolution studies utilizing the nonlinearities of molecular transitions to obtain narrow Doppler free resonances.

I. INTRODUCTION

In our efforts to apply lasers to the studies of fundamental atomic and molecular processes, we passed through a period of devising novel approaches and discovering new effects which were potentially useful in such studies. The field has come to fruition in that the challenges ahead now lie in utilizing the newly discovered effects and novel methods to make detailed observations of processes not possible previously. This paper gives a summary of three recent experiments at the MIT Optical and Infrared Laser Laboratory relating to the use of the nonlinearities of molecular resonances in high resolution studies of matter. Two of these experiments relate to the studies of dynamic behaviors where the nonlinear resonances are used to obtain detailed information on intermolecular force laws and rapid molecular relaxation. The third experiment relates to high resolution studies over an entire molecular rotation-vibration band in which the Zeeman effect due to the weak rotational moments in the ground $^1\Sigma$ electronic state of a molecule is observed. The narrow Doppler free resonances observed over the entire band reveal the anomalous feature of the Zeeman effect, enabling measurements of the small g-factors, their signs and their dependencies on molecular vibrations.

II. STUDIES OF COLLISION BROADENING IN MOLECULES OF SELECTED VELOCITIES IN A ROOM TEMPERATURE GAS

Collision width of a molecular resonance arises from intermolecular forces and their disruptive effects on the interaction of the molecule with the radiation field. As such, the general features of collision broadenings and their behaviors in different transitions have always been considered a treasure chest of information on intermoleculer force laws. In the past, however, the collision widths of resonances have been studied by changing the gas pressure or the gas temperature. These studies suffered from some basic experimental difficulties, greatly limiting the accuracy of the observations; the varying conditions of a gas at different pressures and temperatures and the determination of exact gas pressures, can be listed as the foremost among the difficulties. As a result, the line width measurement accuracies were done, at the best, to within a few percent, severely limiting the wide scope of their usefulness.

The method described here makes possible precise studies of the collision broadening at a non-varying gas condition having a fixed gas pressure and temperature. In the measurements, neither the exact knowledge of the gas pressure nor its temperature is of importance. As a result, compared to the previous studies, measurements can be performed and interpreted at much improved accuracy.

Consider a Doppler broadened absorption line at a low gas pressure where the collision line width, γ, is considerably below the transition's Doppler $1/e$ width, $\Delta\omega_D = \frac{\omega_o}{c} u$, where ω_o is the center frequency of the resonance, c is light velocity, and u is the most probable velocity given by $u = (2kT/m)^{1/2}$. Consider interaction of the resonance line with an incident traveling wave at a frequency ω propagating along the Z-direction. At the wing of the Doppler line, the molecules resonantly interacting with the incident field have a Z-component of velocity given by $v_z^o = c(\omega-\omega_o)/\omega_o$. For these molecules, the applied field frequency, ω, is Doppler shifted to the resonance frequency ω_o. Accordingly, the applied radiation interacts with selected molecules of known velocity component, v_z^o. Consider now the mean kinetic energy of these "selected" molecules, $(K.E.)_s$ at a room temperature:

$$(K.E.)_s = 1/2 \ \overline{mv_x^2} + 1/2 \ \overline{mv_y^2} + 1/2 \ m(v_z^o)^2.$$

Noting that:

$$1/2 \ \overline{mv_x^2} = 1/2 \ \overline{mv_y^2} = 1/2 \ kT,$$

and with the expression for u given above, we can write:

$$(K.E.)_s = kT[1 + (v_z^\circ/u)^2].$$

In terms of the applied field frequency and the Doppler 1/e width, we can write,

$$(K.E.)_s = kT[1 + (\omega-\omega_o)^2/\Delta\omega_D^2].$$

Accordingly, we note that the mean kinetic energy of the selected molecules differs from the $3kT/2$, the mean kinetic energy of the entire gas molecule at temperature T. An effective temperature T_e, can be defined for the selected molecules by writing

$$(K.E.)_s = 3/2\, kT_e.$$

From the above we find

$$T_e = 2/3\, T[1 + (\omega-\omega_o)^2/\Delta\omega_o^2].$$

It is also useful to give an expression for the rms velocity of the selected molecules:

$$v_{rms} = (v_z^{\circ 2} + u^2)^{1/2}.$$

Note that for $T = 300K^\circ$ (a near room temperature gas), and for $\omega = \omega_o$, the effective temperature is $T_e = 200K^\circ$. When ω is tuned, e.g., to $|\omega-\omega_o| = 2\Delta\omega_D$, we have $T_e = 1000K^\circ$.

Consider now the applied field at frequency ω to be sufficiently intense to cause detectable saturation. Let a weak counterpropagating field at a variable frequency ω_p to probe the resonance. Inspection shows that the transmitted probe field versus ω_p manifests a resonance in the region $\omega_p \approx \omega_o - (\omega-\omega_o)$. This resonance is free from Doppler broadening and originates from the selected molecules saturated by the strong field at frequency ω. As in the case of Lamb dip, the width of the resonance is determined by the collision of the selected molecules with the rest of the gas; power broadening and the transit time effect also contribute to the width. The latter two contributions can be made to be small and, if necessary, corrected for.

By varying the frequency of the saturating field, ω, the selected molecules of different v_z° can be observed. Accordingly, measurements of collision broadening versus v_z° (i.e. versus T_e), can be performed in a gas without requiring a change of pressure or gas temperature. This obviates the difficulties of the previous methods and makes precision studies of collision broadening a reality.

In the initial observation[1] of the effect, the line width of the $\nu_2[asQ(8,7)]$ transition in NH_3 was observed, using an N_2O laser oscillating on its P(13) line. The laser output provided the saturating field at the frequency ω. The counterpropagating probe field was

obtained by separating a portion of the laser output with a beam splitter
and, with an acousto-optics modulator, shifting its frequency by the
desired amount, δ, which lie in the radio frequency region. The experimental arrangements were such that for $\delta = 0$, one would observe the
Lamb dip; for δ other than zero, one would observe the resonance at a
given v_z°. Both self broadening, as well as buffer broadening with Xe
were studied. In the initial measurements[1], the dependence of collision
broadening versus v_z° was observed with an accuracy of about six or seven
percent. Subsequently, advantage has been taken of the inherent possibility for high precision studies offered by this method. In the studies,
a resonance for a given $|\omega-\omega_o|$ was observed and compared with the Lamb
dip measured simultaneously at the same gas pressure. This eliminated
the need for a knowledge of the exact gas pressure or a change of its
conditions. The measurements were repeated at a number of fixed frequencies on the wing of the Doppler line. These corresponded to varying
T_e up to 1150° (for which $(v_z^\circ/u) \cong 2.5)$.

For precise line width measurements, it became necessary to make
detailed point by point computer fits of the carefully recorded resonances, with a Lorentzian line shape. Figure 1 gives the observed line

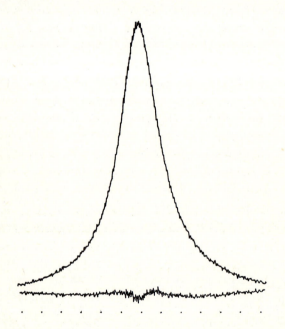

Figure 1. A typical saturation dip recorded for self broadening of
ammonia at p = 36 mTorr. Residual from computer least squares fit to a
Lorentzian is shown at base. (The full width of the resonance is
2.47 MHz.)

shape; it also gives the difference between the observed curve and the least mean square computer fit of the curve to a Lorentzian. This is given here to point out that line widths with much improved accuracy can be obtained, as compared to the standard method which generally relies on a single point half width measurement. The standard methods are perfectly adequate for studies versus pressure or gas temperature; our method, however, can make use of data analysis at much improved accuracy.

In the theoretical interpretation, starting from a collision model, the dependence of the line broadening cross section versus relative velocity of the colliding molecules is estimated. Subsequently, this cross section is averaged over the velocity distributions of the colliding species. Ammonia self broadening has been known to follow a dipole-dipole collision force law, corresponding to $1/r^3$ potential. Using the Anderson collision theory[2], this predicts a collision broadening $\gamma = \mu <v\sigma>$, independent of v. According to this, line width should remain constant at varying v_z°. While our initial less accurate observation verified this, the subsequent detailed studies showed deviation by an amount considerably larger than the accuracy of the more precise data. Detailed results will be shortly submitted for publication elsewhere[3].

Observations were also made on NH_3 broadening by Xenon buffer. For this case, $\gamma = n<v\sigma>$ is found to be strongly dependent on v_z°. It was known from previous studies that collision forces between NH_3 and Xe dominantly arise from an induced-dipole quadrupole interaction corresponding to a $1/r^7$ potential. Experimental observations indicate considerable deviation from this; the explanation in this case seems to lie in the contribution of velocity changing collisions. The details will be given in the forthcoming publication[3] mentioned above.

This work has been done by T. Mattick in collaboration with N. A. Kurnit. A. Sanchez and myself.

III. OBSERVATION OF ADIABATIC RAPID PASSAGE UTILIZING DOPPLER FREE RESONANCES; MEASUREMENTS OF RAPID MOLECULAR RELAXATIONS

I will now give the summary of another recent experiment in which population inversion by Adiabatic Rapid Passage (ARP) is observed for selected molecules of a specific velocity group. The initial experiment is done on selected molecules with $v_z^\circ \approx 0$. For this, the narrow Doppler free resonance at the center of a Doppler broadened absorption line, (the Lamb dip), is observed as a strong saturating wave and a weak counter-propagating probe wave of the same frequency is rapidly tuned through the line center. The method has the advantage of easing the conditions required for ARP, in that the frequency sweep need only be of the order of the homogeneously broadened width of the resonance, much narrower than the Doppler width. Furthermore, since molecules in neighboring velocity groups are probed before, during, and immediately after the ARP, fast relaxation processes can be investigated with relatively slow sweep rates. This in turn reduces the requirements on laser power. In addition, from the discussions given above, we note that for the two counterpropagating waves having different frequencies, relaxation effect can be studied versus molecular velocity.

In the experiment, the observations are made with the $\nu_2 = [asQ(8,7)]$ absorption line of the NH_3 molecules, using an N_2O laser oscillating on its P(13) line. The experimental arrangement consisted of first observing the Lamb dip signal with an intense saturating field and a weak counterpropagating probe field. For this, the transmitted probe wave is detected and displayed on an oscilloscope as the laser frequency is swept across the center of the absorption line. For the intense field weakly saturating the resonance and its frequency swept at a slow rate (corresponding to slow passage,) the observed Lamb dip is a symmetrical resonance with a width $\gamma = 1/T_e$ determined mainly by the collision broadening. As the saturating field intensity is further increased, the observed slow passage Lamb dip signal becomes deeper and broader due to power broadening. (A deeper Lamb dip corresponds to a more heavily saturated line and hence less absorption of the probe field at the peak of the dip.) Power broadening becomes dominant at field intensities where $\beta > \gamma$ with $\beta = \mu E/\hbar$. To observe the ARP effect, the intense saturating field is maintained at a power level where the power broadening dominates the observed width. With this satisfied, the sweep rate is gradually increased. The condition for ARP induced by the intense saturating field is then automatically approached as the sweep rate, $\frac{d\omega}{dt}$, approaches the power broadening parameter, β. For $\frac{d\omega}{dt} > \beta$, the intense field inverts[4] the level populations for molecules in the successive regions of the velocity distributions as they undergo ARP. When the frequency is swept across the center of the absorption line, the weak probe field interacts with the molecules of the selected velocity group with $v_z^\circ \approx 0$ as they are inverted by the ARP caused by the strong field. This leads to an amplification of the probe field in the region of the Lamb dip and an asymmetrical line shape due to relaxation of the inverted molecules returning to equilibrium. Accordingly, for increasing sweep rates, the Lamb dip signal becomes deeper and deeper and breaks into the amplification and appears as an asymmetrical resonance with a tail. Note that in this way, the molecules with v_z° at around zero are probed by the weak field at times before, during and immediately after they are inverted by the ARP effect. Hence, the study of the asymmetry appearing on the Lamb dip versus the sweep rate, gives direct information on rapid relaxation of the molecules returning to equilibrium. For the results and additional details, the reader is referred to a publication which has recently appeared in print.[5]

The work, currently under further scrutiny, is done by S.M. Hamadani, in collaboration with T. Mattick, N. A. Kurnit and myself.

IV. ANOMALOUS ZEEMAN EFFECT IN THE 10.6μ BAND OF CO_2 MOLECULES; THE USE OF DOPPLER FREE RESONANCES OVER AN ENTIRE BAND

Conventional molecular spectroscopy in the infrared, visible and the U.V. regions has vastly benefited by the ability to make observations over an entire emission or absorption band. From this has followed a whole host of distinguished measurements, providing accurate molecular parameters.

Similarly, the Doppler free resonances observed in the entire transitions of a rotation-vibration band, can reveal features not obtainable from the studies of an isolated transition in the band. Comparing a high-J with a low-J transition, for instance, can reveal the presence of small energy correction terms which are generally dependent on the angular momentum quantum number, J. The method can be expected to provide minute details of a variety of novel effects, particularly when applied to the study of causes which result in small splittings of the rotational energy levels.

Observation of narrow Doppler free resonances require operating at very low gas pressures. In cases where the transition studied belongs to a hot band, the absorption coefficient at a low gas pressure is much too weak to allow detection of the Doppler free resonance (e.g. the Lamb dip), via direct observation of the absorption effect. A recent method utilizing a different approach[6] has made possible detection of Doppler free resonances in weakly absorbing transitions interacting with an intense incident radiation in the form of a standing wave. In the case of a hot band, the spontaneously emitted radiation from the upper levels of the hot band to the ground vibrational state of the molecule is collected and focused with a lens on a detector. The detector output is observed as the incident standing wave is tuned in frequency across the line profile of one of the hot band transitions. The Doppler free resonance appears as a resonant change in the intensity of the detected spontaneous emission, as the frequency of the standing wave radiation is swept through the center of the transition. The method, first applied[6] to the 10.6μ and the 9.3μ bands of CO_2, has been used in a variety of experiments.

In the CO_2 experiment, the output of a stable single-mode laser oscillating on a preselected transition of its 10.6μ or 9.3μ band is applied to an external cell containing CO_2 absorbing gas at a low pressure. The change in the intensity of the 4.3μ emission band, (the (001)⟶(000) band), is used to observe the Doppler free resonance, in the transition selected by the oscillating laser line. In a series of experiments to be published[7], highly distortion free and narrow Doppler free resonances are observed. (The limiting half width at the present is about 70 kHz and is mainly caused by power broadening. The observations can be made at gas pressures below a fraction of millitorr where collision broadening is small. At the present, the integration time in the detection system is about a fraction of a second. As in the methane experiments by J. Hall and colleagues[8] at NBS, integrating electronics and frequency stabilizing schemes can be applied to increase the integration time to a few hours, thus improving the detection sensitivity by orders of magnitude. The improved sensitivity makes possible operation at much reduced powers and, consequently, a decrease in the line width down to a very small transit time limit.)

The method described allows observations to be made on a small volume of the gas sample, not exceeding tens of cm^3. This has made possible placing the gas sample in between the pole pieces of an electromagnet which proveds magnetic fields up to approximately ten kilogauss.

Consider the case where the laser E-field is perpendicular to the direction of the applied magnetic field. Further, assume the g-factors to be the same for the lower and upper vibrational levels. In this case, at elevated magnetic fields, an observed Doppler free resonance symmetrically splits into three components. The two side-components arise from the $\Delta M_J = \pm 1$ transitions. (For g>0, the upshifted component belongs to $\Delta M_J = + 1$ and the down-shifted component to $\Delta M_J = - 1$. The reverse is true for g<0.) The middle component is the cross over resonance[9] caused by the Doppler effect and the standing wave nature of the incident radiation. Confining our attention to the two side-components, we note that for J>1, each component consists of superposition of a number of $\Delta M_J = +1$ or -1 transitions.

Anomalous Zeeman effect due to small dependence of the g-factor on the vibrational quantum number causes small splittings of each of the two side components into a number of lines, with each line belonging to a different ΔM_J transition. Unless the line width is less than the size of the splitting, different ΔM_J transitions within each component remain overlapping and unresolved. Even in the face of this, much accurate information on anomalous Zeeman effect can be obtained, if a high-J transition is compared to a low-J transition and the effect is studied both in the P- as well as the R-branch transitions. Figure 2

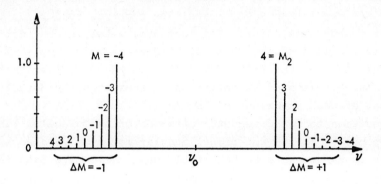

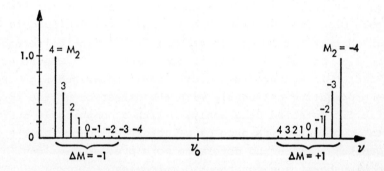

Figure 2. Anomalous Zeeman effect for a P- and R-branch line. The lower level is assumed to be the same for both lines and correspond to $J = 4$. (It is assumed that $\Delta g \ll g$.) The upper spectrum belong to the P-branch line and the lower belong to the R-branch line when the magnitude of the upper g-factor is larger than that of the lower g-factor; otherwise, the reverse would be the case. (The cross over resonances appear at ν_o and are not shown.)

gives an example by displaying the anomalous Zeeman effect for a P-branch and an R-branch transition having the same lower level. (The Figure is given for the lower level with $J = 4$.) Note that, eventhough the spacing between the adjacent ΔM_J transitions is small, the center of gravity of the $\Delta M_J = +1$ lines and that of the $\Delta M_J = -1$ lines each is shifted by an appreciable amount due to the anomalous Zeeman effect — the larger the J value, the larger is this shift. Further we note that the sign of this shift for a P-branch line is different from that of an R-branch line, (see Figure 2.) Inspection shows that for the case where the magnitude of the g-factor is larger for the upper vibrational level than the lower vibrational level, the centers of the two side components are shifted towards each other for a P-branch transition, and shifted away from one another for an R-branch transition. From this it follows that, even for a weak dependence of the g-factor on the vibrational quantum number, the spacing between the two side components can be sizably different for an R (J) transition compared to a P (J) transition, particularly when J is large. For instance, a two percent difference in the g-factor will cause about 50% difference in the splitting of a P (20) line compared to the R (20) line. We therefore see that study of the Zeeman effect over an entire band makes possible detection and measurement of weak anomalous effect.

In addition to the dependence of g-factor on molecular vibration, a weak dependence on the rotational angular momentum is also expected. The effect is expected to be detectable when the entire band is analyzed and a low-J transition is compared with a high-J transition.

The preliminary data analysis for the 10.6µ band of CO_2 gives a value of g_ν = -0.043. As for the anomalous feature of the Zeeman effect, the magnitude of g for the upper vibrational level is found to be larger than that of the lower vibrational level. The difference is found to be 1.8% of the magnitude of the g-factor.[10] The negative sign for the g-factor is measured using circularly polarized standing wave radiations propagating colinearly with the direction of the applied magnetic field. Detailed data analysis with a computer is currently underway to inspect the presence of g-factor dependence on the angular momentum quantum numbers and determine more precise g-values.[11] Measurements have also been made for the N_2O molecules.

This work is performed by M. Kelly, in collaboration with J. Thomas, N. Kurnit and myself.

The above topics will be discussed in more detail at the forthcoming theoretical summer school at Les Houche. Other activities of the MIT Optical and Infrared Laser Laboratory, including the recent developments in the area of extending methods of microwave electronics into the optical regions, will also be covered at Les Houche and at the forthcoming CIRP Conference at Zürich.

REFERENCES

[1] A. T. Mattick, A. Sanchez, N. A. Kurnit, and A. Javan, Appl. Phys. Lett., 13, 675, 1973.

[2] P. W. Anderson, Phys. Rev., 80, 511, 1950. This article lists other references.

[3] A. T. Mattick, N. A. Kurnit, and A. Javan, Jour. of Chem. Phys., to be published.

[4] See, for instance, A. Abragam, The Principles of Nuclear Magnetism, Oxford University Press, 1961.

[5] S. M. Hamadani, A. H. Mattick, N. A. Kurnit, and A. Javan, Appl. Phys. Lett., 87, 21, 1975.

[6] C. Freed and A. Javan, Appl. Phys. Lett., 17, 53, 1970.

[7] M. Kelly, J. Thomas, N. A. Kurnit, and A. Javan, Phys. Rev., to be published.

[8] J. L. Hall, Proceedings of Esfahan Symposium, edited by M. S. Feld, A. Javan, and N. A. Kurnit, Wiley Interscience, 1973.

[9] A. Javan, Proceedings of Esfahan Symposium, edited by M. S. Feld, A. Javan, and N. A. Kurnit, Wiley Interscience, 1973. This is a review article and gives the early developments in laser application to precise spectroscopy.

[10] M. Kelly, J. Thomas, N. A. Kurnit and A. Javan, Appl. Phys. Lett., to be published.

[11] M. Kelly, J. Thomas, N. A. Kurnit and A. Javan, Phys. Rev., to be published.

TITLES AND ABSTRACTS OF POST-DEADLINE PAPERS

SOME COMMENTS ON THE DISSOCIATION OF POLYATOMIC MOLECULES BY INTENSE 10.6μm RADIATION[†]

N. Bloembergen
Pierce Hall, Harvard University, Cambridge, MA 02138

The discussion of the physical processes involved in the dissociation of polyatomic molecules by intense infrared radiation at a vibrational absorption band leads to an estimate of the required electric field amplitude of (1 to 3) × 10^6 volts/cm.

[†]Supported by the Joint Services Electronics Program and the Advanced Research Projects Agency.
Paper submitted to Phys. Rev. Letters.

EXCITATION OF HIGHLY FORBIDDEN TRANSITIONS BY TUNABLE LASERS AND SEARCH FOR PARITY VIOLATION INDUCED BY NEUTRAL CURRENTS

M.A. Bouchiat
Laboratoire de Spectroscopie Hertzienne, Ecole Normale Supérieure, 24, Rue Lhomond, F-75231 Paris Cedex 05

The theories of weak interactions featuring neutral currents, which have recently been detected by high energy neutrino-scattering on nuclei, also predict the existence of a short range electron-nucleus potential which, in general, will violate the invariance under space reflexion but preserve time reversal invariance. In several publications [1] a general analysis of the parity violating phenomena induced by neutral currents in atomic physics has shown that under well defined conditions observable effects can be predicted. The parity violating interaction mixes states of opposite parities. The usual parity selection rules are no longer strictly valid and mixed electric and magnetic dipole transitions become possible. The photons emitted in such transitions by unpolarized atoms will have, on the average, a small circular polarization P_c. The feasibility of an experiment is linked to the following remarks [1a,b]:
i) the effect exhibits a very strong enhancement in heavy atoms and roughly goes like Z^3
ii) recent progress in laser technology allows the excitation of highly forbidden radiative transitions where parity mixing has the best chance to show up.
As an example the M.D. 6S-7S transition of cesium is discussed. The parity violating E.D. amplitude $E_{1z}^{p.v}$ is not very sensitive to the details of the atomic wave function and has been computed in the Weinberg Salam model: $E_{1z} = <6S|7S> = i1.7ea_0 10^{-11}$. Several effects contributing to the M.D. amplitude $M_{1z} = <6S|\mu_z|7S>$ have been discussed [1]; only limits have

been derived, $10^{-5} \mu_B < M_1 < 10^{-4} \mu_B$, leading for $P_c = 2 \, \text{Im}\{E_{1z}/M^*_{1z}\}$ to the limits: $10^{-4} < P_c < 10^{-3}$. However, observation of such a highly forbidden transition brings up different problems already discussed [1]. As a preliminary step in an experimental program designed to detect parity violation in forbidden transitions, the Cs 6S-7S single photon transition has already been studied in the presence of an external static electric field $\vec{E}_O$, used to control the interdiction factor [2]. It is stressed that such an experimental investigation of radiative S-S transitions, which bears some analogy with two-photon transitions, is by itself of great interest independently of the question of parity violation in atomic physics [1c,2]. The direct single photon electric dipole excitation of the 7S state has been observed by illuminating the 6S ground state with a single mode, cw, tunable dye laser and $E_O = 10$ to 1000 volt/cm [2]. The resonance is detected by monitoring the decay fluorescence from the 7S state to the $6P_{1/2}$ state. The spectra obtained exhibit considerable changes in intensity and structure for different relative orientations of $\vec{E}_O$ and the linear polarization of the incident beam. Excellent agreement with theoretical prediction [1c] is obtained. In this preliminary experiment the weakest detectable transition was observed with $E_O = 12$ volt/cm and corresponds to an oscillator strength of 2×10^{-13}. It can be concluded that $|M_{1z}| < 3 \times 10^{-4} \mu_B$ and $|E_{1z}| < 10^{-6} e a_O$.

[1] C. BOUCHIAT, M.A. BOUCHIAT: Phys. Lett 48B, 111 (1974); J. Phys. 35, 899 (1974); and 36, 493 (1975).
[2] M.A. BOUCHIAT, L. POTTIER: J. Phys. Lettres 36, L. 189 (1975).

COLLISIONAL ANGULAR MOMENTUM MIXING IN RYDBERG STATES OF SODIUM

T.F. Gallagher, S.A. Edelstein, and R.M. Hill
Stanford Research Institute, Menlo Park, CA 94025

Collisions with rare gas atoms and nitrogen molecules are observed to produce a lengthening of the fluorescent decay times of excited Na d states. The effect is interpreted as a collisional mixing of the nearly degenerate $\ell \geq 2$ substates so that the observed decay reflects the average lifetime of all substates for which $\ell \geq 2$. The measured values of this decay rate are in excellent agreement with a theoretical prediction for the hydrogen atom. The cross section for the process appears to increase as the geometrical size of the excited Na atom independent of the collision partner, suggesting that the mixing is due to a strong short range interaction.

SPECTROSCOPY OF HIGHLY EXCITED S AND D STATES OF POTASSIUM BY TWO PHOTON ABSORPTION[†]

M.D. Levenson, C.D. Harper, and G.L. Eesley
University of Southern California, Los Angeles, CA

We have excited the even parity Rydberg states of potassium by two photon absorption through a principle quantum number of 26. Detection was by means of the fluorescence from the excited state to the 4P levels. A xenon laser pumped pulsed dye laser was necessary to obtain reasonable signal levels since the two-photon absorption cross sections were calculated to be less than 10^{-28} cm^2/W/cm^2. A multipass configuration employing lenses and corner cubes resulted in intensities at focus of 10 MW/cm^2. The energies of the Rydberg states were determined by scanning the laser frequency while monitoring both the fluorescence output and a Fabry-Perot frequency marker. The results indicate that the quantum defects for these states begin to systematically decrease for n > 20. In a Doppler-free configuration, the resolution of the two photon absorption experiment was 30 MHz, limited by vibrations in the single mode laser cavity structure. The ultimate achievable resolution would be ≈3 MHz for lasers of this type.

[†]Supported by the Research Corporation.

HIGH-RESOLUTION, TWO-PHOTON ABSORPTION SPECTROSCOPY OF HIGHLY-EXCITED D STATES OF Rb ATOMS

Yoshiaki Kato and B.P. Stoicheff
Department of Physics, University of Toronto

Using a single-mode, cw dye laser, we have observed two-photon absorption transitions (without Doppler broadening [1]) from the ground 5s state to highly-excited d states in Rb vapour. Thus far, the transitions 5s-11d up to 5s-30d have been observed, the state 30d being only 134 cm^{-1} below the ionization limit.
The experiment is essentially the same as that described by Hänsch et al. [2] in their study of the 3s-4d two-photon transitions in Na. In Rb, the 5s-nd (n ≥ 11) transitions were detected by monitoring the n'p-5s (n' ≥ 6) fluorescence. A relatively high pressure of Rb (~1.5 × 10^{-3} Torr) was necessary in the present experiment, because the laser wavelengths for the 5s-nd transitions were far off resonance with the strong 5s-5p transitions at 7948.1 and 7800.7 Å (for example, at 6149.4 Å for the 5s-11d transitions), and because of the smaller oscillator strengths (varying as ~$1/n^3$) to the high d states.
All of the spectra were investigated with a laser power of 40 to 50 mW. During each scan, the dye-laser frequency was simultaneously monitored with a stable, 50 cm, confocal Fabry-Perot interferometer. Sufficient signal-to-noise intensity was achieved even with the 5s-30d spectrum to indicate that it will be possible to extend this investigation to higher states. Extension to the lower d states will be made by using a dye laser oscillating at longer wavelengths. All of the observed linewidths are ~20 MHz and are instrument limited.

The hyperfine splittings of the 5s states of Rb^{85} and Rb^{87} are known to be 3035.7 ± 0.15 MHz and 6834.1 ± 0.1 MHz, respectively [3]. Since all of the possible two-photon transitions from a ^2S to a ^2D state are allowed [4], each spectrum consists of 8 discrete components. The intensity distribution of these components shows that the 11d to 30d ^2D states of Rb are not inverted (that is, the $D_{5/2}$ levels lie above the $D_{3/2}$ levels in contrast to the ^2D states of Na [5,6]).
The fine structure intervals were determined from the observed spectra with a typical accuracy of 15 MHz. Within this accuracy, no differences were observed in these intervals for Rb^{85} and Rb^{87}. Their dependence on the quantum number n and effective quantum number n* reveals that the fine structure intervals of the d states of Rb do not vary as n^{-3} nor n^{*-3}. The isotope shift δ of the ground state was also found from the relative positions of the Rb^{85} and Rb^{87} transitions: the separation (566 ± 6 MHz) of two components together with the known hyperfine splittings give an isotope shift of 160 ± 12 MHz, in agreement with the calculated value of 150 MHz arising from the mass shift [7].
The effect of pressure on the two-photon transitions was studied using the 5s-14d transitions. Although no increase in linewidth was observed up to a pressure of $\sim 10^{-2}$ Torr, the signal intensity saturated at $p \sim 5 \times 10^{-3}$ Torr, indicating collisional quenching of the excited d state. Further investigations including transitions to higher states are in progress.

[1] L.S. VASILEVIKO, V.P. CHEBOTAYEV, A.V. SHISHAEV: JETP Lett. 12, 113 (1970).
[2] T.S. HÄNSCH, K.C. HARVEY, G. MEISEL, A.L. SCHAWLOW: Opt. Commun. 11, 50 (1974).
[3] P. KUSCH, H. TAUB: Phys. Rev. 75, 1477 (1949).
[4] B. CAGNAC, G. GRYNBERG, F. BIRABEN: J. de Phys. 34, 845 (1973).
[5] C. MOORE: "Atomic Energy Levels", NSRDS-NBS (U.S.A.) 1971.
[6] C. FABRE, M. GROSS, S. HAROCHE: Opt. Commun. 13, 393 (1975).
[7] H.G. KUHN: Atomic Spectra (Academic Press, New York 1962) p. 370.

TWO-PHOTON MOLECULAR ELECTRONIC SPECTROSCOPY IN THE GAS PHASE

L. Wunsch, H.J. Neusser, and E.W. Schlag
Institut für Physikalische und Theoretische Chemie, Technische Universität, D-8 München

In this paper we wish to study the problems associated with the measurement of two-photon molecular spectra in polyatomic gases. Molecular electronic spectra have been one of our prime sources of information about the structure of molecules. This information has been derived from one-photon absorption experiments in the gas phase, under medium to high resolution. Many molecular states, however, are by reasons of parity as well as symmetry not accessible to one-photon experiments. Hence for complete understanding of molecular states, and their behaviour, two-photon experiments appear to be required.
In order to obtain sufficient resolution, and absence of falsification due to medium effects, only experimentation in the gas phase will be considered. In contrast to the growing two-photon literature on condensed media, the polyatomic gas phase has just now become amenable to study. The difficulty lies in the 5-8 order reduction in particle density which is characteristic of the gas phase relative to the condensed phase. Two-photon spectroscopy of atomic systems, often carried

out with low peak powers, usually employs near resonant intermediate states which strongly enhance the transition. This is usually not possible in molecular electronic transitions where the very interest lies in the first few electronic transitions.

As our particular example we chose benzene as the archtype of polyatomic molecules for which perhaps most theoretical information is also available as a basis for theoretical calculations. The first transition is $^1A_{1g} \rightarrow {}^1B_{2u}$ which is, however, parity forbidden in two-photon absorption. The two-photon absorption is also symmetry forbidden, as indeed is the one-photon absorption. Hence any transition can only be allowed by virtue of vibronic effects. These effects, however, would be of particular interest also in two-photon spectra as new vibrational states could thus be populated.

Experiments with a 1 MW nitrogen laser pumped dye laser were successful in our laboratory and have been reported in preliminary form [1,2]. A very rich two-photon molecular spectrum has been observed for the first time, allowing for the assignment of new states, hitherto unknown in benzene. The intensity is shown to be sufficient to pick up states from vibrationally excited ground states even though they are down by 3 orders of magnitude due to the Boltzmann factor. These hot bands are essential in the unequivocal assignment of the newly observed states in the absorption spectrum. The intensity is also sufficient to study gaseous systems in the low pressure limit, 0.1-1 Torr, and hence enables one also to measure the characteristic lifetimes of these newly prepared states.

We here have demonstrated how entirely new molecular states can be in fact observed by this technique. We have obtained a rich, well resolved strong molecular electronic two-photon spectrum in the gas phase which allowed for the first assignment of a polyatomic molecule in a gas phase two-photon spectrum. We hope this shows that it is now readily possible to exploit a valuable complement to molecular electronic spectroscopy: two-photon molecular spectroscopy.

[1] L. WUNSCH, H.J. NEUSSER, E.W. SCHLAG: Chem. Phys. Lett. **31**, 433 (1975).
[2] L. WUNSCH, H.J. NEUSSER, E.W. SCHLAG: Chem. Phys. Lett. **32**, 210 (1975).

TWO-PHOTON LASER ISOTOPE SEPARATION OF ATOMIC URANIUM - SPECTROSCOPIC STUDIES, EXCITED STATE LIFETIMES, AND PHOTOIONIZATION CROSS SECTIONS

G.S. Janes, I. Itzkan, C.T. Pike, R.H. Levy[†] and L. Levin[††]
Avco Everett Research Laboratory, Inc., Everett, MAS 02149

One of the more attractive Laser Isotope Separation schemes [1] involves the use of atomic uranium vapor. This paper describes experimental studies at low particle density which demonstrate selective two-step excitation and ionization of atomic uranium vapor and, additionally, measure a number of relevant parameters such as photoionization cross sections, and state lifetimes. The apparatus should also prove to be a useful tool for the spectroscopic study of other atomic vapors as well.

[†]Present Address: Exxon Nuclear Co., Inc., Bellevue, Washington.
[††]Present Address: P.O. Box 9001, Beer-Cheva, Israel.
This work was supported by Jersey Nuclear-Avco Isotopes, Inc., Bellevue, Washington.

In this experiment, uranium atoms are excited and ionized by light from short pulse tuned lasers. The ions are detected and analyzed by a mass spectrometer. Uramium vapor is produced at the bottom of the vacuum tank by electrom beam evaporation. Only a small fraction of the vapor passes through a hole in the top of the shield and reaches the sensitive region of the mass spectrometer. The tuned laser light is provided by either a pair of pulsed dye laser systems or else by a single pulsed dye laser system for excitation followed by a pulsed nitrogen laser (3371 Å) for ionization. The exciter consists of two synchronized nitrogen lasers which transversely pump a dye oscillator and a dye amplifier. The dye oscillator is basically an Avco "Dial-A-Line"[†] laser with some tuning modifications [2]. The wavelength and linewidth of this dye laser are determined by the grating used as the rear reflector of the oscillator and two Fabry-Perot etalons. This combination of elements gave an oscillator linewidth of about 0.02 Å. A lens focuses the oscillator output into the amplifier. The wavelength of the exciter laser is monitored with both a standard 1.25 meter spectrometer and a Fabry-Perot spectrometer. The ionizer and exciter beams are combined with a dichroic mirror prior to entering the vacuum tank through a window. A lens is used to focus this light into the sensitive region of the quadrupole mass spectrometer where it intercepts the uranium atoms. The total number of ions per pulse produced by two-step laser photoionization was small; however, the time of production was known precisely. By counting single ions, using digital logic and multiplexing the mass spectrometer between U^{238} and U^{235}, it was possible to not only measure isotope ratios as a function of exciter wavelength, but also to estimate and correct for background effects and for ions produced by single photons. The raw data without background correction yielded a 30% enrichment. We have measured excited state lifetimes by observing yields as a function of the delay between the two laser pulses. These results are given in the table below. In addition, for an excitation wavelength of 4266.325 Å the variation of two-step photoionization efficiency was measured as a function of the wavelength of the ionizing laser. The maximum yield at an ionizing wavelength of 3609 Å corresponded to a cross section of $2 \times 10^{-17} \Gamma_L/\Gamma$ cm^2, where Γ is the linewidth of the optical transition and Γ_L is the laser linewidth. Finally, the ionization potential of uranium was determined to be 6.187 ± .002 ev.

Table: Lifetimes of some excited states with large isotope shifts

Energy [cm^{-1}]	Exciting Wavelength in Air [Å]	J	Isotope Shift	Lifetime [ns]
23 572	4241.1	6	-.076	60
23 433	4266.3	5	-.057	155
23 212	4306.8	5	-.080	170
22 862	4372.8	6	-.061	135
22 583	4426.9	6	-.055	130
22 056	4532.6	6	-.081	300

[1] R.H. LEVY, G.S. JANES: "Method of and Apparatus for the Separation of Isotopes", United States Patent 3,772,519 (Nov. 13, 1973).
[2] I. ITZKAN, F.W. CUNNINGHAM: J. Quant. Electr. QE-8, 101 (1972).

[†]Registered Trademark.

ISOTOPE SEPARATION IN THE SOLID STATE[†]

David S. King and Robin M. Hochstrasser
Department of Chemistry and Laboratory for Research on the Structure of Matter, University of Pennsylvania, Philadelphia, PA 19174

The very efficient unimolecular photolysis of s-tetrazine is utilized to study isotopic selectivity and photochemistry in molecular crystals at 4.2 K to 1.6 K. A dye laser excitation spectrum of the $^1B_{3u}$ $(n\pi^*)$ origin of tetrazine in benzene at 1.6 K consists of four peaks with instrument limited linewidths ca 0.8 cm^{-1}, and the intensity ratios expected from natural abundances for $^{12}C_2{}^{14}N_4H_2$, $^{12}C_2{}^{15}N^{14}N_3H_2$, $^{13}C^{12}C^{14}N_4H_2$, and $^{12}C_2{}^{14}N_4DH$, respectively. The vibrational frequencies of normal, ^{13}C-, and ^{15}N- tetrazine were obtained in the excited state in excitation monitoring a specific isotopic species fluorescence; and in the ground state in fluorescence following a selective excitation. Excitation into the vibronic manifold of a given isotopic composition of tetrazine in benzene at 1.6 K results in the enhancement of the rate of photodissociation of only that particular isotopic species--leaving all other species unaffected. There is negligible isotopic scrambling due to the rapid depopulation of the excited state via the photoreaction k' ~ 2 × 10^9 s. Enrichments of 10^4-fold for ^{13}C- and ^{15}N- tetrazine were obtained after irradiation of the mixed crystal system at about 5801 Å. Infrared spectra at 4.2 K, taken immediately following irradiation into either the $^1B_{3u}$ $(n\pi^*)$ or $^3B_{3u}$ $(n\pi^*)$ manifold identifies the quantitative presence of HCN, and fractional distillation identifies N$_2$. Although isotopic composition has negligible effect on the photoprocess kinetics, the rate for the photoprocess from the triplet manifold is down by a factor of 10^5 consistent with a spin orbit coupling mechanism. The non-appearance of any intermediate at 4.2 K could conceivably be caused by a local thermal process utilizing the energy released in same first step, but we expect aromatic lattices to have very high thermal conductivity at 4.2 K. It has become apparent that molecular mixed crystals at 4.2 K to 1.6 K provide an excellent system to both study photoreactions and their radical intermediates, and to prepare and trap radicals and other molecules of a desired isotopic composition at high purity levels.

[†]This research was supported by the National Institutes of Health and by the NSF-MRL program through LRSM at the University of Pennsylvania. To be published in J. Am. Chem. Soc.

SATURATED DISPERSION BY LASER BEAM DEVIATION IN A SATURATED MEDIUM

B. Couillaud, A. Ducasse
Laboratoire de Spectroscopie Moléculaire, Université de Bordeaux I, F-33405 Talence, France

A narrow band laser beam sent through iodine vapor changes the susceptibility of a velocity group of molecules, those which can absorb at the frequency of the saturating light. A weak beam (probe beam) propagating in a nearly opposite direction to the saturating beam will interact with it if both light waves interact with the same molecules:

that is those with essentially vanishing axial velocity if the two
beams have the same frequency corresponding to the frequency of a resonant transition. In addition to the well-known saturated absorption
phenomenon, this interaction is characterised by geometrical effects
upon the probe beam. We define geometrical effects as effects which do
not affect the total intensity of the beam. They can be schematically
resumed as deviation, displacement and distortion due to the lens-like
medium produced by the saturating beam. Hence if we record the total
intensity of the probe beam with a detector whose sensitive area is
larger than the beam section, we will obtain a pure saturated absorption signal. In another way if we use a detector whose sensitive area
is smaller than the section of the beam, the spatially non-uniform intensity in the beam (Gaussian beam) makes it sensitive to the variations of intensity as well as to the geometrical effects.
An experiment was performed with a freerunning cw dye laser and iodine
vapor. The experimental set-up had basically the geometry proposed by
Hänsch [1] for iodine experiments with an argon laser, but in order to
obtain the maximum of deviation for the probe beam, the crossing point
for the beams was located at one end of the iodine cell. When the
laser was continuously tuned across a resonance, the signal was a
superposition of an absorption curve corresponding to the saturated absorption signal and a dispersion curve which describes the refractive
index variation associated to it. This last signal was called saturated dispersion and observed with an interferometric set-up by Borde,
et al. [2]. The predominance of the deviation upon the distortion (the
displacement was quite negligible in our experimental conditions) was
illustrated by the change of sign of the dispersion signal for particular arrangements of the beams in the iodine cell and by the evolution
of this signal with respect to the position of the detector in the
probe beam.
A second experiment was then performed which allowed separate recording
of the saturated absorption and dispersion signals at the same time.
The experimental set-up differed essentially from the previous one as
follows: to assure a better frequency stability the cw dye laser was
locked to a transmission fringe of a high fineness optical cavity. A
reference beam monitored by a diode D_3 was used to increase the signal
over noise ratio by a differential method. The probe beam was divided
into two beams. One was focused with a short focal lens on the sensitive surface of a detector D_1, the other was set on a detector D_2 which
monitored only a part of its intensity. After the difference between
D_1 and D_3 and phase detection the signal of saturated absorption was
recorded on the first channel of a recorder. The signal D_2-D_3 gave,
after phase detection, the sum of the saturated absorption plus the
saturated dispersion. The signal of saturated dispersion was obtained
from a new difference between the outputs of the two phase detections,
and was plotted on the second channel of the recorder.
Three components of the hyperfine structure of an iodine line were investigated. The lineshape of the saturated absorption and saturated
dispersion signals was numerically calculated and found in great agreement with the experimental data.

[1] T.W. HÄNSCH, M.D. LEVENSON, A.L. SCHAWLOW: Phys. Rev. Lett. 26, 946 (1971).
[2] C. BORDE, G. CAMY, B. DECOMPS, L. POTTIER: C.R. Acad. Sc., t. 277, 381 (1973).

PROGRESS IN SATURATED DISPERSION SPECTROSCOPY OF IODINE

C. Bordé, G. Camy, and B. Decomps
Laboratoire de Physique des Lasers, Université de Paris-Nord, F-93206 Saint-Denis France

By making use of the ring interferometer described in [1] it has been possible to obtain very narrow unmodulated dispersion signals in iodine. The peak to peak width of these signals (corresponding to the full width at half maximum for saturated absorption) is currently of the order of 600 kHz. Half of that width is still due to the residual frequency jitter of the argon laser which is slaved on the side of a transmission fringe of a confocal Fabry-Perot. The other half results from collision broadening, natural width and transit time broadening. The saturated dispersion curves have been used as error signals to stabilize the frequency of commercial argon lasers without any frequency modulation of the lasers. The sensitivity of the ring interferometer method with crossed polarizations has been analyzed and the applicability of the Kramers-Kronig relations to saturation experiments has been discussed [2]. These are shown to be valid in the limit of an infinite Doppler width (compared with the homogeneous width) and of small saturation parameters.

[1] C. BORDÉ, G. CAMY, B. DECOMPS, L. POTTIER: C.R. Acad. Sc. (Paris) <u>277B</u>, 381 (1973).
[2] D.R.M.E. Contract Report No. 7234293 (1974).

MAGNETIC OCTUPOLE INTERACTION IN I_2

K.H. Casleton, L.A. Hackel, and S. Ezekiel
Research Laboratory of Electronics, Massachusetts Institute of Technology, Cambridge, MA 02139

We have observed magnetic octupole and scalar spin-spin interactions in the optical spectrum of I_2 at 5145 Å. The lines excited are the hyperfine transitions on the P(13) 43-0 $B^3\Pi - X^1\Sigma$ line in I_2^{127}. The experimental set-up is similar to that described earlier [1], where two single frequency 5145 Å argon ion lasers were individually stabilized to hyperfine transitions excited in two independent molecular beams of I_2. The line positions were determined with a precision of one part in 10^{11} (5 kHz) using a heterodyne technique employing two argon ion lasers individually stabilized to I_2 hyperfine lines excited in molecular beams.
In order to describe the hyperfine structure accurately it was necessary to construct a Hamiltonian which included the following interactions

$$H_{HFS} = H_{NEQ} + H_{SR} + H_{TSS} + H_{SSS} + H_{NMO}$$

The first three terms of the Hamiltonian have been included in previous analyses [2] of iodine hyperfine spectra. These are, respectively, the nuclear electric quadrupole, the magnetic spin rotation and tensor spin-spin interactions. Whereas previously it was sufficient to calculate the quadrupole energy to second order, in the present work we also considered effects of third-order contributions to the interaction by including matrix elements off diagonal by $J' = J \pm 4$. This expanded the energy matrix to 105 × 105. However, by sorting the states with common F a block diagonal matrix is obtained in which the largest submatrix is only 11 × 11.

The fourth term in the Hamiltonian is the scalar part of the nuclear spin-spin interaction which results from the indirect electron coupled spin-spin interaction. This term has the form $\vec{I}_1 \cdot \vec{I}_2$, $\vec{I}$ being the iodine nuclear spin, and was first discussed by Ramsey and Purcell [3]. The last term of the Hamiltonian, the magnetic octupole interaction, results from the coupling of the nuclear magnetic octupole moment with the third derivative of the molecular vector potential. The form of the matrix elements has been discussed by Casimir and Karremann [4], and Svidzinskii [5].

The coupling strengths associated with every term in the Hamiltonian were varied in a least-squares computer program to obtain the best fit to the data. The results showed that the inclusion of the tensor nuclear spin-spin interaction dramatically reduced the standard deviation of the fit to 12.2 kHz. The addition of the scalar nuclear spin-spin term improved the fit to 8.5 kHz and the nuclear magnetic octupole interaction improved the fit further to 6.5 kHz. Extending the quadrupole calculations to third order improved the fit slightly to 6.3 kHz. A statistical F test [6] was performed to determine the validity of including the scalar spin-spin and magnetic octupole interactions in the theoretical model. The results of the F test indicated a better than 99% confidence level that the improvement in the fit for each of these terms was indeed real and not just statistical.

The high resolution available in the present experiment has allowed the precise determination of both ground and excited state quadrupole coupling constants as well as differences between upper and lower state constants for spin-rotation interaction, tensor and scalar spin-spin interactions and the magnetic octupole interaction. These are, respectively,

$$eQq' = -554,094 \pm 13 \text{ kHz}$$
$$eQq'' = -2,448,025 \pm 10 \text{ kHz}$$
$$C' - C'' = 186.71 \pm 0.10 \text{ kHz}$$
$$D_t' - D_t'' = -100.5 \pm 1.0 \text{ kHz}$$
$$D_s' - D_s'' = -2.72 \pm 1.0 \text{ kHz}$$
$$\Omega m' - \Omega m'' = -2.17 \pm 0.70 \text{ kHz}$$

This precise value for the ground state quadrupole coupling constant agrees very well with a previous estimate [7] of -2452 ± 40 MHz, while it is about 10% larger than the value found for crystalline I_2 [8]. Using $Q = -0.79 \times 10^{-24}$ cm^2 for the iodine quadrupole moment [9], the field gradients for the two states are then $eq' = 0.465 \times 10^7$ and $eq'' = 2.05 \times 10^7$ dyne/cm. The ratio of these two field gradients is $q''/q' = 4.4$ indicating that the distribution of charges which contribute to the electric field gradient is much more spherical for the excited electronic state than for the ground state of I_2. Alternatively, in terms of the Townes-Dailey model, this indicates there is little unbalanced p electron character near the iodine atoms in the excited state.

Schwartz's value [10] for the corrected nuclear octupole moment of I^{127} is $\Omega = 0.181 \mu_N \times 10^{-24}$ cm^2. From this we can calculate $m' - m'' = -79.4$ dyne/cm $\cdot \mu_N$, where μ_N is the nuclear magneton and m' and m'' represent, respectively, the upper state and lower state values of the divergence of the electron magnetization near the iodine nuclear site. From the negative sign, we see that again this ground state electronic property has a larger variation than the excited state quantity.

With an order of magnitude or more improvement in the data, the individual constants m' and m'' could probably be determined, thereby giving more quantitative information about the electronic structure of iodine. A further increase in the precision should also permit the investigation of the nuclear hexadecapole interaction in iodine. We have already calculated matrix elements for this interaction and have estimated the coupling constants based on the discussion of Wang [11]. Preliminary attempts to fit the data by including the hexadecapole term showed that the effect of this interaction is indeed small. The shift in line positions is expected to be on the order of a few hundred hertz so that an improvement in the data of one to two orders of magnitude would be required to observe the effect of the hexadecapole.

This work was sponsored by the Air Force Office of Scientific Research.
[1] D.G. YOUMANS, L.A. HACKEL, S. EZEKIEL: J. App. Phys. 44, 2319 (1973).
[2] P.R. BUNKER, G.R. HANES: Chem. Phys. Lett. 28, 377 (1974).
[3] N.F. RAMSEY, E.M. PURCELL: Phys. Rev. 85, 143 (1952).
[4] H.B.G. CASIMIR, G. KARREMAN: Physica 9, 494 (1942).
[5] K.K. SVIDZINSKII: Soviet Maser Research (Consultant's Bureau/Plenum, New York 1964) pp. 88-148.
[6] P.R. BEVINGTON: Data Reduction and Error Analysis for the Physical Sciences (McGraw-Hill, New York 1969) p. 200.
[7] M.S. SOREM, T.W. HÄNSCH, A.L. SCHAWLOW: Chem. Phys. Lett. 17, 300 (1972).
[8] R.V. POUND: Phys. Rev. 82, 343 (1951).
[9] W. GORDY, R.L. COOK: Microwave Molecular Spectra (Wiley-Interscience, New York 1970) Appendix V.
[10] C. SCHWARTZ: Phys. Rev. 105, 173 (1957).
[11] T.-C. WANG: Phys. Rev. 99, 566 (1955).

HIGH-RESOLUTION RAMAN SPECTROSCOPY WITH A TUNABLE LASER

B. Bölger
Philips Research Laboratories, Eindhoven, The Netherlands

In the usual method of Raman spectroscopy a fixed frequency laser is used for illumination of the sample, and the scattered light is analyzed by a double monochromator. The acceptance of the system is small especially at large resolutions. Detection at a fixed wavelength allows constructions with large acceptance and resolution. We have done a feasibility experiment using this principle. A pulsed tunable dye laser, bandwidth 0.1 cm^{-1} with 10 ns pulses of 1 kW, pumped by a N_2 laser, illuminated the sample. The wavelength was so chosen that the spontaneously scattered Raman light was near 455 nm. This light was measured through the fluorescence (at 850 nm) it induced in Cs vapour by the narrow band absorption of Cs at 455 nm. Stray light and excitation light were efficiently suppressed by filters. Additional suppression was obtained by using the fluorescence delay due to lifetime and trapping effects in a gated signal detection. Despite the low average exciting power of 1 mW and the high resolution, limited by the dye laser, of 0.1 cm^{-1}, the sensitivity is presently comparable to the conventional method using 1 W excitation and 1 cm^{-1} resolution.

TIME DEPENDENCE OF THE THIRD-HARMONIC GENERATION IN Rb-Xe MIXTURES

H. Puell, Physik-Department der Technischen Universität D-8000 München, F.R. Germany
C.R. Vidal, Max-Planck-Institut für Extraterrestrische Physik, D-Garching, F.R. Germany

Third-harmonic generation in a phase-matched Rb-Xe mixture [1] was investigated with incident light powers of up to 200 MW. Experiments were performed with two different pulse durations (7 and 300 ps, generated by a Nd:glass mode-locked laser system) to study the saturation effects at high input powers. The nonlinear Rb-vapor at a pressure of 1 Torr was prepared in a concentric heat pipe with an effective length of 28 cm. Adding Xe-gas to the system optimum phase-matching occurred at a Rb-Xe pressure ratio of 1:372. A slight asymmetry observed in the phase-matching curve (increase of the side maxima at the low pressure side ($\Delta k > 0$) and a corresponding decrease on the other side) was found to be in agreement with theoretical calculations considering the density gradients at the end of the Rb-vapor column.

Focusing the fundamental light beam with a confocal parameter of 47 into the Rb-Xe mixture, the conversion efficiency for third-harmonic generation was measured for input intensities ranging from 10^9 up to $2 \cdot 10^{11}$ W/cm^2. Over several orders of magnitude the third-harmonic energy increased with the third power of the input energy, as expected from small signal theory. Numerical calculations taking into account the radial and temporal intensity distribution of the fundamental light beam as well as focusing effects show good quantitative agreement with the experiments for light intensities below 10^{10} W/cm^2 (300 ps pulses) and $5 \cdot 10^{10}$ W/cm^2 (7 ps pulses), respectively. At higher input powers the observed energy conversion was considerably smaller than the theoretically expected values (0.3% instead of 2% for 300 ps pulses at $5 \cdot 10^{10}$ W/cm^2, and 3% instead of 15% for 7 ps pulses at $2 \cdot 10^{11}$ W/cm^2).

In order to explain this discrepancy the intensity dependent change of the refractive index of Rb due to the second-order Kerr-effect was included, which may destroy the phase-matching condition at high light intensities. Good agreement of the modified theory with the experimental results was obtained for the 300 ps pulses, but for the 7 ps pulses an even lower energy conversion (0.6%) was predicted than experimentally observed. It appears that there is an additional mechanism which effectively reduces the Kerr constant $\chi^3(\omega)$ at very high intensities. This was further supported by the fact that our experiments showed no evidence for self-focusing, whereas from theoretical estimates (taking only the value for $\chi^3(\omega)$) a very distinct onset of self-focusing at $5 \cdot 10^{10}$ W/cm^2 is expected.

Evaluating the nonlinear susceptibilities of Rb up to the third order with the density matrix formalism, the following effects appear to modify the effective Kerr constant: Including the time dependence of the envelope of the electric field E one finds an excited state population N_1 due to single photon absorption proportional to $E^2/(\omega - \omega_0)^2$, which is independent of the linewidth Γ_{10} of the relevant transition. In our case this term (which may be also evaluated from the adiabatic following model [2]) dominates the usual first-order term $N_1 \propto \Gamma_{10} \int E^2 dt/(\omega - \omega_0)^2$.

Similarly, in case of two-photon absorption the excited state population is proportional to the usual term $\Gamma_{10} \int E^4 dt/(2\omega - \omega_0)^2$ and an additional transient term $E^4/(2\omega - \omega_0)^2$. With these effects we computed for our experimental conditions ($2 \cdot 10^{11}$ W/cm^2) excited state populations in the 5P and the 4D level of Rb of the order of 4% and 12%, respectively. The contribution of the excited states to the refractive

index, Kerr-effect, and third-harmonic generation can then no longer be neglected. Our computations show indeed that the Kerr constant of some of the excited states under consideration have even the opposite sign to that of the groundstate and, hence, give rise to a significant cancellation.
In calculating the nonlinear susceptibilities one also has to take into account that in case of the 7 ps pulses only the m = ±1/2 sublevels will be excited, whereas in case of the 300 ps pulses (which last long compared to the collisional dephasing time $T_2 \sim 50$ ps) essentially all the m-sublevels will be equally populated. Incorporating the excited 5P and 4D level populations into our calculation for the third-harmonic generation we got a quantitative agreement with our experimental results within a factor of 2. Further improvement may be achieved considering some of the F levels which may noticeably be populated from the 4D level.

[†]Present address: Max-Planck-Institut für Extraterrestrische Physik, D-Garching, F.R. Germany.
[1] Such a system was investigated extensively by R.B. MILES and S.E. HARRIS: IEEE J. Quant. Electr. QE-9, 470 (1973).
[2] D. GRISCHKOWSKY: Phys. Rev. A7, 2096 (1973); M.D. CRISP: Phys. Rev. A8, 2128 (1973).

GENERATION OF TUNABLE COHERENT RADIATION AT 1460 Å IN MAGNESIUM

Stephen C. Wallace and G. Zdasiuk
Department of Physics, University of Toronto

Coherent radiation in the spectral region 1400-1600 Å (a range of 9800 cm^{-1}) has been generated by four wave mixing in Mg vapour. Compared to strontium, efficiencies in Mg are increased by 10^3 and wavelength tunability is superior because of the absence of a multiplicity of autoionizing levels at the harmonic wavelength. Power conversion efficiencies as high as 0.1% have been obtained at 1436 Å with a dye laser power of 41 kW, with no evidence of saturation.

NON-OPTICAL OBSERVATION OF ZERO-FIELD LEVEL CROSSING EFFECTS IN A SODIUM BEAM

Jean-Louis Picqué
Laboratoire Aimé Cotton, C.N.R.S. II, Bât. 505, F-91405 Orsay, France

Intense coherent laser radiation can induce high-order multipole moments in the two atomic states connected by an optical transition [1]. Because of the coherent oscillation or optical nutation of the atoms between these states, the hertzian coherences can be coupled to the populations of the Zeeman sublevels [2]. This was shown previously using optical detection, with the so-called "saturation resonances" [3]. It has also allowed us to observe zero-field level crossing effects on the atoms themselves, in a Rabi-type atomic beam apparatus. A cw dye laser was tuned to the D_1 absorption line in a sodium beam. We observed, in particular, narrow resonances associated with the optical pumping of the ground state of the atoms [4]. In our experiment, the observation of such resonances required one more interaction with the laser field than in a fluorescence experiment [5] and two more interactions than in a forward scattering experiment [6], and thus corresponded to higher-order nonlinear effects.

The atomic beam apparatus is similar to those known from conventional magnetic resonance experiments [7]. It has been previously used for a spectroscopic study of the sodium resonance lines with a frequency-swept dye laser [8]. In the central C region, between the two-pole inhomogeneous magnetic fields A and B, the dye-laser light irradiated the sodium beam at right angle. The stray magnetic field was compensated over the interaction volume to within about 10 mG, with the use of a set of three orthogonal Helmholtz coils. The scanning magnetic field was parallel to the field in the A and B magnets. The laser radiation was linearly σ-polarized. A small portion of the laser output was removed to lock the single-mode laser frequency to one hfs component of the D_1 line, by monitoring the fluorescence from the beam. The laser frequency was thus stabilized to better than 1 MHz. The intensity available in the main beam under these conditions was of the order of 10 mW/mm^2.

The geometry of the atomic beam apparatus was adjusted so that only the atoms transferred in the C region from a state with m_J = +1/2 (in high magnetic field) to a state with m_J = -1/2 could be detected. The laser was tuned to one of the two hfs transitions arising from the F = 2 ground level. Thus, the level crossing signal consisted in the modification with magnetic field scanning of the population of the F = 2, m_F = -2 sublevel (since the effect involves stimulated emission, the level F = 1 could not contribute to this signal).

In a typical experiment the atomic beam signal as a function of magnetic field was recorded. The laser frequency was stabilized to the line $^2S_{1/2}$, F = 2 ↔ $^2P_{1/2}$, F = 2. The signal exhibited three resonances. The broadest one arose from the variation of the alignment induced by the linearly polarized radiation in the excited state. This resonance corresponds to the classical Hanle effect, here broadened by the strong light intensity, observed in fluorescence experiments. Its width (tens of Gauss) was related to the lifetime (∼10^{-8} s) of the excited sodium 3P state. The two narrow resonances can be attributed to the alignment of quadrupole moment (Δm_F = 2 coherence) and to the hexadecapole moment (Δm_F = 4 coherence, destroyed by a weaker magnetic field) generated in the ground state [6]. Their width (tens of mG) is related to the transit time (∼10^{-5} s) of the atoms through the laser beam.

[1] M. DUCLOY, M.P. GORZA, B. DECOMPS: Opt. Commun. 8, 21 (1973).
[2] C. COHEN-TANNOUDJI: In Atomic Physics IV, ed. by G. zu Putlitz, E.W. Weber and A. Winnacker (Plenum Press, New York 1975).
[3] M. DUCLOY: Opt. Commun. 3, 205 (1971).
[4] J. DUPONT-ROC, S. HAROCHE, C. COHEN-TANNOUDJI: Phys. Lett. 28A, 638 (1969).
[5] R. SCHIEDER, H. WALTHER: Z. Phys. 270, 55 (1974).
[6] W. GAWLIK, J. KOWALSKI, R. NEUMANN, F. TRÄGER: Opt. Commun. 12, 400 (1974).
[7] N.F. RAMSEY: Molecular Beams (Clarendon Press, Oxford 1956).
[8] H.T. DUONG, P. JACQUINOT, S. LIBERMAN, J.L. PICQUÉ, J. PINARD, J.L. VIALLE: Opt. Commun. 7, 371 (1973).

DRESSED ATOM PICTURE OF HIGH INTENSITY GAS LASER[†]

Paul R. Berman and Jehuda Ziegler
Physics Department, New York University, New York, N.Y. 10003

When an atom is placed in a strong optical field, its energy levels are effectively split by the field. This phenomenon may be interpreted as an ac Stark effect and the atom which has been modified by the strong field can be termed a "dressed atom" [1]. It is then of interest to examine the interaction of this dressed atom with additional fields (usually weak) that can serve to probe the dressed atom's structure. However, this type of approach breaks down when one considers the interaction of atoms or molecules with two strong fields. If both fields are strong, the concept of one being the dressing field and one the probe field is not particularly useful. Nevertheless, it is still possible to interpret the interaction of two strong fields with atoms or molecules using a "dressed atom" approach.

To illustrate this technique, we consider the interaction of a two-level system with a strong standing wave field (equivalent to two strong traveling wave fields) such as is encountered in the theory of a high intensity single-mode gas laser [2]. To apply the dressed atom approach, we somewhat arbitrarily break up the atom-field interaction into a dressing field plus probe field interaction. The "dressed atoms" consist of atoms interacting with the oppositely directed traveling waves in the rate equation approximation (neglect of coupling between the fields except insofar as they share the same available population of atoms) while the "probe" interaction is represented by the difference between the true atom-field interaction and that of the rate equation approximation. With this approach, we can semiquantitatively predict which velocity subsets of atoms will experience saturation resonances. This resonance structure has been previously discovered in numerical solutions to the high intensity laser problem [2], and our predictions are in very good agreement with the actual position of the resonances. As such, the dressed atom approach is a useful method for obtaining some physical intuition in problems involving the interaction of two strong fields with atoms or molecules.

The resonances discussed above are not directly measurable in the laser output. However, the resonances can be investigated by subjecting the atoms in this standing wave field to an additional probe field [3]. Alternatively, one can study the resonances by considering the mathematically equivalent problem of a single traveling wave interacting with an atom, but without using the rotating-wave or resonance approximation. In that case the resonances mentioned above can be viewed as representing multiphoton interactions of the field with the atoms.

[†]Supported by the U.S. Army Research Office.
[1] C. COHEN-TANNOUDJI, S. HAROCHE: J. Physique $\underline{30}$, 153 (1969);
S. HAROCHE: Ann. Phys. (Paris) $\underline{6}$, 189 (1971);
C. COHEN-TANNOUDJI, J. Physique $\underline{32}$, C5a-11 (1971).
[2] S. STENHOLM, W.E. LAMB, JR.: Phys. Rev. $\underline{181}$, 618 (1969);
B.J. FELDMAN, M.S. FELD: Phys. Rev. $\underline{A1}$, 1375 (1970);
H.K. HOLT: Phys. Rev. $\underline{A2}$, 233 (1970).
[3] B.J. FELDMAN, M.S. FELD: Phys. Rev. $\underline{A6}$, 899 (1972).

NONLINEAR RESONANT PHOTOIONIZATION IN MOLECULAR IODINE

F.W. Dalby, G. Petty, and C. Tai
Department of Physics, University of British Columbia, Vancouver, B.C.

Strong photoionization spectra was observed in molecular iodine following laser excitation [1]. From the dependence upon laser power, and the vibrational analysis one can assign the observed resonances to 2 photon transitions to a previously unobserved state of g symmetry, followed by absorption of a third photon and photoionization. In view of the sensitivity and simplicity of the resonant photoionization technique it should have further applications [2].

[1] G. PETTY, C. TAI, F.W. DALBY: Phys. Rev. Lett. 34, 1207 (1975).
[2] See, e.g., P.M. JOHNSON, M. BERMAN, D. ZAKHEIM: J. Chem. Phys. 62, 2500 (1975); P.M. JOHNSON: J. Chem. Phys. 62, 4562 (1975).

INFRARED - X-RAY DOUBLE RESONANCE STUDY OF $2P_{3/2} - 2S_{1/2}$ SPLITTING IN HYDROGENIC FLUORINE

H.W. Kugel[†]
Rutgers, The State University, New Brunswick, NJ, USA
M. Leventhal, D.E. Murnick*, C.K.N. Patel and O.R. Wood, II
Bell Telephone Laboratories, Incorporated, Murray Hill and Holmdel, NJ, USA

We report a measurement of Lamb shift (S) in $^{19}F^{8+}$ obtained by the observation of the Lyman - α radiation at 826 eV induced via resonant absorption of infrared laser radiation at 2382.52 cm^{-1} produced by pulsed HBr laser beam incident upon a high energy beam of $^{19}F^8$ atoms in the metastable $2S_{1/2}$ state. The fixed frequency HBr laser is tuned through the $2S_{1/2} - 2P_{3/2}$ ($\Delta E - S$, where ΔE is the fine structure splitting and S is the Lamb shift) resonance using the Doppler shift in the reference frame of the particle beam which has a velocity $v \approx 0.085c$. As ΔE can be calculated to high accuracy since it is insensitive to radiative corrections, the measured $\Delta E - S$ splitting is used to obtain a Lamb shift $(2S_{1/2} - 2P_{1/2})$ value of $S = 3339 \pm 35$ GHz. The series expansion calculation of $S = 3349$ GHz and the closed form calculations of $S = 3360$ GHz and $S = 3342$ GHz can not be distinguished by our present measurements. However, an anticipated improvement in the precision of measurements will allow discrimination between these various calculations and provide a crucial check on the QED calculations.

[†]Supported in part by the NSF
*Associate of the Graduate Faculty, Rutgers University

STARK IONIZATION OF HIGH-LYING RYDBERG STATES OF SODIUM*

Theodore W. Ducas, Richard R. Freeman, Michael G. Littman, Myron L. Zimmerman and Daniel Kleppner
Research Laboratory of Electronics and Department of Physics, Massachusetts Institute of Technology, Cambridge, MA 02139, USA

We have used stepwise excitation in an atomic beam to excite slow-moving atoms to pure high-lying quantum states at densities low enough to eliminate collisional effects. The atoms were detected with high efficiency by Stark ionization. Results are given of a study of the threshold field for ionization of s-states of sodium with principal quantum number n from 26 to 37.

The sodium atoms in an atomic beam were excited stepwise by two pulsed dye lasers pumped by a common nitrogen laser. The first dye laser was tuned to the D_1 line (5890 Å), while the second laser (∼4100 Å) caused transitions from the p-state to high-lying s or d states. The highly excited atoms were detected by direct ionization in an applied Stark field. The laser beams intersected the atomic beam between electric field plates. A pulsed ionizing field was applied after laser excitation, and the resulting ions were observed with a channel electron multiplier. In addition to avoiding the problem of signal loss due to long radiative lifetime, this method provides close to 100% detection efficiency and very low background. The approximate ionization field required for s and d stated with principal quantum number n was $(16n^4)^{-1}$ a.u. (∼390 V/cm for n=30). Resolved s and d levels up to n=60 have been observed.

We have studied the ionization probability as a function of electric field for levels n=26 to n=37. For each s and d state, a greater value of applied field was required to ionize the atoms, than that obtained from the simple result $E_{crit} = (16n^{*4})^{-1}$, where n* is the effective quantum number. This difference is attributed to the Stark effect at ionization. The present problem has generated great interest over the years as it represents the extreme case of distortion of a free atom by an electric field [1]. For s levels, where the onset of ionization is a sharp function of applied field, we could derive values for the Stark shift at ionization. A simple semi-empirical analysis gives $\Delta W(Stark) = 5.6 \times 10^{-5}$ a.u. for the 30s level.

We have also used the fact that optical selection rules for stepwise two-photon processes are strongly affected by nuclear coupling in the intermediate state. The first laser pulse creates a coherent superposition state of the $P_{3/2}$ level hyperfine states since it is short compared to $\frac{2\pi}{\Delta\omega_i}$, where the $\Delta\omega_i$'s are the characteristic hyperfine splittings of the $P_{3/2}$ level. The time evolution of this superposition state can be probed by means of resonant absorption from the second pulsed (∼4100 Å) laser having a variable delay with respect to the pulse from laser 1. If both lasers are circularly polarized in the same sense, for example, these oscillations can be monitored by measuring the population of a high $ns_{1/2}$ state as a function of the delay. This enables one to measure the hyperfine structure in the intermediate state, and provides an example for a general spectroscopic technique.

One aspect of this phenomenon was used to excite selectively high-lying d states. If the lower laser pulses occur in rapid succession the selection rules for dipole transition are those for the case of no nuclear spin. Excitation of high-lying $ns_{1/2}$ levels is then suppressed if both lasers are circularly polarized in the same sense.

* Work supported by U.S. Air Force Office of Scientific Research (Contract F-44620-72-C-0057).

[1] For early work on this problem see H. Bethe, E. Salpeter; <u>Quantum Mechanics of One and Two Electron Atoms</u>, Academic Press, New York (1957)

Lecture Notes in Physics

Bisher erschienen/Already published

Vol. 1: J. C. Erdmann, Wärmeleitung in Kristallen, theoretische Grundlagen und fortgeschrittenene experimentelle Methoden. 1969.

Vol. 2: K. Hepp, Théorie de la renormalisation. 1969.

Vol. 3: A. Martin, Scattering Theory: Unitarity, Analyticity and Crossing. 1969.

Vol. 4: G. Ludwig, Deutung des Begriffs physikalische Theorie und axiomatische Grundlegung der Hilbertraumstruktur der Quantenmechanik durch Hauptsätze des Messens. 1970. Vergriffen.

Vol. 5: M. Schaaf, The Reduction of the Product of Two Irreducible Unitary Representations of the Proper Orthochronous Quantummechanical Poincaré Group. 1970.

Vol. 6: Group Representations in Mathematics and Physics. Edited by V. Bargmann. 1970.

Vol. 7: R. Balescu, J. L. Lebowitz, I. Prigogine, P. Résibois, Z. W. Salsburg, Lectures in Statistical Physics. 1971.

Vol. 8: Proceedings of the Second International Conference on Numerical Methods in Fluid Dynamics. Edited by M. Holt. 1971. Out of print.

Vol. 9: D. W. Robinson, The Thermodynamic Pressure in Quantum Statistical Mechanics. 1971.

Vol. 10: J. M. Stewart, Non-Equilibrium Relativistic Kinetic Theory. 1971.

Vol. 11: O. Steinmann, Perturbation Expansions in Axiomatic Field Theory. 1971.

Vol. 12: Statistical Models and Turbulence. Edited by C. Van Atta and M. Rosenblatt. Reprint of the First Edition 1975.

Vol. 13: M. Ryan, Hamiltonian Cosmology. 1972.

Vol. 14: Methods of Local and Global Differential Geometry in General Relativity. Edited by D. Farnsworth, J. Fink, J. Porter and A. Thompson. 1972.

Vol. 15: M. Fierz, Vorlesungen zur Entwicklungsgeschichte der Mechanik. 1972.

Vol. 16: H.-O. Georgii, Phasenübergang 1. Art bei Gittergasmodellen. 1972.

Vol. 17: Strong Interaction Physics. Edited by W. Rühl and A. Vancura. 1973.

Vol. 18: Proceedings of the Third International Conference on Numerical Methods in Fluid Mechanics, Vol. I. Edited by H. Cabannes and R. Temam. 1973.

Vol. 19: Proceedings of the Third International Conference on Numerical Methods in Fluid Mechanics, Vol. II. Edited by H. Cabannes and R. Temam. 1973.

Vol. 20: Statistical Mechanics and Mathematical Problems. Edited by A. Lenard. 1973.

Vol. 21: Optimization and Stability Problems in Continuum Mechanics. Edited by P. K. C. Wang. 1973

Vol. 22: Proceedings of the Europhysics Study Conference on Intermediate Processes in Nuclear Reactions. Edited by N. Cindro, P. Kulišić and Th. Mayer-Kuckuk. 1973.

Vol. 23: Nuclear Structure Physics. Proceedings of the Minerva Symposium on Physics. Edited by U. Smilansky, I. Talmi, and H. A. Weidenmüller. 1973.

Vol. 24: R. F. Snipes, Statistical Mechanical Theory of the Electrolytic Transport of Non-electrolytes. 1973.

Vol. 25: Constructive Quantum Field Theory. The 1973 "Ettore Majorana" International School of Mathematical Physics. Edited by G. Velo and A. Wightman. 1973.

Vol. 26: A. Hubert, Theorie der Domänenwände in geordneten Medien. 1974.

Vol. 27: R. Kh. Zeytounian, Notes sur les Ecoulements Rotationnels de Fluides Parfaits. 1974.

Vol. 28: Lectures in Statistical Physics. Edited by W. C. Schieve and J. S. Turner. 1974.

Vol. 29: Foundations of Quantum Mechanics and Ordered Linear Spaces. Advanced Study Institute Held in Marburg 1973. Edited by A. Hartkämper and H. Neumann. 1974.

Vol. 30: Polarization Nuclear Physics. Proceedings of a Meeting held at Ebermannstadt October 1-5, 1973. Edited by D. Fick. 1974.

Vol. 31: Transport Phenomena. Sitges International School of Statistical Mechanics, June 1974. Edited by G. Kirczenow and J. Marro.

Vol. 32: Particles, Quantum Fields and Statistical Mechanics. Proceedings of the 1973 Summer Institute in Theoretical Physics held at the Centro de Investigacion y de Estudios Avanzados del IPN – Mexico City. Edited by M. Alexanian and A. Zepeda. 1975.

Vol. 33: Classical and Quantum Mechanical Aspects of Heavy Ion Collisions. Symposium held at the Max-Planck-Institut für Kernphysik, Heidelberg, Germany, October 2-5, 1974. Edited by H. L. Harney, P. Braun-Munzinger and C. K. Gelbke. 1975.

Vol. 34: One-Dimensional Conductors, GPS Summer School Proceedings, 1974. Edited by H. G. Schuster. 1975.

Vol. 35: Proceedings of the Fourth International Conference on Numerical Methods in Fluid Dynamics. June 24-28, 1974, University of Colorado. Edited by R. D. Richtmyer. 1975.

Vol. 36: R. Gatignol, Théorie Cinétique des Gaz à Répartition Discrète de Vitesses. 1975.

Vol. 37: Trends in Elementary Particle Theory. Proceedings 1974. Edited by H. Rollnik and K. Dietz. 1975.

Vol. 38: Dynamical Systems, Theory and Applications. Proceedings 1974. Edited by J. Moser. 1975.

Vol. 39: International Symposium on Mathematical Problems in Theoretical Physics. Proceedings 1975. Edited by H. Araki. 1975.

Vol. 40: Effective Interactions and Operators in Nuclei. Proceedings 1975. Edited by B. R. Barrett. 1975.

Vol. 41: Progress in Numerical Fluid Dynamics. Proceedings 1974. Edited by H. J. Wirz. 1975.

Vol. 42: H II Regions and Related Topics. Proceedings 1975. Edited by T. L. Wilson and D. Downes. 1975.

Vol. 43: Laser Spectroscopy. Proceedings of the Second International Conference, Megève, June 23-27, 1975. Edited by S. Haroche, J. C. Pebay-Peyroula, T. W. Hänsch, and S. E. Harris. 1975.

TITLES OF RELATED INTEREST

LASER SPECTROSCOPY
W. Demtröder
Springer-Verlag, Berlin, Heidelberg New York
2nd enlarged edition
1973, Pp.III+106

TOPICS IN APPLIED PHYSICS

DYE LASERS, Vol.1
F.P. Schäfer (editor)

F.P. Schäfer: Principles of Dye Laser Operation

B.B. Snavely: Continuous-Wave Dye Lasers

C.V. Shank, E.P. Ippen: Mode-Locking of Dye Lasers

K.H. Drexhage: Structure and Properties of Laser Dyes

T.W. Hänsch: Applications of Dye Lasers

1973, Pp.XI+285

LASER SPECTROSCOPY OF ATOMS AND MOLECULES, Vol.2
H. Walther (editor)

H. Walther: Atomic and Molecular Spectroscopy with Lasers

E.D. Hinkley, K.W. Nill, F.A. Blum: Infrared Spectroscopy with Tunable Lasers

K. Shimoda: Double-Resonance Spectroscopy of Molecules

J.M. Cherlow, S.P. Porto: Laser Raman Spectroscopy of Gases

B. Decomps, M. Dumont, M. Ducloy: Linear and Nonlinear Phenomena in Laser Optical Pumping

K. Evenson, F.R. Petersen: Laser Frequency Measurements, the Speed of Light and the Meter

1975

LIGHT SCATTERING IN SOLIDS, Vol.8
M. Cardona (editor)

M. Cardona: Introduction

A. Pinczuk, E. Burstein: Raman Scattering in Semiconductors

R.M. Martin, L.M. Falicov: Resonance Raman Scattering

M.V. Klein: Electronic Raman Scattering

M.H. Brodsky: Raman Scattering in Amorphous Semiconductors

A.S. Pine: Brillouin Scattering in Semiconductors

Y.R. Shen: Stimulated Raman Scattering

1975, Pp.ca.360

FORTHCOMING TITLES

HIGH RESOLUTION LASER SPECTROSCOPY
K. Shimoda (editor)

K. Shimoda: Introduction

K. Shimoda: Line Broadening and Narrowing Effects

P. Jacquinot: Atomic Beam Spectroscopy

V.S. Letokhov: Saturation Spectroscopy

J.L. Hall: Recent Studies on Very High Resolution Spectroscopy

V.P. Chebotayev: Three-Level Laser Spectroscopy

S. Haroche: Quantum Beat Spectroscopy

N. Bloembergen, M.D. Levenson: Doppler-Free Two-Photon Spectroscopy

LASER MONITORING OF THE ATMOSPHERE
E.D. Hinkley (editor)

S.H. Melfi: Remote Sensing for Air Quality Management

V.E. Zuev: Laser Transmission of the Atmosphere

R.H.T. Collis, P.B. Russell: Lidar Measurement of Particles and Gases by Elastic Backscattering and Differential Absorption

H. Inaba: Detection of Atoms and Molecules by Raman Scattering and Resonance Fluorescence

E.D. Hinkley, R.T. Ku, P.L. Kelley: Molecular Pollutant Detection by Differential Absorption

R.T. Menzies: Laser Heterodyne Detection Techniques

TOPICS IN MODERN PHYSICS

BEAM-FOIL SPECTROSCOPY
S. Bashkin (editor)

S. Bashkin: Introduction

S. Bashkin: Instrumentation

I. Martinson: Wavelengths Measurements and Level Analysis

L. Curtis: Lifetime Measurements

I. Sellin: Autoionizing Levels

H. Marrus: Studies of H-Like and He-Like Ions of High Z

W. Whaling, L. Heroux: Applications to Astrophysics

O. Sinanoglu: Fundamental Calculation of Level Lifetimes

W. Wiese: Systematic Effects in Z-Dependence of Oscillator Strengths

J. Macek, D.J. Burns: Coherence, Alignment, and Orientation Phenomena